Biology of Copper Complexes

Experimental Biology and Medicine

Biology of Copper Complexes, edited by *John R. J. Sorenson*, 1987

Epstein–Barr Virus and Human Disease, edited by *P. H. Levine, D. V. Ablashi, M. Nonoyama, G. R. Pearson, and R. Glaser*, 1987

Methods of Protein Sequence Analysis • 1986, edited by *Kenneth A. Walsh*, 1987

Regulation of Immune Gene Expression, edited by *Marc Feldmann and Andrew McMichael*, 1986

Biological Methylation and Drug Design, edited by *Ronald T. Borchardt, Cyrus R. Creveling, and Per Magne Ueland*, 1986

Retroviruses and Human Pathology, edited by *Robert C. Gallo, Dominique Stehelin, and Oliviero E. Varnier*, 1985

Directed Drug Delivery, edited by *Ronald T. Borchardt, Arnold Repta, and Valentino Stella*, 1985

Immune Regulation, edited by *Marc Feldmann and N. A. Mitchison*, 1985

Human T Cell Clones, edited by *Marc Feldmann, Jonathan R. Lamb, and James N. Woody*, 1985

Inositol and Phosphoinositides, edited by *John E. Bleasdale, Joseph Eichberg, and George Hauser*, 1985

Growth, Cancer, and the Cell Cycle, edited by *Philip Skehan and Susan J. Friedman*, 1984

Ir Genes, edited by *Carl W. Pierce, Susan E. Cullen, Judith A. Kapp, Benjamin D. Schwartz, and Donald C. Shreffler*, 1983

Methods in Protein Sequence Analysis, edited by *Marshall Elzinga*, 1982

Inflammatory Diseases and Copper, edited by *John R. J. Sorenson*, 1982

Membrane Fluidity, edited by *Morris Kates and Arnis Kuksis*, 1980

Biology of Copper Complexes

Edited by

John R. J. Sorenson

University of Arkansas for Medical Sciences, Little Rock, Arkansas

Humana Press • Clifton, New Jersey

Crescent Manor
PO Box 2148
Clifton, NJ 07015

Library of Congress Cataloging-in-Publication Data:

Biology of copper complexes.

(Experimental biology and medicine)
Based on a symposium held in 1986.
Includes index.
1. Copper—Physiological effect—Congresses.
2. Coordination compounds—Physiological effect—Congresses.
3. Copper—Metabolism—Congresses.
4. Coordination compounds—Metabolism—Congresses.
5. Copper—Therapeutic use—Congresses.
6. Coordination compounds—Therapeutic use—Congresses.
I. Sorenson, John R. J. II. Series: Experimental biology and medicine (Clifton, NJ)
[DNLM: 1. Copper—metabolism—congresses. 2. Copper—therapeutic use—congresses
QU 130 B61442 1986]

QP 535.C9B57 1987 612'.01524 87-17047
ISBN 0-89603-123-3

Preface

In 1928, it was discovered that copper was essential for normal human metabolism. Ten years later, 1938, it was observed that patients with rheumatoid arthritis had a higher than normal serum copper concentration, which returned to normal with remission of this disease. Thirteen years later, it was found that copper complexes were effective in treating arthritic diseases. The first report that copper complexes had antiinflammatory activity in an animal model of inflammation appeared twenty-two years after the discovery of essentiality. In 1976, it was suggested that the active forms of the antiarthritic drugs are their copper complexes formed in vivo. This suggestion was confirmed and extended in the interim with over 1000 recent publications, and many of these were addressed in the proceedings of our first symposium, published in 1982.

The present symposium was organized to present new normal physiological, nutritional, and biochemical aspects of essential metalloelement metabolism as well as variations in metabolism associated with disease states. In addition new data concerning antimicrobial, antiinflammatory, antiulcer, anticancer, anticarcinogenic, analgesic, and radioprotectant activities of copper complexes were presented. These activities are consistent with the notion that they represent the facilitation of normal copper-dependent metabolic processes in disease states. The presentations and interactive discussions that followed are contained in these proceedings.

John R. J. Sorenson

Dedication

These proceedings are dedicated to those who made this truly memorable scientific and social Arkansas experience possible. They include: those whose persistent pioneering essential-metalloelement research efforts now provide important background information in support of new and exciting approaches to the understanding and treatment of chronic diseases; those who persist in providing new knowledge in these areas; Dr. Harry P. Ward, Chancellor of the University of Arkansas for Medical Sciences, and Dr. L. D. Milne, Dean of the College of Pharmacy, for the academic environment they provide and their personal and financial support; the sponsors,

International Copper Research Association
Arkansas Catfish Farmers Association
Arkansas Forestry Association
Arkansas Poultry Federation
Arkansas Rice Festival
Coca-Cola Bottling Company of Arkansas
Dean Foods
Mr. Floyd G. Villines, Little Rock City Director
Kellogg Company
Little Rock Convention Center
Little Rock Regional Airport
Mericon Industries, Incorporated
Post Winery
Silbernagel Company, Incorporated
Skaggs Alpha-Beta Stores
Smith Whetstones, Inc.

and the participants for their persistent efforts and support in organizing this meeting, as well as their free exchange of knowledge.

Special thanks are also due to Mrs. Stacie A. Hronas for manuscript corrections and retyping as well as transcribing and inserting recorded proceeding discussions, Mrs. Catherine M. Vaughan for her assistance in these proceedings, and Mrs. Mary Vick for her efforts in premeeting organization and coordinating activities during the meeting.

John R. J. Sorenson

Contents

Physiologic Aspects of Copper Complexes

Session Chairperson: ***Peter Lipsky***

Biochemical Aspects of Copper Complexes

Session Chairperson: ***Edward D. Harris***

Copper Complexes in Essential Metalloelement Nutriture and Disease States

Session Chairperson: **Robert J. Cousins**

Biological Aspects of Copper and Iron in Inflammation

Session Chairperson: **Carl L. Keen**

Antiinflammatory Activities of Copper Complexes

Session Chairperson: ***Dean P. Loven***

Analgesic and Antimicrobial Activities of Copper Complexes

Session Chairperson: ***Ayalla Barnea***

Anticancer Activities of Copper Complexes and Iron Chelators

Session Chairperson: ***Miera Fields***

Anticarcinogenic and Radioprotectant Activities of Copper Complexes

Session Chairperson: ***Loren Pickart***

Other Biological Effects of Complexes and Complexing Agents

Session Chairperson: ***Gary T. Elliott***

Poster Presentations

PARTICIPANTS

Adolfs, M. J. P., Institute of Pharmacology, Erasmus Universiteit Rotterdam, P.O. Box 1738, 3000 DR Rotterdam, The Netherlands

Aihara, Hironaka, Research Center, Taisho Pharmaceutical Co., Ltd., 1-403, Yoshio-cho, Ohmiya, Saitama 330, Japan

Ali, Fadia, Smith Kline and French Laboratories, 709 Swedeland Road, Swedeland, PA 19479

Anderson, Alfred W., Veterinary Medicine, 19383 Ave 300 Exeter, CA 93221

Antholine, William E., National Biomedical ESR Center, Medical College of Wisconsin, 8701 Watertown Plank Road, Milwaukee, WI 53226

Avila, M. A., Instituto de Oncologia "Angel H. Roffo", Avenida San Martin 5481, Buenos Aires, Argentina

Badawi, Abdel Fattah M., Department of Chemistry, Centre of Science and Mathematics, Taif, Saudi Arabia

Baker, Max L., Department of Radiology, College of Medicine, University of Arkansas for Medical Sciences, 4301 W. Markham, Little Rock, AR 72205

Ball, Frank, 4 Atlantic St., Charleston, SC 29401

Barber, Ellen F., Food Science and Human Nutrition Department, University of Florida, Gainesville, FL 32611

Barnea, Ayalla, Department of OB/GYN and Physiology, University of Texas Health Sciences Center at Dallas, 5323 Harry Hines Blvd., Dallas, TX 75235

Barnett, John B., Department of Microbiology and Immunology, College of Medicine, University of Arkansas for Medical Sciences, 4301 W. Markham, Little Rock, AR 72205

Basker, Kala, University of Texas Health Sciences Center at Dallas, 5323 Harry Hines Blvd., Dallas, TX 75235

van Batenburg, M., Institute of Pharmacology, Erasmus Universiteit Rotterdam, P.O. Box 1738, 3000 DR Rotterdam, The Netherlands

Bates, George W., Department of Biochemistry & Biophysics, Texas A & M University, College Station, TX 77843

Benson, R. Wayne, National Center for Toxicological Research, Jefferson, AR 72079

Bhathena, Sam J., Carbohydrate Nutrition Laboratory, Bldg. 307 Rm. 311, Beltsville, MD 20705

Bonta, Ivan L., Institute of Pharmacology, Erasmus Universiteit Rotterdam, P.O. Box 1738, 3000 DR Rotterdam, The Netherlands

Carlson, M. P., Veterinary Diagnostic Center, University of Nebraska, Lincoln, NE 68508

Chang, Louis W., Department of Pathology, College of Medicine, University of Arkansas for Medical Sciences, 4301 W. Markham, Little Rock, AR 72205
Chen, Harold, College of Medicine, Louisiana State University, 1501 Kings Highway, P.O. Box 33932, Shreveport, LA 71130
Chidambaram, M.V., Department of Biochemistry & Biophysics, Texas A & M University, College Station, TX 77843
Clegg, Michael S., Department of Nutrition, University of California at Davis, Davis, CA 95616
Clement, John R., College of Medicine, Louisiana State University, 1501 Kings Highway, P.O. Box 33932, Shreveport, LA 71130
Cousins, Robert J., Food Science and Human Nutrition Department, University of Florida, Gainesville, FL 32611
Crispens, Charles G., Jr., Department of Biology, The University of Alabama in Birmingham, Birmingham, AL 35294
Crook, James E., Oak Ridge Associated Universities, Medical and Health Division, Oak Ridge, TN 37831-0117
Cypher, George, International Copper Research Association, Inc., 708 Third Ave., New York, NY 10017
Dameron, Charles T., Department of Biochemistry and Biophysics, Texas A & M University, College Station, TX 77843
DiSilvestro, Robert A., Department of Food Science & Human Nutrition, Purdue University, West Lafayette, IN 47907
Doster, A.R., Veterinary Diagnostic Center, University of Nebraska, Lincoln, NE 68508
Doughty, Darrell, Richardson-Vicks, Inc., One Far Mill Crossing, Shelton, CT 06484
Dubick, Michael, Department of Nutrition, University of California at Davis, Davis, CA 95616
Egner, Patricia A., Department of Environment Health Sciences, Johns Hopkins University School of Hygiene and Public Health, 615 N. Wolfe St., Baltimore, MD 21205
Elliott, Gary T., Department of Biopharmaceutical Sciences, College of Pharmacy, University of Arkansas for Medical Sciences, 4301 W. Markham, Little Rock, AR 72205
Elliott, Graham R., Institute of Pharmacology, Erasmus Universiteit Rotterdam, P.O. Box 1738, 3000 DR Rotterdam, The Netherlands
El-Shazely, Rafaat M., Chemistry Department, Faculty of Science, Mansoura University, P.O.Box 21, Mansoura, Egypt
Famaey, Jean-Pierre, Department of Rheumatology, Hopital Universitaire Saint-Pierre, University of Brussels, 322 Rue Haute, 1000 Brussels, Belgium

Ferrell, Fay, Department of Nutrition, University of California at Davis, Davis, CA 95616
Fields, Meira, USDA, ARS, Bldg. 307 Rm. 312, Beltsville, MD 20705
Foly, J.J., Smith Kline and French Laboratories, 709 Swedeland Road, Swedeland, PA 19479
Fontaine, Jeanine, Laboratory of Pharmacology, Institute of Pharmacy, University of Brussels, 322 Rue Haute, 1000 Brussels, Belgium
Fuchs, Alicia Graciela, Instituto de Oncologia "Angel H. Roffo", Avenida San Martin 5481, Buenos Aires, Argentina
Gaisser, H.-D., Department of Pharmacochemistry, Vrije Universiteit, de Boelelaan 1083, 1081 HV Amsterdam, The Netherlands
Gasparian, Vardan K., Institute of Biochemistry of Armenian Academy of Sciences, 375044, Yerevan-44, USSR
Gibson, Thomas R., Molecular Biology Institute and Department of Biology, San Diego State University, San Diego, CA 92182
Glembotski, Christopher C., Molecular Biology Institute and Department of Biology, San Diego State University, San Diego, CA 92182
Gohar, M.A., Department of Chemistry, Faculty of Science, Alexandria University, Egypt
van der Goot, Hendrikus, Department of Pharmacochemistry, Vrije Universiteit, de Boelelaan 1083, 1081 HV Amsterdam, The Netherlands
Granger, D. Neil, Department of Physiology, College of Medicine, University of South Alabama, Mobile, AL 36688
Greenaway, Frederick, T., Department of Chemistry, Clark University, Worcester, MA 01610
Grisham, Matthew B., Department of Physiology, College of Medicine, University of South Alabama, Mobile, AL 36688
Hafez, A.H., Department of Animal Hygiene, Faculty of Veterinary Medicine, Assiut University, Assiut Egypt
Harris, Edward D., Department of Biochemistry & Biophysics, HEEP Building, Texas A & M University, College Station, TX 77843-2128
Hartter, Darryl, University of Texas Health Sciences Center at Dallas, 5323 Harry Hines Blvd., Dallas, TX 75235
Hashimoto, Sanae, Research Center, Taisho Pharmaceutical Co., Ltd., 1-403, Yoshio-cho, Ohmiya, Saitama 330, Japan
Hernandez, Lucrecia A., Department of Physiology, College of Medicine, University of South Alabama, Mobile, AL 36688

Hopper, John H., Kellogg Company, Hampton, AR 71744
Hough, Jr., Aubrey, Department of Pathology, College of Medicine, Univeristy of Arkansas for Medical Sciences, 4301 W. Markham, Little Rock, AR 72205
Hunter, Glenn C., Department of Surgery, Veterans Administration Medical Center, Martinez, CA 94553
Hyde, James S., National Biomedical ESR Center, Medical College of Wisconsin, 8701 Watertown Plank Road, Milwaukee, WI 53226
Iwakawa, Mayumi, University of Texas System Cancer Center, Texas Medical Center, Houston, TX 77030
Johnson, Jerre L., Veterinary Science Laboratory, University of Nebraska, R 4, Box 46B, North Platte, NE 69101
Kasemeier, Scott L., Department of Biopharmaceutical Sciences, College of Pharmacy, University of Arkansas for Medical Sciences, 4301 W. Markham, Little Rock, AR 72205
Keen, Carl L., Department of Nutrition, University of California at Davis, Davis, CA 95616
Kennedy, F. Scott, College of Medicine, Louisiana State University, 1501 Kings Highway, P.O. Box 3392, Shreveport, LA 71130
Kensler, Thomas W., Johns Hopkins University, 615 North Wolfe St., Baltimore, MD 21205
Kishore, Vimal, Xavier University of Louisiana, College of Pharmacy, New Orleans, LA 70125
Krishnaswamy, S., Department of Biochemistry, University of Vermont, Burlington, VT 05405
Kyle, Joseph, Department of Chemistry, Chicago State University, 95th St. at King Drive, Chicago, IL 60628
Leipold, H.W., Department of Pathology, Kansas State University, Manhatten, KS 66502
Lipsky, Peter E., Rheumatic Disease Unit, University of Texas Health Science Center at Dallas, 5323 Harry Hines Blvd., Dallas, TX 75235
Loven, Dean P., Division of Radiation Biology and Oncology, East Carolina University, Greenville, NC 27858-4354
de Lustig, E.S., Instituto de Oncologia "Angel H. Roffo", Avenida San Martin 5481, Buenos Aires, Argentina
Madhu, Cherukury, Department of Pharmacology, University of Kansas Medical Center, 39th & Rainbow, Kansas City, KS 66103
Melethil, Srikumaran, M3-206A School of Medicine and Pharmacy, University of Missouri-Kansas City, 2411 Holmes, Kansas City, MO 64108-2792

Mohan, Govind, Department of Pharmacology, S. N. Medical College, Agra, India

Mostafa, Mohsen M., Chemistry Department, Faculty of Science, Mansoura University, P.O. Box 21, Mansoura, Egypt

Musa, S.A., Department of Poultry Diseases, Faculty of Veterinary Medicine, Assiut University, Assiut, Egypt

Mylroie, Augusta A., Department of Chemistry, Chicago State University, 95th St. at King Drive, Chicago, IL 60628

Nalbandyan, Robert M., Institute of Biochemistry of Armenian Academy of Sciences, 375044, Yerevan-44, USSR

Nassar, Ahmed Yasien, Department of Biochemistry, Faculty of Medicine, Assiut University, Assiut, Egypt

Neve, J., Laboratory of Analytical Chemistry and Toxicology, Institute of Pharmacy, University of Brussels, 322 Rue Haute, 1000 Brussels, Belgium

Nuccio, M., Department of Biochemistry & Biophysics, Texas A & M University, College Station, TX 77843

O'Neal, Hubert R., Hochem Enterprises, Inc., 11202 Pecan Creek Drive, Houston, TX 77043

Okuyama, Shigeru, Research Center, Taisho Pharmaceutical Co., Ltd., 1-403, Yoshio-cho, Ohmiya, Saitama 330, Japan

Parashar, R.K., Department of Chemistry, Agra University, Agra 282004 U.P., India

Pelle, Edward, Research and Development, Estte Lauder Company, 125 Pinelawn Road, Melville, NY 11747

Peretz, A, Department of Rheumatology, Hopital Universitaire Saint-Pierre, University of Brussels, 322 Rue Haute, 1000 Brussels, Belgium

Petering, David H., Department of Chemistry, University of Wisconsin-Milwaukee, Milwaukee, WI 53201

Pickart, Loren, Bioheal, Inc., 15232 S.E. 48th Drive, Bellevue, WA 98006

Recant, Lillian, Diabetes Research Laboratory, V.A. Medical Center, Washington, DC 20422

Reed, Gregory A., Department of Pharmacology, University of Kansas Medical Center, 39th & Rainbow, Kansas City, KS 66103

Reiners Jr., John, J., University of Texas, System Cancer Center, Science Park Research Division, Smithville, TX. 78957

Risk, Martin, Department of Obstetrics and Gynecology, School of Medicine, University of South Dakota, Sioux Falls, SD 57101

Roberts, Dean W., National Center for Toxicological Research, Jefferson, AR 72079

Salari, Hamid, Department of Biopharmaceutical Sciences, College of Pharmacy, University of Arkansas for Medical Sciences, 4301 W. Markham, Little Rock, AR 72205

Sarau, H.M., Smith Kline and French Laboratories, 709 Swedeland Road, Swedeland, PA 19479

Scaife, Rhetta, Department of Natural Sciences, University of Arkansas at Pine Bluff, Pine Bluff, AR 71601

Schmidt, Sophie, Department of Obstetrics and Gynecology, School of Medicine, University of South Dakota, Sioux Falls, SD 57101

Schneider, N.R., Veterinary Diagnostic Center, University of Nebraska, Lincoln, NE 68508

Seasley, Ross, 371 Marylhurst Dr., Dayton, OH 45459

Shallaby, Abd El-Hammid M., Chemistry Department, Faculty of Science, Mansoura University, P.O. Box 21, Mansoura, Egypt

Sharma, R.C., Department of Chemistry, Agra University, Agra 282004 U.P., India

Shetty, Bhasker V., School of Medicine and Pharmacy, University of Missouri-Kansas City, 2411 Holmes, Kansas City, MO 64108-2792

Simmons, Jerry Department of Laboratory Medicine, School of Medicine, V.A. Medical Center, Sioux Falls, SD 57101

Smit, H., Department of Pharmacochemistry, Vrije Universiteit, de Boelelaan 1083, 1081 HV Amsterdam, The Netherlands

Smith, Jr., J. Cecil, USDA, ARS, Vitamin and Mineral Nutrition Lab, Bldg 307. Rm. 117, Beltsville, MD 20705

Soderberg, Lee S.F., Department of Microbiology and Immunology, Univeristy of Arkansas for Medical Sciences, 4301 W. Markham, Little Rock, AR 72205

Sorenson, John R.J., Department of Biopharmaceutical Sciences, College of Pharmacy, University of Arkansas for Medical Sciences, 4301 W. Markham, Little Rock, AR 72205

Subczynski, Witold K., Department of Biophysics, Institute of Molecular Biology, Jagiellonian University, Krakow, Poland

Summan, Abdul Hadi M., Faculty of Science, Umm-Al-Qura University, Holy Makkah, Saudi Arabia

Taffe, B.G., Department of Environment Health Sciences, Johns Hopkins University School of Hygiene and Public Health, 615 N. Wolfe Street, Baltimore, MD 21205

Teeters, Winnie, Environmental Protection Agency, 401 M Street SW, Washington, DC 20460

Thomas, Edwin L., Department of Biochemistry, St. Jude's Children's Research Hospital, 332 N. Lauderdale, Memphis, TN 38101

Timmerman, H., Department of Pharmacochemistry, Vrije Universiteit, de Boelelaan 1083, 1081 HV Amsterdam, The Netherlands

Torregrosa, Doris, Department of Pathology, College of Medicine, University of Arkansas for Medical Sciences, Little Rock, AR 72205

Umbles, Constance, Department of Chemistry, Chicago State University, 95th St. at King Drive, Chicago, IL 60628

Vertongen, F., Department of Medicinal Chemistry, Hopital Universitaire, Saint-Pierre University of Brussels, 322 Rue Haute, 1000 Brussels, University of Brussels, Belgium

Williams, Darryl M., College of Medicine, Louisiana State University, 1501 Kings Highway, P.O. Box 33932, Shreveport, LA 71130

Willingham, William M., Department of Natural Sciences, University of Arkansas at Pine Bluff, Pine Bluff, AR 71601

Young O'Gara, Coleen, Department of Chemistry, Clark University, Worcester, MA 01610

de Zwart, M.A.H., Department of Pharmacochemistry, Vrije Universiteit, de Boelelaan 1083, 1081 HV Amsterdam, The Netherlands

Physiologic Aspects of Copper Complexes

A PHYSIOLOGICAL BASIS FOR PHARMACOLOGICAL ACTIVITIES OF COPPER COMPLEXES: AN HYPOTHESIS

John R.J. Sorenson

Department of Biopharmaceutical Sciences

College of Pharmacy, University of Arkansas for Medical Sciences, 4301 W. Markham, Little Rock, Arkansas 72205.

Copper is recognized as an essential metalloelement like sodium, potassium, magnesium, calcium, iron, zinc, chromium, vanadium, and manganese[1]. Like essential amino acids, essential fatty acids, and essential cofactors (vitamins), essential metalloelements are required for normal metabolic processes but can not be synthesized in vivo and daily dietary intake and absorption are required. The adult body contains between 1.4 and 2.1 mg of copper per kilogram of body weight. The infant body contains 3 times this amount which is consistent with the fact that their metabolic needs are that much greater than adults. All tissues of the body need copper for normal metabolism but some tissues have greater metabolic needs than others and tissue contents of copper reflect this fact. Amounts of copper found in various body tissues and fluids of individuals are shown in Table I[2,3].

The amount of copper in each tissue correlates with the number and kind of metabolic processes requiring copper in that tissue. In this regard, it is of interest to point out that brain and heart contain more copper than all other tissues except the liver which is the major copper storage organ. Gall bladder and bile also contain a large amount of copper and this has been attributed to their suggested role in excretion. However, bile may also

TABLE 1. MEAN CONCENTRATION[a] OF Cu IN TISSUES AND FLUIDS

Tissue	Cu	Tissue	Cu
Adrenal	210	Milk	
Aorta	97	colostrum	0.35-0.50 ug/ml
Bile	547	mature	0.20-0.50 ug/ml
Blood (total)	1.01 ug/ml	Muscle	85
erythrocytes	0.98 ug/ml	Nails	23 ug/g
plasma	1.12 ug/ml	Omentum	190
serum	1.19 ug/ml	Ovary	130
Bone	25 ug/g	Pancreas	150
Brain	370	Pancreatic Fluid	105
Breast	6 ug/g	Placenta	4 ug/g
Cerebrospinal Fluid	0.22 ug/g	Prostate	110
Diaphragm	150	Saliva	0.08 ug/ml
Esophagus	140	Skin	120
Gall Bladder	750	Spleen	93
Hair	19 ug/g	Stomach	230
Heart	350	Sweat	0.55 ug/ml
Intestine		Testes	95
Duodenum	300	Thymus	4 ug/g
jejunum	250	Thyroid	100
ileum	280	Tongue	4.6 ug/g
cecum	220	Tooth	
signoid colon	230	dentine	2 ug/g
rectum	180	enamel	10 ug/g
Kidney	270	Trachea	65
Larynx	59	Urinary Bladder	120
Liver	680	Urine	0.04 ug/ml
Lung	130	Uterus	110
Lymph Node	60		

[a] ug/g of tissue ash or as shown

contain a mobile form of copper suitable for intestinal reabsorption and the gall bladder may also serve as a storage tissue. Kidney copper content is also relatively large which may reflect a conservatory role since urinary copper is very small. Remaining tissues have lesser amounts of copper because of their relatively lower metabolic activity but it is just as important for normal metabolism in these tissues as it is in all others.

Although bile may serve as the major excretory vehicle for excess copper, significant but lesser amounts are lost via hair, stratum corneum, finger- and toe-nails, sweat, and urine as end-products of metabolism. These losses point out the need for compensating daily intake and absorption to replenish this essential metalloelement.

Ionic copper has a particularily high affinity for other molecules (Ligands) capable of bonding with it. A consequence of this is that all measurable copper in biological systems exists as complexes or chelates composed of copper bonded to organic components of these systems. Calculated amounts of ionic copper suggested to be present in biological systems (10^{-18}M in plasma[4]) are too small to be measured using the most sensitive instrumentation available. As a result, measurable tissue copper contents reflect contents of copper complexes and these complexes account for the absorption, distribution, and biologically active forms of copper in vivo.

Recognized and recently suggested copper-dependent enzymes[6-13] are listed in Table II. Cytochrome c Oxidase is required by all cells to produce energy needed to drive energy requiring biochemical reactions. Cytosolic and extracellular superoxide dismutases are required to prevent cellular destruction associated with accumulation of superoxide (O_2-), which is thought to play a role in causing pathologic changes associated with affected tissues in chronic diseases such as arthritides, seizures, diabetics, and cancers[14]. Tyrosinase is the enzyme required for synthesis of dihydroxyphenylalanine (DOPA) and its subsequent transformations to melanin, required for pigmentation. Dopamine-β-Hydroxylase is required for conversion of dopamine to norepinephrine, which is then converted to epinephrine. Lysyl Oxidase is required for cross-linking of collagen and elastin in maintenance and repair of all connective tissues. Amine Oxidases are required for removal of primary amines which are no longer needed for hormonal activity i.e. norepinephrine or dopamine or other amines which are toxic metabolic end-products. Ceruloplasmin, the principal copper-containing component of plasma, is now suggested to be multifunctional having: transport function, serum antioxidant (SOD-like) activity, serum amine oxidase activity, cellular growth promoting activity, ferroxidase activity in mobilizing stored iron [Fe(II)] for hemoglobin synthesis, and angiogenic activity required for vascularization. Clotting factor V is now suggested to be copper-dependent. The peptidyl α-Amidating Monooxygenases are a new class of copper-dependent enzymes suggested to be required for syntheses of neuroendocrine peptides from their precursor glycyl-terminus peptides. In addition, copper-dependent processes appear to be required for modulation of prostaglandin syntheses, lysosomal membrane stabilization, and modulation of histaminic activity[15].

Ingested copper complexes follow the pathway presented in Figure 1. One of a large number of possible copper complexes that might be found in foods and/or beverages (CuL_2), following ingestion and digestion, would give rise

Table II. RECOGNIZED COPPER-DEPENDENT ENZYMES AND THEIR CHEMICAL FUNCTION

Enzyme	Function
Cytochrome c Oxidase	Reduction of Oxygen: $O_2 \xrightarrow{H+,e-} HO_2 \xrightarrow{H+,e-} H_2O_2 \xrightarrow{H+,e-} H_2O + HO\cdot \xrightarrow{H+,e-} H_2O$
Superoxide Dismutase	Disproportionation of Superoxide in prevention of its accumulation: $2O_2^- + 2H^+ \longrightarrow O_2 + H_2O_2$
Tyrosinase	Hydroxylation of tyrosine in melanin synthesis: 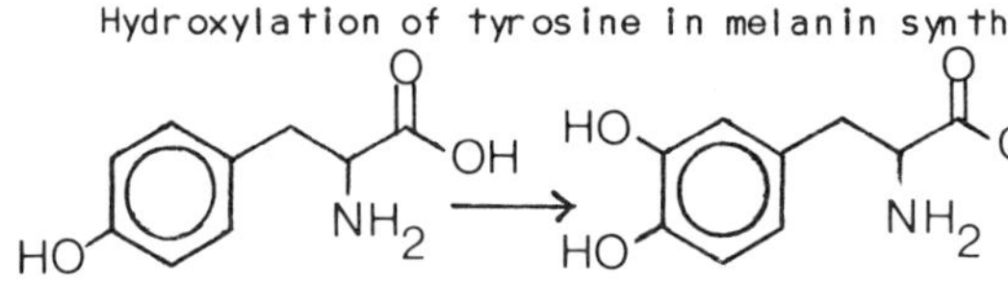 DOPA
Dopamine-B-Hydroxylase and Extremely Acidic Copper-Containing Protein	Hydroxylation of dopamine in catecholamine synthesis: 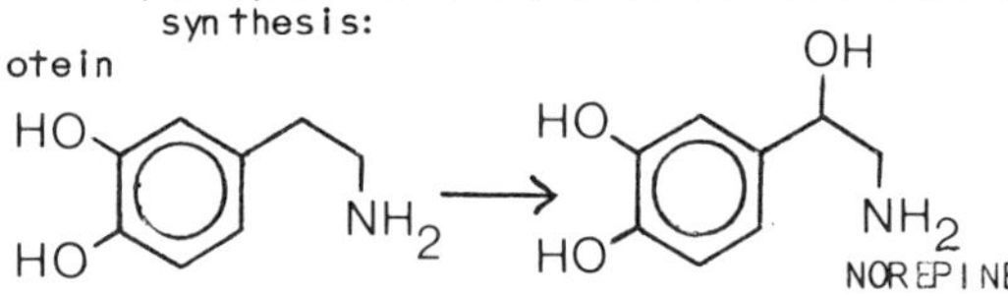 NOREPINEPHRINE
Lysyl Oxidase	Oxidation of terminal amino group of lysyl amino acids in procollagen and proelastin to an aldehyde group: $-CH_2-CH_2-NH_2 \longrightarrow -CH_2-\overset{\displaystyle O}{\overset{\|}{C}}H$
Amine Oxidases	Oxidation of primary amines to aldehydes in catecholamine and other primary amine metabolism: $R-CH_2-NH_2 \longrightarrow R-\overset{\displaystyle O}{\overset{\|}{C}}H$
Ceruloplasmin	Mobilization and utilization of stored iron: Ferroxidase $Fe(II) \longrightarrow Fe(III)$ "Angiogenin" - stimulates endothelial cell division and vascularization
Factor V	Blood clotting factor
Peptidyl α-Amidating Monooxygenases	Synthesis of neuroendocrine peptides (hypothalamic releasing factors, melanocyte stimulating hormones from anterior pituitary, oxytocin and vasopressin from the posterior pituitary, gastrin from stomach, and cholecystokinin from the small intestine): $\text{Peptidyl}-\underset{H}{N}-CH_2\overset{O}{\overset{\|}{C}}OH \xrightarrow[\text{Vit C}]{} \text{peptidyl}-NH_2 + H\overset{O}{\overset{\|}{C}}\overset{O}{\overset{\|}{C}}OH$

to the formation of other copper complexes as a result of exchange with ligands (L) in the enzyme digest (amino acids, fatty acids, amines, etc.) or tertiary complexes of the original complex. It is important to remember that gastric digestion is enzyme catalyzed and not acid catalyzed and, since the gastric pH is not likely to be less than 3[16], some of the originally ingested complex may remain intact. It is also worth noting that pharmacologic doses of copper complexes have an antisecretory effect and thus prevent lowering of gastric pH below 6.0.

Additional complexes such as Transcuprein[5], which may be a tertiary complex of an absorbed complex, or a tertiary albumin complex may also be formed with or without ligand exchange following systemic absorption. Again,

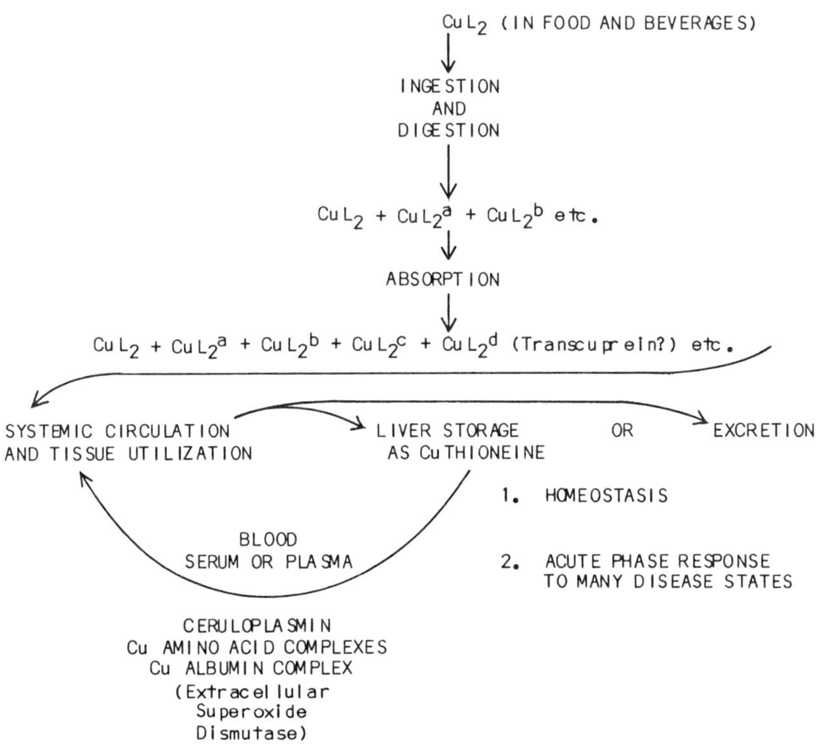

Figure 1. Distribution of copper complexes following ingestion, digestion, and absorption.

some of the original complex may remain intact. These copper complexes then undergo systemic circulation to all tissues and are: 1) utilized by tissues following ligand exchange with apoenzymes and apoproteins to form metalloenzymes and metalloproteins, 2) stored in the liver following ligand exchange with thionine to form copper-thioneine, or 3) excreted in the event tissue needs have been met and stores replenished. Since copper storage has never been reported in any normal population and copper absorption must vary in these populations, efficient homeostatic mechanisms must regulate retention and excretion of varying amounts of absorbed copper.

Also shown in Figure 1, copper-thioneine stored copper is released from the liver as complexed forms; ceruloplasmin, copper amino acid complexes, and a copper albumin complex into blood to meet normal metabolic needs. This homeostatic release of copper complexes from the liver meets normal copper-dependent physiologic requirements of body tissues which includes de novo synthesis of copper-dependent enzymes.

The essentiality of copper is now understood based upon its recognized need for activation of copper-dependent enzymes. Complexed forms of copper also facilitate absorption, tissue distribution, and tissue utilization. In the non-disease state these forms of copper account for the physiologic regulation of copper-dependent homeostatic processes. Since copper is needed for normal metabolism and prevention of disease, great care should be taken to assure that dietary intake provides required amounts of copper. Unfortunately, many or nearly all modern diets studied[17] do not supply required amounts of this and other essential metalloelements. It is then reasonable to suggest that marginal or deficient intakes or impaired absorption may lead to decreased enzyme activity and manifestations of acute disease in the short term as well as manifestations of chronic disease in the long term.

Larger than normal quantities of blood copper complexes are associated with various disease states[15] listed in Table III. The elevation usually found in early disease is generally two to three times the normal concentration during the active disease phase. With disease remission plasma copper concentration returns to normal.

The central question concerning this observation is whether the elevation is a cause of the disease or a response to the disease. Many interpret an elevated blood copper concentration as a cause of the respective disease. It is equally plausible that this elevation is a response to the respective disease. Since there is no evidence that elevation of normal blood copper-containing components cause any of these diseases, the hypothesis that elevation of plasma copper-containing components is a general physiologic response to various diseases merits serious consideration. The fact that small molecular weight copper complexes have pharmacologic activity in animal models of various disease states (Table III) is offered as evidence that the associated increase in blood copper-containing components is one of the physiologic responses which mediate remission when remission occurs. If it can be established that elevation in plasma copper containing components is a general physiologic response to disease states, then facilitating this response with copper complexes can be viewed as a physiologic approach to disease treatment and copper complexes may be useful in prevention of disease.

Table III. Disease States Associated With Elevated Concentrations of Plasma or Serum Copper Complexes and Pharmacological Effects of Copper Complexes.

The Concentration of Plasma Copper Complexes is Elevated In:	Copper Complexes Have:
Inflammation	Antiinflammatory Activity
Infection	Antimicrobial Activity
Fever	Antipyretic Activity
Epilepsy	Anticonvulsant Activity
Cancers	Anticancer Activity
Carcinogenesis	Anticarcinogenic (and Antimutagenic) Activity
Pain	Analgesic Activity

References

1. Underwood, EJ. Trace Elements in Human and Animal Nutrition, 3rd ed. Academic Press, New York, p. 57, 1977.
2. Tipton, IH; Cook MJ. Health Phys. 1963, 1, 103.
3. Iyengar, GV; Kollmer, WE; Bowen, HJM. In: The Elemental Composition of Human Tissues and Body Fluids, Springer, New York, 1978.
4. May, PM; Linder, PW; Williams DR. Experientia 1976, 32, 1492.
5. Weiss, KC; Linder, MC. Am. J. Physiol. 1985, 249, E77.
6. Frieden, E. Clin. Physiol. Biochem. 1986, 4, 11.
7. O'Dell, B.. In: Trace Elements in Human Health and Disease, Vol. I, Zinc and Copper (eds. A.S. Prasad and D. Oberleas) Academic Press, New York, 1976, p 391.
8. McCord, JM; Fridovich, I. J. Biol. Chem. 1969, 244, 6049.
9. Fridovich, I. Arch. Biochem. Biophys. 1986, 247, 1.
10. Hamilton, GA. Copper Proteins (eds. T.G. Spiro) Wiley, New York, 1981, p 193.
11. Anonymous. Nutr. Revs. 1985, 43, 117.
12. Marklund, S; Holme, E; Hellner, L. Clin. Chem. Acta 1982, 126, 41.
13. Marklund, S. Proc. Natl. Acad. Sci. 1982, 79, 7634.
14. Sorenson, JRJ; Oberley, LW; Crouch, RK; Kensler, TW. Oxygen Radicals in Chemistry and Biology (eds. W. Bors, M. Saran, and D. Tait) Walter de Gruyter and Co., Berlin 1984, p 821.
15. Sorenson, JRJ. In: Metal Ions in Biological Systems-14 (ed. H. Sigel), Marcel Dekker, New York, 1976, p 77.
16. Davenport, HC. Physiology of the Digestive Tract, 5th ed. Year Book Medical Publishers, Chicago, 1982, p 133.
17. Sorenson, JRJ; Oberley, LW; Kishore V; Leuthauser, SWC; Oberley, TD; Pezeshk, A. Inorg. Chim. Acta 1984, 91, 285.

DISCUSSION

HARRIS: To what extent can we reutilize Cu from spent Cu-containing enzymes?

SORENSON: Intuitively, reutilization is not 100% efficient and the net loss accounts for our recognized daily dietary requirement. Part of bile Cu is probably reabsorbed and the kidney may have a conservatory role but there is still a daily net loss.

SMITH: Please clarify your statements concerning re-absorption of biliary Cu from the GI tract. Is Cu reabsorbed to a large extent or is less being secreted into bile? Secondly, we don't have much urinary Cu to conserve do we? Urine contains about 15 to 40 ug/l at the most.

SORENSON: What I am suggesting is that some of the Cu in bile is an intestinal absorbable form and only that part which is not reabsorbed is excreted in feces. With regard to the kidney, what I am suggesting is that urine Cu content is very low because the kidney has a role in conserving Cu. This suggestion accounts for the low urinary Cu content and high kidney Cu-thioneine content.

RISK: We're trying to track down the significance of elevated plasma Cu during gestation and labor and I'm interested in your suggestion that there may be a multitude of Cu complexes, each with its own physiological role. We suspect that there may be some other molecule different from ceruloplasmin that's involved. I would appreciate your comments on that.

SORENSON: It's likely that specific complexes have specific biochemical effects in specific tissues. However, there may be a number of complexes with the appropriate physicochemical properties that facilitate physiologic distribution of Cu to specific tissues or cells.

RISK: Are you suggesting that there are different Cu complexes responsible for different responses such as antimicrobial activity, antiflammatory activity, etc?

SORENSON: Yes, those responses may be produced by different complexes or the same or similar complexes depending upon their physicochemical properties. Not all Cu complexes have the same pharmacological effects in models of disease.

MELETHIL: Shouldn't we be looking at other elements in conjunction with changes in Cu? Just by focusing on Cu alone are we guilty of tunnel vision?

SORENSON: Yes we should and we do. If you poll the rest of the participants in this conference you will find that they understand that Cu is only one of perhaps 12 essential metalloelements. I learned a very long time ago that I was going to have a lot of difficulty teaching that Cu-dependent processes were important in responding to disease and there was no chance that I was going to be able to teach, in the beginning, that Zn, Mn, Fe, Mg, and other essential metalloelement-dependent processes were also important in responding to disease states. Copper complexes are the theme of this conference because we can do some rather remarkable things with them even though they represent only one of many responses, in the physiologic sense, that occur to correct whatever lesion has occurred. Copper is no more important than Na, K, Zn, Mg, Mn, Fe, or the other essential metalloelements. Generally we are educated to understand something about Na, K, Fe, and Ca and that's it. This conference and published proceedings will partially correct the deficit with regard to understanding Cu metabolism in response to disease states.

LIPSKY: A central tenant of your hypothesis is that these various Cu complexes exert a very important host defense role and the reason that they are elevated in plasma is to protect the host against tumors, infection, or what have you. Now in this regard, it's interesting to look at a clinical situation. Probably one situation that has been studied most is chronic inflammatory disease, rheumatoid arthritis, which is characterized by markedly and chronically elevated levels of ceruloplasmin and Cu at the inflammatory site as well as in plasma. Empirically, it's interesting that a class of pharmacologic agents, the so called disease modifying agents, has been developed in a search for drugs which actually have the capacity to lower so called acute phase reactants including ceruloplasmin. Now, would you think that is logical based on your hypothesis or would these be, perhaps, agents that might have some negative influence on this response to the disease process?

SORENSON: Yes, these drugs, such as penicillamine, are likely to have a negative influence on the physiologic response to these diseases in the long term. They may have a beneficial effect in the short term due to mobilization. However, persistent loss is likely to exacerbate these diseases.

Data should be gathered to determine whether or not they have a positive or negative influence in this patient population. There would be a firm basis for claiming that they are useful if patients feel better, they're more active, much more functional, and they complain less of their disease. However, based upon the recognized removal of Cu and Zn associated with penicillamine therapy, it is likely that the response to disease is markedly impaired. It seems to me that it is possible to determine whether or not the patient is really benefited as a result of using these drugs. This question should be held open and data gotten with existing patient populations who are being treated with these drugs. Penicillamine actually makes inflammation worse in animal models of inflammation and it has no antiinflammatory activity. Penicillamine may also exacerbate an existing problem associated with diet. It has been recently suggested that no more than 25% of the U.S. population has an adequate daily dietary Cu intake. It is also known that dietary intakes of Zn, Ca, and Fe are less than recommended. In addition, patients who do not feel well do not eat well. These observations along with the recognized depleting effect of penicillamine may impede the required essential metalloelement-dependent responses needed to overcome arthritic disease processes.

CHIDAMBARAM: Is Cu(II) or Cu(I) absorbed duodenally? Is there a reductive process required for its absorption as there is with Fe?

SORENSON: I don't know. I do recall a recent report concerning preliminary attempts to identify Cu complexes in bile based on molecular weight but I do not recall any particular point being made concerning oxidation state or duodenal or subsequent intestinal absorption.

SMITH: The dogma is that it's Cu(II) that is absorbed based upon the observation that vitamin C and other reducing agents decrease absorption by reducing Cu(II) to Cu(I). However, there are not a lot of supporting data. We have submitted samples of intestinal tissue for electron spin resonance (ESR) determination of oxidation state but it's been hard to find an interested ESR spectroscopist. I would hope that we can discuss the valence state of Cu and also what compounds are in bile.

SORENSON: Okay, but I don't think I can add much to the discussion. The oxidation state of Cu in any biological system depends upon the redox potential of that system and those ligands that are bonded to Cu atoms. They determine the redox potential of Cu complexes. An interested electrochemist would be most helpful in this regard.

JOHNSON: What happens to Fe and Cu levels in deficient animals and humans?

SORENSON: A very large physiologic increase in plasma Cu is associated with active acute disease states. In chronic disease, levels of Cu are much lower and this may be evidence of depleted stores. Fe as well as Zn levels fall in acute and chronic disease as a physiologic response and these decreases never go below one-half of the normal level unless there is Zn or Fe deficiency. It is well known that inflammation is made much worse by feeding animals a Cu-deficient diet. Even though plasma Cu is markedly depressed in a Cu-deficient animal, plasma Cu will increase with an inflammatory insult but the increase will be much less than that observed for a similarily treated animal ingesting a diet containing adequate Cu. This small increase may not be sufficient since inflammation is always worse in these animals.

JOHNSON: Does the Cu level return to near normal?

SORENSON: No, it remains elevated for the duration of active disease. The experience with regard to inflammatory disease, rheumatoid arthritis for example, is that an individual will enter into an acute active disease state and then that will subside with a period of remission. Cu is elevated with active disease and returns to normal with remission. Then there will be another flare of the disease and that will subside with a rise and return to normal plasma Cu content. Periods between remission get shorter and shorter and finally that person has symptoms of Definate Rheumatoid Disease which is a chronic systemic ongoing disease and only slightly elevated plasma Cu levels. This slight elevation may be due to depleted Cu stores resulting in an inability to continue to physiologically respond to disease.

CROOK: Can you perhaps tell us how sex and age might modify, if at all, what you've talked about so far with regard to pre- and post-menopausal females?

SORENSON: Yes, women appear to have a slightly higher plasma Cu level and more active Cu metabolism than men. Estrogens are known to increase plasma copper levels in a physiologic regulatory fashion. It is occasionally mentioned that testosterone and androgens have a role in physiologic regulation of Cu in men but definitive data documenting as well as explaining, in a mechanistic sense, these regulations are lacking. It is my understanding that there is an age related decrease in tissue Cu levels. The elderly have problems with apetite as well as economics that may cause them to not eat as well as younger individuals. Recognizing a need to pay attention to all essential metalloelement metabolism in order to maintain normal metabolic processes may improve normal sexual function, the quality of life, and perhaps increase longevity in both pre- and post-menopausal women as well as in male populations.

FAMAEY: Chronic rheumatoid arthritics have near normal plasma Cu levels. Is there also an important effect of treatment on these patients? In our experience patients who are worse receive many more drugs than others and we have the impression that drugs used to treat these patients are influencing levels of all trace elements. We can measure effects on Zn, Cu, and Se metabolism in normal rats, without any kind of disease, when they are treated with drugs used to treat rheumatic patients.

SORENSON: Yes, antiarthritic drugs do affect Zn and Cu metabolism. Later we will show that Cu complexes of antiarthritic drugs are more effective antiinflammatory agents and they are less toxic.

FAMAEY: In our chronic patients we still have a lower

level of Zn but we have a normal level of Cu so it is our impression that Zn is more sensitive in rheumatoid arthritis than Cu. When we treat our normal rats with steroids, Zn is decreased and Cu is less affected. So we think chronic disease patients given steroids, or non-steroidal antiarthritic drugs for long periods, should be studied to determine what effect these drugs have on trace element metabolism. It's probably also important to know what influence these drugs have on trace element levels in normal animals.

SORENSON: The low level of Cu in your chronic patients may be evidence of Cu inadequacy in these patients. I agree that studying essential metalloelement metabolism is an important immediate need with regard to providing these patients with much more beneficial therapy.

REGULATION OF CERULOPLASMIN SYNTHESIS BY RETINOIC ACID AND INTERLEUKIN-1

Robert J. Cousins and Ellen F. Barber

Food Science & Human Nutrition Department

University of Florida, Gainesville, Florida 32611

Research conducted over a period of fifty years, both in a clinical setting with human subjects, as well as in experiments with animals, has shown that plasma ceruloplasmin levels increase during periods of stress, inflammatory disease and chronic infection. However, there is little known about the factors which influence ceruloplasmin synthesis. The physiological role for the increase in circulating levels of this acute phase protein is equally obscure. The plethora of hormonal changes that precede and/or accompany these conditions suggest a variety of signals contribute to the phenomenon.

Ceruloplasmin accounts for 80-90% of the total plasma copper in most species. It is a glycoprotein of about 132-kDa containing 6 Cu atoms per molecule and 7-8% carbohydrate. Three major polypeptide fragments of 67, 50 and 19-kDa may represent specific domains.[1-3]

The protein has a variety of potential functions. Most of these would encompass a role in host defense. The oxidation of Fe^{2+} for iron mobilization from tissues as a transferrin complex is an example.[3] It has been suggested that the hypoferremic state caused by inflammation or infection produces a compensatory increase in plasma ceruloplasmin to aid in iron mobilization from the reticuloendothelial system which allows plasma iron to return to normal levels during the recovery period.[4] Copper transport to tissues and subsequent uptake by specific receptors has been proposed by Harris and associates as a physiological role for ceruloplasmin.[5] Clearly during the cascade of cellular events related to inflammation, there may be a need to deliver extra copper to specific tissues or cells. Ceruloplasmin may also function as an antioxidant. Goldstein et al. have shown that ceruloplasmin can serve as a superoxide radical scavenger.[6] Since O_2^-, H_2O_2 and •OH are produced in pathogenic processes such as inflammation, increased plasma ceruloplasmin levels could have a beneficial effect by acting as a free radical scavenger.[7] An anti-inflammatory action of injected ceruloplasmin has been demonstrated by Denko and correlated to elevated serum ceruloplasmin levels.[8]

Our strategy for studying ceruloplasmin gene expression is based principally on three features: 1) the protein is glycosylated, therefore, it may respond to changes in vitamin A status and administration of vitamin A and related analogues, 2) vitamin A analogues may regulate the expression of specific genes and 3) interleukin-1 seems to induce other acute phase proteins and may induce ceruloplasmin via regulated gene expression.

The induction of ceruloplasmin by vitamin A analogues was examined in a series of experiments with rats of both deficient and adequate vitamin A status. Vitamin A appears to have a role in reducing susceptibility to infection.[9] Since ceruloplasmin may function during the infectious process to regulate iron metabolism,[4] a homeostatic link to vitamin A status is plausible.

Vitamin A deficiency may depress ceruloplasmin synthesis and/or secretion. The synthesis of a number of glycoproteins, e.g. α_{2u}-globulin[10] and α_1-macroglobulin[11] are affected by vitamin A status. This response could relate to the finding that retinoids, viz. retinoic acid, regulate genes for alkaline phosphatase in prostatic adenocarinoma cells,[12] plasminogen activator in teratocarcinoma stem cells,[13] transglutaminase in CHO cells,[14] and transferrin in sertoli cells.[15] In addition, retinol and retinoic acid administration produce dramatic changes in total mRNA abundance in some tissues including liver.[16] Alternatively, this deficiency has been shown to reduce glycosylation[17] which would reduce the secretion of glycoproteins from liver (e.g. glycosylation of

α_{2_u}-globulin).[10] Since ceruloplasmin is a glycoprotein, its secretion could be reduced by low vitamin A status, thus lowering the circulating levels of the protein. The intracellular ceruloplasmin content of rat liver was reduced by vitamin A, although not significantly, in one study where plasma levels, which are more important, were not measured.[18]

We have approached this problem in two ways. The first was to examine the effect of dietary vitamin A depletion and repletion on the synthesis and serum levels of ceruloplasmin. As shown in Table 1, p-phenylenediamine (pPD) oxidase activity following repletion with 13-cis-retinoic acid for 3 days returned the activity to normal limits. In addition, ceruloplasmin synthesis was measured by immunoprecipitation of ^{3}H-ceruloplasmin in serum and expressed as a % of total ^{3}H-precipitable with TCA. Incorporation of ^{3}H-leucine into ceruloplasmin was increased, when vitamin A stores were repleted with retinoic acid. No change in the total serum protein concentration was detected. These studies were repeated many times and repletion with retinoic acid or retinyl acetate always increased ceruloplasmin levels. This suggests that vitamin A may alter expression of the ceruloplasmin gene. Other factors may also contribute to increasing serum ceruloplasmin levels. Combinations of epinephrine + dexamethasone and glucagon + dexamethasone injections also produced increases comparable to those produced by retinoic acid.[19] Multiple doses of retinoic acid (i.p.) given daily for 4 days elevated serum ceruloplasmin levels to 4-fold that found in untreated normal rats[19] or 2-fold when given (p.o.) to rats with the early inflammatory phase of adjuvant arthritis.[20] Serum copper concentration fairly closely followed ceruloplasmin oxidase activity in most studies, but the relationship deviated as the latter increased in response to multiple doses of 13-cis-retinoic acid.

An adequate supply of dietary copper exerts a permissive effect on the stimulation of ceruloplasmin oxidase activity by retinoic acid. Feeding rats a purified diet (AIN-76a formulation) containing <1 mg Cu/kg resulted in a reduction of the serum Cu levels to 0.03 μg/ml. Ceruloplasmin pPD oxidase activity was virtually undetectable (ca. 0.12 $\mu mole \cdot min^{-1} \cdot L^{-1}$). Retinoic acid administration did not increase activity, in the copper

Table 1
Effect of Vitamin A Deficiency and Repletion on Synthesis and Serum Levels of Ceruloplasmin

Retinoic Acid	Days of Repletion 1	3	5
	^{3}H-Leucine Incorporation[a] (Ceruloplasmin c.p.m./Total Protein c.p.m.)		
-	3.0 ± 0.2	3.1 ± 0.5	3.6 ± 0.4
+	3.6 ± 1.0	4.5 ± 0.5*	5.2 ± 1.0*
	pPD Oxidase Activity[b]		
-	22.1 ± 4.1	24.0 ± 3.1	24.3 ± 4.7
+	24.0 ± 3.8	32.2 ± 5.1*	36.2 ± 6.2*

Rats were fed a vitamin A deficient diet for 9 weeks. Repletion was with 100 µg 13-cis-retinoic acid (p.o.) given daily for up to 5 days. Pulse labeling was with 100 µCi ^{3}H-leucine given to each rat 2 hr before they were killed. Aliquots of serum were mixed with rabbit anti-rat ceruloplasmin serum and immunoprecipitated as described previously.[21] The immunoprecipitate was passed through a sucrose gradient to remove unincorporated ^{3}H-leucine. An aliquot of serum was mixed with 10% TCA to precipitate total ^{3}H-labeled serum proteins. Results are expressed as % ^{3}H incorporated into ceruloplasmin vs total proteins. b) Ceruloplasmin p-phenylenediamine oxidase activity measured as described previously.[20] Each value is the mean ± SD (n=3-5). *Significantly different from nontreated control at P<.05.

deficient rats, however. If copper (125 μg Cu/100 g as Cu acetate) was administered 8 hr after retinoic acid, then oxidase activity was similar to that in rats fed adequate amounts of copper. Activity was increased to 28 and 37 (μmole pPD oxidized • min^{-1} • L^{-1}) in copper deficient rats given copper and copper plus retinoic acid, respectively.[19]

As discussed above, there is considerable evidence that vitamin A, retinoic acid and related metabolites influence expression of certain genes. Therefore we used actinomycin D, an inhibitor of transcription, to examine its effect on induction of ceruloplasmin by retinoic acid. We observed that actinomycin D when given alone did not influence pPD oxidase activity, but the stimulation caused by retinoic acid was totally inhibited. In copper deficient rats that were supplemented with copper, actinomycin D also completely inhibited the induction of pPD oxidase activity by retinoic acid. These observations suggest that retinoic acid stimulates ceruloplasmin related pPD oxidase activity in plasma via differential regulation of ceruloplasmin gene expression. This could be a direct effect of this vitamin A metabolite or an indirect effect related to differentiation of interleukin-1-producing cells, e.g. macrophages.[22] Although it is less likely, inhibition of secretion of the holoprotein from the liver by actinomycin D could also explain the results.

Induction of ceruloplasmin pPD oxidase activity by leukocytic endogenous mediator (LEM) was initially demonstrated in 1972 by Pekarek et al.[23] A considerable body of literature in the past decade has demonstrated the significance of LEM-like molecules in initiating a variety of host defense functions. The initial protein derived from leukocytes which is responsible for the induction of hepatic acute phase protein synthesis and other events in the cascade of events of the acute phase response has been designated as interleukin-1 (IL-1). Detailed reviews on interleukin-1 have described the multiplicity of effects involved.[24,25] The interleukin-1 preparation used in these studies was isolated from rabbit peritoneal macrophages followed the procedure described by Keenan et al.[26] Briefly, it involved peritoneal infusion of 0.2% glycogen into rabbits. Twenty hr later the peritoneal exudate was removed, filtered and centrifuged at 900xg. The macrophages were resuspended and cultured for 24 hr in phosphate-buffered saline. After centrifugation at 3000xg, the interleukin-1-containing supernatant was removed and stored at -20°C. The crude interleukin-1 preparation was bioassayed based upon a depression of the serum zinc concentration as described previously.[27] A 40-50% reduction in this concentration was produced within 7 hr of administration with 1.0 ml (1.1 mg protein) of the crude preparation. No change in serum copper was observed within this time frame. High performance liquid chromatography on a Superose-12 column demonstrated the interleukin-1 preparation contained one major protein peak and a number of minor peaks. The major peak was reduced upon heating the preparation for 30 min at 90°C. The latter destroyed its biological activity. We have used this preparation in lieu of more highly refined material because the basics of interleukin-1 terminology and biochemistry are in an evolutionary phase. For example, it is not clear whether interleukin-1 or a related protein, hepatocyte stimulating factor, actually induces acute phase protein synthesis.[28]

Two doses of the interleukin-1 preparation given 8 hr apart were found to produce a greater stimulation of ceruloplasmin oxidase activity (204 ± 10% above controls) compared to one dose (142 ± 20% above controls) by 24 hr after the first dose. Insulin given alone or with dexamethasone in addition to interleukin-1 produced a slight, but not significant ($P<.05$) increase in activity compared to the two interleukin-1 doses alone.

Dietary copper had a significant influence on the response to interleukin-1. For these experiments, purified diets based on the AIN-76a formulation containing 0.6 mg Cu/kg (deficient) or 6.0 mg Cu/kg (sufficient) were fed to rats for 3-4 weeks. The deficient diet reduced serum copper levels to ca. 4% of normal levels. In the sufficient group, interleukin-1 injections increased serum pPD oxidase activity significantly ($P<.05$) by 24 hrs later, while copper injections (150 μg as copper acetate) given at the same time as interleukin-1 did not. These pPD oxidase activities were ($\mu mole \cdot min^{-1} \cdot L^{-1}$), 30± 3 (controls), 38 ± 5 (+Cu), 59 ± 7 (+IL-1) and 47 ± 4 (IL-1 + Cu). In contrast, in the copper deficient rats serum pPD oxidase activity was only stimulated with supplemental copper: 1.2 ± 1.0 (controls), 28 ± 5 (+ Cu), 0.07 ± 0.01 (+IL-1) and 33 ± 5 (IL-1 + Cu). Clearly, an adequate supply of copper is needed for interleukin-1 to induce ceruloplasmin oxidase activity.

The evidence in support of the inducibility of ceruloplasmin was sufficient for us to examine the effects of interleukin-1 on the rate of ceruloplasmin synthesis using pulse labeling techniques. The 2 hr pulse with ^{3}H-leucine (100 µCi/rat) was chosen because this gives maximum incorporation into secreted proteins.[29] For these experiments, rats were fed either copper deficient or copper sufficient diets as described above. Interleukin-1 was given at 0 and 8 hr of the 24 hr time course. As shown in Table 2, the rate of ceruloplasmin synthesis in copper sufficient rats was maximally stimulated at 12 hr after the first interleukin-1 dose (0 hr). Thereafter, it decreased and by 24 hr it was the same as the rate observed in the controls. In rats fed the copper deficient diet, the maximal rate of ceruloplasmin synthesis was observed 18 hr after the first interleukin-1 injection. These results suggest that the induction of ceruloplasmin protein is independent of dietary copper status. Supplementation of copper in the form of an injection as described above did not further increase the rate of synthesis, but did increase pPD oxidase activity. The antibody used in these immunoprecipitation studies appears to precipitate apoceruloplasmin as well as the holoprotein.[21] This explains why interleukin-1 stimulates the rate of synthesis even in copper deficient rats that exhibit nearly undetectable serum pPD oxidase activity. Actinomycin D administration blocked the induction of ceruloplasmin synthesis (Table 3).

The pulse labeling methods used in the present experiments suggest that interleukin-1 may regulate expression of the ceruloplasmin gene. Actinomycin D blocked the stimulation of ceruloplasmin-related pPD oxidase activity of plasma induced by retinoic acid and interleukin-1 and the induction of synthesis produced by interleukin-1. Both agents have been demonstrated to alter gene expression. The antioxidant properties of ceruloplasmin in plasma may serve as a defense mechanism against the generation of oxygen radicals by macrophages and neutrophils required for destruction of invading pathogenic microorganisms. An increase in ceruloplasmin would act as a scavenger of these free radicals. It has been argued that superoxide dismutase is a much more efficient free radical scavenger than ceruloplasmin. However, ceruloplasmin is an extracellular protein and thus could act in plasma whereas superoxide dismutase is

Table 2
Relative Rate of Ceruloplasmin Synthesis in Copper Deficient and Copper Sufficient Rats Injected with Interleukin-1

Hours After Interleukin-1	^{3}H-Leucine Incorporation % Ceruloplasmin c.p.m./ Total Protein c.p.m.	
	Copper Sufficient	Copper Deficient
2	2.9 ± 0.3	3.8 ± 1.1
6	5.6 ± 0.3*	5.8 ± 1.9*
12	7.0 ± 0.2*	6.2 ± 1.7*
18	5.0 ± 0.2*	9.0 ± 0.4*
24	4.1 ± 0.4	5.8 ± 0.8

Rats were fed copper deficient (0.6 mg Cu/kg) or copper sufficient (6.0 mg Cu/kg) diet for 3-4 weeks. Interleukin-1 preparation was given at 0 and 8 hr of the time course. Pulse labeling was as described in Table 1. Each value is the mean ± SD (n=4). *Significantly different from respective 2 hr rate at P<.05.

Table 3
Effect of Actinomycin D on Rate of Ceruloplasmin Synthesis

Treatment	^{3}H-Leucine Incorporation % Ceruloplasmin c.p.m./ Total Protein c.p.m.
Control (PBS)	5.1 ± 0.1
Actinomycin D	2.6 ± 0.2*
Interleukin-1	6.3 ± 0.4*
Interleukin-1 + Actinomycin D	2.3 ± 0.4*

Rats were fed a copper sufficient diet and treated with interleukin-1 as described in Table 2. Actinomycin D (0.8 mg/kg) was given 2 hr before the first interleukin-1 injection. Pulse labeling was as described in Table 1. Each value is the mean ± SD (n=3-4). Significantly different from control rate at p<.05.

intracellular. Plasma copper levels increase during exercise and pregnancy.[1] This probably is directly related to increased plasma interleukin-1 levels associated with these physiological states.[25] Furthermore, the inducible nature of ceruloplasmin has considerable potential related to development of therapeutic strategies to optimize it's important role(s) in host defense.

ACKNOWLEDGEMENTS

This work was supported by NIH Grant AM 31127 and Boston Family Endowment Funds of the Institute of Food and Agricultural Sciences (University of Florida). The authors wish to thank Mavis R. Swerdel for her help with the vitamin A deficiency studies and Dawn Mendoza for typing the manuscript. The 13-cis-retinoic acid used in these experiments was generously provided by Hoffman LaRoche.

REFERENCES

1. Cousins, R.J. Physiol. Rev. 1985, 65, 238.

2. Takahashi, N.T: Ortel, T.L: Putnam, F.W. Proc. Natl. Acad. Sci. U.S.A. 1984, 81, 390.

3. Frieden, E. In: Biological Roles of Copper; Elsevier/North Holland; New York, 1980; p, 93.

4. Beaumier, D.L: Caldwell, M.A: Holbein, B.L. Infect. Immun. 1984, 46, 489.

5. Stevens, M.D: DiSilvestro, R.A: Harris, E.D. Biochemistry, 1984, 23, 261.

6. Goldstein, I.M: Kaplan, H.B: Edelson, H.S: Weissman, G. J. Biol. Chem. 1979, 254, 4040.

7. Aust, S.A: Morehouse, L.A: Thomas, C.E. J. Free Radical Biol. Med. 1985, 1, 3.

8. Denko, C.W. Agents Actions 1979, 9, 333.

9. McLaren, D.S. In: Present Knowledge in Nutrition, 5th Ed.; Nutrition Foundation Inc; Washington, D.C, 1984; p, 202.

10. Haars, L.J: Pitot, H.C. J. Biol. Chem. 1979, 19, 9401.

11. Kiorpes, T.C: Anderson, R.S: Wolf, G. J. Nutr. 1981, 111, 2059.

12. Reese, D.H: Fiorention, G.J: Claflin, A.J: Malinin, T.I: Politano, V.A. Biochem. Biophys. Res. Commu. 1981, 102, 315.

13. Strickland, S: Mahdavi, V. Cell. 1978, 15, 393.

14. Strickland, S: Smith, K.K: Marotti, K.R. Cell. 1980, 21, 347.

15. Skinner, M.K: Griswold, M.D. Biol. Reprod. 1982, 27, 211.

16. Omori, M: Chytil, F. J. Biol. Chem. 1982, 257, 14370.

17. Wolf, G. Physiol. Rev. 1984, 64, 873.

18. Rask, L: Valtersson, C: Anundi, H: Kvist, S: Eriksson, U: Dallner, G: Peterson, P.A. Exper. Cell. Res. 1983, 143, 91.

19. Barber, E.F: Cousins, R.J. 1986, unpublished results.

20. Cousins, R.J: Swerdel, M.R. Proc. Soc. Exp. Biol. Med. 1985, 179, 168.

21. Weiner, A.L: Cousins, R.J. Biochem. J. 1983, 212, 297.

22. Moore, W.T: Murtaugh, M.P: Davies, P.J.A. J. Biol. Chem. 1984, 259, 12794.

23. Pekarek, R.S: Powanda, M.C: Wannemacher, R.W. Proc. Soc. Exp. Biol. Med. 1972, 141, 1029.

24. Dinarello, C.A. N. Eng. J. Med. 1984, 311, 1413.

25. Dinarello, C.A. Rev. Infect. Dis. 1984, 6, 51.

26. Keenan, R.A: Moldauen, L.L: Sakamoto, A: Blackburn, G.L: Bistrian, B.R. J. Surg. Res. 1982, 33, 151.

27. DiSilvestro, R.A: Cousins, R.J. Am. J. Physiol, 1984, 247, E436.

28. Woloski, B.M.R.N.J: Fuller, G.M. Proc. Natl. Acad. Sci. U.S.A. 1985, 82, 1443.

29. Schnerbert, G: Howlett, G: Nagashima, M: Millership, A: Martin, H: Urban, J: Kother, L. J. Biol. Chem. 1982, 257, 10271.

DISCUSSION

SMITH: In regard to your findings that acute phase reactant interleukin-1 (IL-1) affects ceruloplasmin (Cp) synthesis which is not affected by Cu status, would you give your opinion of our use of oxidase activity to assess Cu status rather than Cu concentration? We find that oxidase activity falls markedly to nearly zero in Cu-deficient animals, within a week, while there is some Cp present.

COUSINS: The oxidase activity, which is Cu-dependent, is very important for assessing status but it might be a nice adjunct to determine total Cu level. Our results indicate that something like IL-1 will induce synthesis of the apoprotein which then needs Cu to produce an active protein.

HARRIS: Your work has cleared up some important problems in terms of Cu deficiency. I've always questioned whether Cu is an oblicatory agent for Cp synthesis and you've shown that is not necessarily the case and that the synthesis of the apoprotein is regulated at a hormonal level. Normally, vitamin A is not thought of as an inducer but you've shown us that it has this potential. Will vitamin A form a stable complex with Cu? Do you find immunochemically reactive Cp induced in a deficient animal with vitamin A?

COUSINS: We have done post-labeling experiments with retinoic acid induction in a repletion experiment and yes, there is an increase in the immunoreactive material. One reason we were interested in these retinoic acid experiments was that there is no good model protein for those in the vitamin A field to use. There are some but it's not like the vitamin D story where Ca Binding Protein has been very effectively used for almost 20 years to explain important aspects of the effect of vitamin D and the calcitriol-induced gene expressions in that system. We thought the Cp system might help in the vitamin A area. It's a non-stored form of the vitamin and we have to give therapeutic levels, levels needed to produce a reduction in adjuvant arthritis, to get this response.

(Editor's note: Vitamin A is an irritant and this may have a role in accounting for these observations.)

FIELDS: You've shown in one of your previous slides that

Cp has a lot of functions. Did you attempt to measure functional Cp activity when Cp oxidase activity was not measurable?

COUSINS: No we haven't. Oxidase activity is determined as a correlary to our other measurements of system regulation.

SORENSON: Were you suggesting that Cu wasn't needed for Cp synthesis?

COUSINS: Yes.

SORENSON: Couldn't Cu be derived from another tissue?

COUSINS: Perhaps it could be but oxidase activity is fairly low so that would argue against it. Apparently Cu has to be there in sufficient amounts only at the right time.

SORENSON: With regard to the scheme showing the appearance of Cu and the decline of Zn in serum with stress, infection, or trauma, I noticed that you had the decline in plasma Zn lasting for less than a day. Is there some evidence for that?

COUSINS: That's a hypothetical slide developed with data from a number of studies both with human subjects as well as with rats. It fits in very nicely to the scheme that we are most familiar with in rats. Twenty-four hours after different acute insults the depression in serum Zn has returned to normal levels, with 6, 7, or 8 hours being the times of maximum depression.

SORENSON: What is the insult to the animal?

COUSINS: It can be any number of things; serious stress, administration of a hormonal agent, IL-1, etc. That depression, in rats at least, always seems to have the same approximate chronology. It's lowest at about 7 hours. That is why we use the depression of serum Zn at 7 hours after IL-1 administration as the method of bioassay.

SORENSON: In polyarthritis we find a persistent depression of plasma Zn, a chronic event. Is polyarthritis different from those other treatments?

COUSINS: Yes, it is probably a continuous stimulus. If administration of IL-1 were continued, the response would last longer than it does with a single administration. We've recently completed some experiments which will be published in American Journal of Physiology using dibutryl cAMP to induce metallothioneine. It does clearly act as an inducer at the transcriptional level. When it is given repeatedly it produces a somewhat longer response, and we plan to use that model for prolonged inductions because it's one that is fairly amenable to that type of chronic treatment without many other side effects. The depression in plasma Zn will persist if the stimulus is there.

SORENSON: Is there a recognized truly accurate measure for Cp which is precise and specific, without interferences?

COUSINS: It depends on whether or not you are talking about the apoprotein or the physiologically active holoprotein.

DISILVESTRO: PPD oxidase is an accurate assay if there is not a lot of hemolysis. The Fe in red samples is going to give ridiculously high readings. Samples that have very little hemolysis give pretty good specific assay data, particularly if azide sensitivity is determined. If azide wipes it all out then it is pretty specific.

COUSINS: When we followed PPD oxidase activity over a period of a number of days in the retinoic acid study, the serum Cu level plateaued while oxidase activity tended to continue to increase. That could be a reflection of some problem with PPD oxidase, but under appropriate conditions it's not a bad indicator.

LIPSKY: On your diagram you used an arrow to suggest that IL-1 was activating a promoter for the Cp gene. Do you have any evidence for that?

COUSINS: I was suggesting that the effect was a cellular membrane mediated effect. There's no other agent known to signal this response.

LIPSKY: It's known that in chronic polyarthritis there are persistent high levels of IL-1 at the inflammatory site and some researchers find it in plasma. Are you aware of any mechanism by which target tissues like the liver might be desensitized to IL-1 or would you suggest that it's a question of depleting the animal of Cu and ending up with apoceruloplasmin in plasma?

<u>COUSINS</u>: The only comments I can make are of a general nature. IL-1 doesn't induce just Cp. It induces all the acute phase proteins, α_2-macroglobulin and so on. We have to assume that there is some commonality in terms of what the actual genetic inducer is at the level of transciption. For that reason I suggest that Cp has evolved by the same mechanism during evolution as syntheses of other acute phase proteins. I am suggesting that that is what we will find in the end. It's one of a class of proteins synthesized in response to the same signal.

ROLE OF CERULOPLASMIN IN COPPER TRANSPORT

Edward D. Harris

Department of Biochemistry and Biophysics
and the Texas Agricultural Experiment Station,
Texas A&M University

ABSTRACT

Copper in a biological setting is primarily an essential metal cofactor for numerous biocatalysts. Transport systems that specifically handle copper are known to exist but have not been fully characterized. We are left, therefore, with an incomplete understanding as to how copper becomes incorporated into cells and into the cellular enzymes. In this paper we discuss the evidence that ceruloplasmin, the major copper-binding globulin, is a transporter of copper to extrahepatic tissues. Much of the evidence rests with the discovery of receptors for ceruloplasmin in cell membranes of numerous tissues and the impact this may have on a transport mechanism.

INTRODUCTION

Ironically, thirty-eight years after ceruloplasmin's discovery and isolation from pig serum[1], there is still basic disagreement as the ways this protein serves vertebrate systems. No more strongly contested is its assumed role in copper transport. Here, the issue is whether an active serum enzyme (132,000 dal) has the dual function of transporting copper to extrahepatic cells. Opposers state that the 6 or 7 copper atoms in ceruloplasmin are so tightly bound that they preclude free and rapid exchange with cellular components. Contrarily, the advocates of the transport hypothesis point to the close correlation of ceruloplasmin with the activity of copper-dependent enzymes in cells, which in some cases, seems to suggest a preference for ceruloplasmin as the copper source. With these basic points still unresolved, we sought more direct evidence for a transport function. Our goal was to determine whether those tissues that seem to require ceruloplasmin for copper-enzyme biosynthesis may in fact contain membrane receptors that recognize and bind the plasma protein. Such specific binding, it could be argued, could signify the presence of portals that regulate copper uptake via a ceruloplasmin-mediated mechanism. These studies led to the discovery of high affinity binding sites for ceruloplasmin in membranes fragments from heart and aortic tissue[2]. Here, we review additional properties of ceruloplasmin receptors in mammalian systems.

The Ceruloplasmin Hypothesis of Copper Transport

Copper in blood is in part considered copper in transit, destined for uptake by cells and incorporation into cellular enzymes. Ceruloplasmin represents the bulk of serum copper, estimated in humans to be 90-95% of the total copper in serum[3]. This value, obtained by measuring 235 normal human subjects, has been questioned[4] but generally found for humans and many animal species[5]. With 90-95% of the copper bound to ceruloplasmin, less than 10% is associated with other components of the serum, mainly albumin.

Early speculation has given rise to supporting evidence, much of it inference, that ceruloplasmin provides copper to cellular enzymes. Marceau and Aspin used partially purified ceruloplasmin to show radiolabeled copper incorporation into numerous animal organs[6] and organ-specific enzymes[7]. Copper derived from ceruloplasmin engaged the mitochondrial complex and appeared to label the cytochrome oxidase of that organelle. Importantly, albumin-bound copper was not incorporated into the enzyme. Hsieh and Frieden showed that ceruloplasmin injected intravenously into copper-deficient rats surpassed free copper, histidine-copper, and copper-albumin in restoring oxidase activity to cytochrome oxidase in numerous tissues, mainly heart[8]. It has since been established that rat organs and affixing tumors take up ceruloplasmin copper preferentially over other copper sources[9].

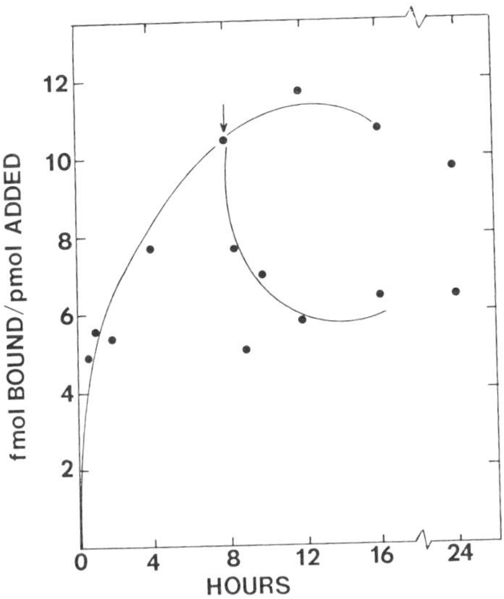

Fig. 1 Time-course and Reversibility of ^{125}I-Ceruloplasmin binding to Heart Membranes. The arrow indicates the point where a 1000 pmole excess of unlabeled ceruloplasmin was added.

Further, ceruloplasmin levels in serum have been linked with cytochrome oxidase activity in human leukocytes[10], superoxide dismutase in liver[7], and lysyl oxidase in aorta[11]. All these studies indirectly support a copper transport function for ceruloplasmin.

Suffice to say that the strongest evidence to date linking ceruloplasmin with copper transport is the discovery of receptors for the native protein in membranes of cells and tissues. This laboratory confirmed the presence of receptors in heart and aorta membranes, established their specificity, and gave general insight into the binding reaction[2]. Receptors for ceruloplasmin have since been reported in rat hepatic endothelial cells[12], human RBCs[13], granulocytes and lymphocytes[14]. Evidence has been obtained that gold-labeled ceruloplasmin binds to endothelial cells (at 37°C) and becomes internalized via a system of coated pits and vesicles[15]. Gold-labeled albumin from bovine serum does not bind and apparently cannot be internalized.

TABLE 1

Binding of ^{125}I-Ceruloplasmin and ^{125}I-Transferrin to Membrane Fragments

Membrane	1000 pmol excess	fmoles bound / pmoles added: Total	Specific
Transferrin (human)			
Pig Heart	-	8.2 ± 0.8	
Pig Heart	+	8.6 ± 0.7	0
Rabbit Reticulocyte	-	10.8 ± 0.8	
Rabbit Reticulocyte	+	7.9 ± 1.1	2.9
Ceruloplasmin (porcine)			
Pig Heart	-	12.4 ± 0.3	
Pig Heart	+	6.3 ± 0.5	6.1
Pig Heart (KCl-treated)	-	10.8 ± 1.4	
Pig Heart (KCl-treated)	+	5.3 ± 0.3	5.4
Rabbit Reticulocyte	-	6.9 ± 0.4	
Rabbit Reticulocyte	+	7.5 ± 3.0	0

14.7 pmol of transferrin and 14.4 pmol of ceruloplasmin were used. Binding assays were run for 16 hr, 4°C with 150 µg of membrane protein. Specific binding represents the difference between total binding and binding observed in the presence of excess unlabeled ceruloplasmin (+).

Properties of the Porcine Receptor for Ceruloplasmin

Binding of ceruloplasmin to the pig heart membranes is shown in Fig. 1. Typically seen is a twelve hour period to reach binding equilibrium. A rapid displacement of the bound ^{125}I-ceruloplasmin from the membrane occurs when a 1000 pmol excess of unlabeled ceruloplasmin is added at or before equilibrium, showing clearly the reversibility of the reaction.

The specificity of the binding is seen in Table 1. In the study, ^{125}I-transferrin was prepared to measure

TABLE 2

Specificity of Binding of Ceruloplasmin to Pig Heart Membranes

Cold Protein Added	fmoles bound / pmoles added Total	Specific
None	10.17 ± 2.02	-
Ceruloplasmin (porcine)	6.59 ± 0.94	3.58
Transferrin (human)	11.06 ± 0.40	0
Ovalbumin	12.12 ± 3.32	0
Albumin (bovine)	15.04 ± 0.31	0
Cytochrome C (equine)	16.69 ± 6.63	0
None + $CaCl_2$ (0.5mM)	10.01 ± 0.11	-
Ceruloplasmin (porcine) +$CaCl_2$ (0.5 mM)	5.18 ± 1.43	4.80

comparative binding. Note that ^{125}I-transferrin showed little or no binding to pig heart membranes under conditions that the ^{125}I-ceruloplasmin did. The labeled transferin, however, did bind effectively to rabbit reticulocyte membranes whereas pig ceruloplasmin showed no tendency to bind to these membranes.

Some of the pig heart membrane preparations were treated with 0.3 M KCl (to dissociate fibrous proteins) prior to the assay. This treatment did not appear to deter binding, suggesting the receptor is firmly attached to the lipid bilayer. Various proteins were also tested for ability to displace the ^{125}I-ceruloplasmin from the receptor (Table 2). Only pig ceruloplasmin was successful. Both albumin and cytochrome C from non-porcine sources appeared to increase the total binding. Also seen in the table are the effects of $CaCl_2$ on the binding. At 0.5 mM, Ca^{2+} neither inhibited nor promoted binding suggesting that divalent cations, calcium in particular, are not needed for maximum binding of ceruloplasmin to the receptor.

It is noteworthy that human transferrin showed no affinity to bind to the pig heart membranes. This could be due to the failure of the heart cells to express transferrin receptors or a failure of pig receptor to recognize a human protein. Then, too, ceruloplasmin showed no affinity to bind to reticulocyte membranes whereas transferrin did. Apparently membrane receptors for ceruloplasmin are not present in young blood cells, but instead are expressed when the reticulocytes mature to erythrocytes. Also note-worthy is that KCl treatment of the heart membranes did not deter binding. Ceruloplasmin has been shown to form a strong complex with cytochrome C in vitro[16]. There is evidence for a type of cytochrome C in plasma membranes[17] suggesting that this protein could be the one engaging ceruloplasmin. Like its counterpart in mitochondria, the plasma membrane cytochrome C is dislodged by high salt treatment[18]. Since the salt-treated pig heart membranes were capable of binding ceruloplasmin, a salt-dissociable protein cannot be the membrane binding component. Failure of equine cytochrome C to prevent ceruloplasmin from binding is further evidence that the receptor is not this protein.

Although the role of ceruloplasmin as a supplier of intracellular copper remains uncertain, the data reported here would tend to support the hypothesis that heart membranes can engage ceruloplasmin in a manner that is both specific and reversible for this protein. Whether this means there are receptors for ceruloplasmin must remain conjectural until such substances can be isolated and fully characterized. That work is currently underway in the laboratory. In this regard Barnes and Frieden succeeded in characterizing a 60,000 dalton component from detergent-treated human erythrocyte ghosts that appeared to represent a receptor for ceruloplasmin[13]. It is not clear whether the 60,000 dalton component is the intact receptor or a subunit of a larger oligomeric complex.

In conclusion, the multicomplex properties of ceruloplasmin are underscored by its versatility as a copper transporter, a catalyst for oxidation-reduction reactions, a serum antioxidant. The focus of this paper is on the transport function of ceruloplasmin. Membranes from avian, mammalian, and human tissue sources bind ceruloplasmin. This would seem to imply the existance of receptors for the native protein in many vertebrate cell membranes. That ceruloplasmin and its receptor constitute components of an extraheptic copper transport-delivery system is by no means certain, but a viable hypothesis to guide future work in copper transport.

REFERENCES

1. Holmberg, C; Laurell, C. Acta Chem. Scand. 1948, 2, 550.

2. Stevens, M; DiSilvestro, R: Harris, E. Biochemistry 1984, 23, 261.

3. Cartwright, G; Wintrobe, M. Am. J. Clin. Nutr. 1964, 14, 224.

4. Wirth, P; Linder, M. J. Nat. Can. Inst. 1985, 75, 1985.

5. Bingley, J; Dick, A. Clin. Chim. Acta 1969, 25, 480.

6. Marceau, N; Aspin, N. Am. J. Physiol. 1972, 222, 106.

7. Marceau, N; Aspin, N. Biochim. Biophys. Acta 1973, 293, 351.

8. Hsieh, H; Frieden, E. Biochem. Biophys. Res. Commun. 1975, 67, 1326.

9. Campbell, C; Brown, R: Linder, M. Biochim. Biophys. Acta 1981, 678, 27.

10. Shokeir, M; Schreffler, D. Proc. Nat. Acad. Sci. (USA) 1969, 62, 867.

11. Harris, E; DiSilvestro, R. Proc. Soc. Exp. Biol. Med. 1981, 166, 528.

12. Kataoka, M; Tavassoli, M. Exp. Cell Res. 1984, 155, 232.

13. Barnes, G; Frieden, E. Biochem. Biophys. Res. Commun. 1984, 125, 157.

14. Kataoka, M; Tavassoli, M. Exp. Hematol. 1985, 13, 806.

15. Kataoka, M; Tavassoli, M. J. Ultrastruct. Res. 1985, 90, 194.

16. Derkachev, E; Shaposhnikov, A. Biokhimiya 1971, 37, 174.

17. Sottocasa, G; Kuylenstierna, B; Ernstern, L; Bergstrand, A. J. Cell Biol. 1967, 32, 415.

18. Jacobs, E; Sanadi, D. J. Biol. Chem. 1960, 235, 531.

ACKNOWLEDGEMENTS

Funding for the work was provided in part through NIH Grant AM35920 and project No. H-6621 of the Texas Agricultural Experiment Station. I thank Mr. Robert Fadal, Mr. Robert Emmick and Mr. Timothy Schmidt for technical assistance and Mr. Michael Stevens for valuable consultations.

DISCUSSION

<u>COUSINS</u>: Have you considered using synovial cells obtained during acute inflammation? They might be receptive to ceruloplasmin (Cp).

<u>HARRIS</u>: Not yet. I'm concerned with the fact that most of our evidence is based on membrane fragments which only provide binding information. We have looked at lymphocytes and found an indication of specific binding. Tavassoli has mentioned the Cp receptor on lymphocytes so lymphocytes might be one cell type to study. Of course, transferrin has been studied extensively with lymphocytes and leukemic cells. We also intend to look at endothelial cells which are becoming more and more prominent in this regard.

<u>RISK</u>: Is the pH of the lysosome sufficient for rapid removal of Cu or is degradation of the protein required?

<u>HARRIS</u>: If it gets to the lysosome that's going to be an important question. The transferrin story suggests a situation where, apparently, it never gets to the lysosome. This protein enters the endosome and the pH gradient there causes the dissociation of Fe from transferrin and transferrin is not degraded. It's simply recycled back to the cell membrane. There could be a similar mechanism for transfer of Cu from Cp.

<u>RISK</u>: Do we know the pH of the endosome?

<u>HARRIS</u>: It is acidic, perhaps about two units or so below neutrality.

<u>SORENSON</u>: The dilemma that you posed at the outset is interesting. This protein has enzyme activity and it's also suggested to be a transport form. The supporting evidence for Cp being a transport protein is not all that strong. Experiments done to suggest transport measured incorporation of radio-labeled Cu from Cp into cytochrome C of Cu-deficient animals. In the light of Linder's suggestion that Transcuprein has transport function and the possibility that albumin and amino acids may have transport function, is Cp really a necessary transport form of Cu?

HARRIS: Before we can answer that question we must first of all decide on criteria that we would accept as firm hard evidence for a transport protein. If a putative Cu transport agent actually has the facility to transport Cu through a cell membrane and it is eventually incorporated into a biologically functional component, that is both transport and incorporation. Using these criteria it may be correct to suggest that Cp has not as yet been demonstrated to be a Cu transport protein. Using these same criteria may call into question results for other proteins suggested to have transport function.

LIPSKY: Does this receptor bind the apoprotein?

HARRIS: I don't know. If sialic acid is removed from Cp, binding is not impeded.

ELLIOTT (Graham): Is it possible to repeat your experiments with Cu depleted Cp?

HARRIS: Yes but we have not done that yet.

ELLIOTT (Graham): If you just incubate a tissue with Cu, do you get normal uptake of Cu?

HARRIS: That's the problem. Yes it does.

ELLIOTT (Graham): There's no necessity for a carrier mechanism?

HARRIS: That's correct. In an in vitro study it is possible to flood the tissue with a non-biological concentration of Cu and show an increase in tissue content. The question we must address is: How is this regulated in vivo?

KEEN: Do you have evidence that there is regulation of receptors? Is there an increase in receptors in Cu-deficient animals or are there fewer receptors in tissues from newborn rats before Cp emerges?

HARRIS: That's what I was eluding to when I mentioned regulation. We plan to study Cu status of the animal as a factor in regulating the emergence of this receptor.

<u>KRISHNASWAMY</u>: Have you looked to see if your Cp binds to synthetic phospholipid vesicles? Is it just lipid binding as opposed to receptor binding?

<u>HARRIS</u>: Phospholipid binding would be non-specific as opposed to specific high-affinity binding.

<u>FIELDS</u>: Do hormones affect binding?

<u>HARRIS</u>: We have not, as yet, studied regulation.

<u>CROOK</u>: What was the status of the one human heart you used? Was it just a young individual who died a traumatic death?

<u>HARRIS</u>: Yes. It was from a 6 week old child that died as a result of some sort of trauma.

<u>CROOK</u>: Do you know if it was a genetic defect?

<u>HARRIS</u>: No. We wanted to get tissue from a Menkes' child but that has not been possible as yet.

<u>CROOK</u>: Were your pigs minature swine?

<u>HARRIS</u>: No, they were mature slaughter house animals at market weight.

<u>RISK</u>: With regard to the heart tissue, was the difference in binding due to differences in composition of the binding receptor or could it be due to degenerative changes?

<u>HARRIS</u>: We analyzed both tissues as soon as we obtained them from their respective sources and we took all necessary precautions to avoid degenerative changes.

<u>RISK</u>: Were differences then due to species?

<u>HARRIS</u>: Yes, the implication is that we measured species and tissue specific binding of Cp.

Copper and the Regulation of Superoxide Dismutase Activity in Aortic Tissue

Charles T. Dameron and Edward D. Harris

Department of Biochemistry and Biophysics
and the Texas Agricultural Experiment Station,
Texas A & M University

ABSTRACT

Understanding how cuproenzymes are regulated in connective tissues is vital to understanding the role copper and copper complexes play in ameliorating chronic inflammations. The rapidly growing chick aorta is an excellent model system to test the effects of copper availability on cuproenzyme production because chicks are readily made deficient, copper metabolism in aorta has been partially characterized through metabolic studies with lysyl oxidase, and chick aortas can be cultured in serum free media. The intent of this presentation is to describe some of the copper regulatory mechanisms being investigated for the aortic copper-zinc superoxide dismutase in chick.

INTRODUCTION

Decreases in tissue copper are frequently observed in chronic inflammations, atherosclerosis, and connective tissue defects but it has not been determined whether low copper is a cause or effect of the disease[1-5]. Copper deficiencies, however, have been directly linked to connective tissue impairments. Gross symptoms of a copper deficiency such as spontaneous fractures, thinned cortices, wide epiphyses in bones, and reduced tensil strength and a tendency to spontaneous rupture in blood vessels are readily observed in afflicted animals. Connective tissue defects in copper deficiencies are associated with increased levels of procollagen and proelastin[6-17]. Of particular interest to our laboratory has been the effects of copper deficiency on the developing chick aorta because in aorta some of the underlying biochemical causes of the lesions are well understood.

Our interest in the young chicks as a model system stems from their susceptibility to copper deficiency and proneness, when deficient, to aortic rupture. Chicks reared on copper deficient diets (< 1 ppm copper) show leg weakness, poor feathering, and decreased weight gain and eventually succumb to aortic rupture[14-17]. The increases in procollagen and proelastin in the connective tissues is firmly linked to copper through the extracellular cuproenzyme lysyl oxidase (LO)[18-23]. Lysyl oxidase oxidizes the ε-amino group of select lysine residues in collagen and elastin to form aldehydes which then react non-enzymatically to form crosslinks through aldol condensations and Schiff base formation[24]. Copper-fed chicks maintain a high level of LO while those fed a deficient diet have a dramatic reduction in activity[18,25] with the activity decreasing to nondetectable levels in 10 days[25]. Equally dramatic is the return of lysyl oxidase activity in deficient chicks that have been given a single i.p. injection of $CuSO_4$ (1 mg Cu/kg). The return, following a 2 hr lag period, is biphasic with the initial 4 fold rise peaking 6 hrs after receiving the copper and a larger sustained increase in activity occurring between 10 and 24 hrs[25]. The regulation of LO by copper prompted us to look more closely at copper metabolism in aortic tissue in the hopes of learning the mechanism of activation.

COPPER INCORPORATION INTO AORTIC ENZYMES

Pulse labeling and chromatography studies showed that in copper-deficient aorta newly absorbed copper seems to be sequestered preferentially by a 30,000 dalton protein[26]. In one key study copper sulfate enriched with ^{67}Cu was injected i.p. into a group of 11-day-old copper deficient cockerels. Aortas were excised 24 hours later, cleaned of fascia and blood, and homogenized. The homogenate was centrifuged at 100,000 x g for 60 min at 4°. A 5 ml aliquot of the supernant was chromatographed on a Sephadex

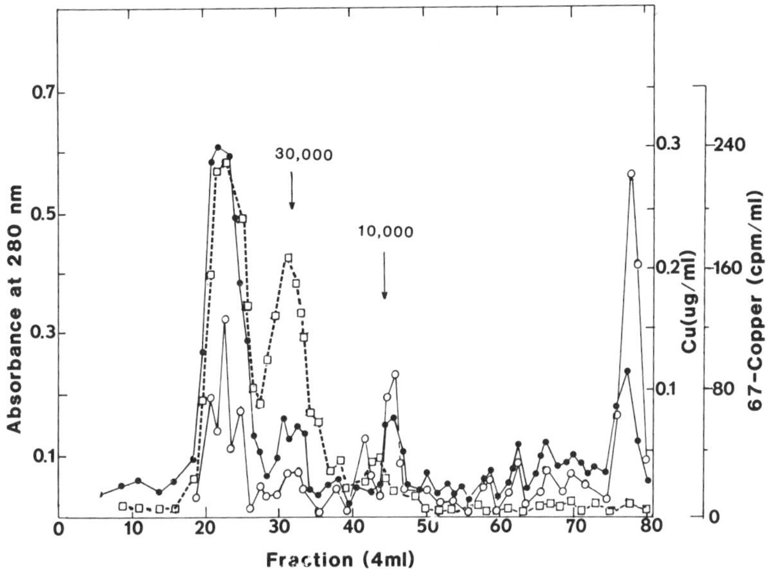

Figure 1. ^{67}Cu incorporation into copper deficient chick aorta. Soluble components of chick aortas were resolved on Sephadex G-75. ○, Copper; ●, A280; □, ^{67}Cu.

G-75 column. The resulting chromatogram is shown in Figure 1[26]. Radiolabel was incorporated nonuniformly into 3 macromolecular peaks and into a region designated "free" copper (fraction 55 and above). The void volume peak received the highest amount of ^{67}Cu. However, a component with an estimated molecular weight of 30,000 daltons showed the highest specific activity (microcuries per micromole copper), suggesting this protein(s) was among the first in aorta to bind copper. In subsequent chromatographic studies we determined that the copper binding protein(s) coeluted with cyanide sensitive superoxide dismutase activity and zinc suggesting that the newly absorbed copper was incorporated into copper-zinc superoxide dismutase (CuZn-SOD). Furthermore, the deficient animals had less of the aortic copper binding component and less superoxide dismutase activity than copper fed chicks. These studies support the suggestion by Marceau and Aspin[27,28] that CuZn-SOD is a major recipient for copper in tissues.

REGULATION OF CUZN-SOD ACTIVITY

The possibility that CuZn-SOD, like LO, could be regulated by copper prompted us to measure the aortic CuZn-SOD activity in copper depletion and repletion experiments. Initially a longitudinal study, similar to the LO study described earlier, was performed in which chicks were reared from hatching on copper sufficient and deficient diets. Figure 2 shows a plot of the total superoxide dismutase activity (cytosolic and mitochondrial) versus days on the diet. An unusual superoxide dismutase activity pattern was observed in the developing aortas of chicks. Newly hatched chicks had very low total aortic SOD activity, but the level increased 3-fold in 2 days, before falling to adult levels. The increase and fall to steady-state levels required 10 to 12 days. A copper-deficient diet (<1mg Cu/kg body weight) did not alter the initial rise but did promote a more rapid decline in and ultimately, within 10-12 days, the loss of detectable total superoxide dismutase activity.

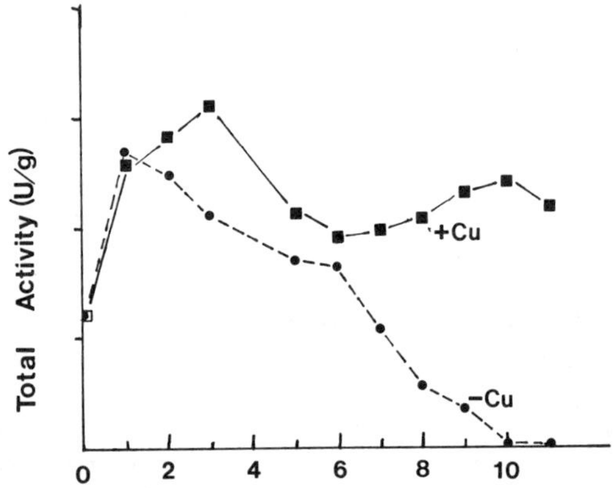

Figure 2. Aortic SOD levels in the developing chick. Total aortic SOD activity was meassured in chicks fed copper-sufficient,■, and copper-deficient,●, diets from hatching for 15 days.

Aortic CuZn-SOD activity returned when deficient chicks were given a copper sufficient diet. The activity increased 3 fold in 4 days after a 1 day lag period. A more controlled and faster method of raising the CuZn-SOD levels, as discussed earlier for LO, is to inject trace amounts of $CuSO_4$ intraperitoneally. Figure 3a, a plot of CuZn-SOD activity versus hours after a single i.p. 1 mg Cu/kg injection, shows that the CuZn-SOD activity in copper deficient chicks was raised in 6 hr to the same level achieved at 3 days on the diet. The CuZn-SOD activation, as observed with LO, showed a 2 hr inital lag period before the more substantial, but steady increase was seen.

In an attempt to further control and characterize the activation response more defined in vitro experiments were performed. In these experiments aortas from 12 day old deficient chicks were removed aseptically and placed directly into petri dishes containing oxygenated, HEPES buffered, serum-free Waymouth media maintained at 37°. After the aortas were cleaned they were placed in screw-capped culture tubes containing copper or saline supplemented-medium (1 ml per aorta) and aerated with 100% oxygen for 10 sec before the tubes were sealed. The tubes were incubated for varying times in a shakerbath maintained at 37°. Results of these studies are shown in Figure 3b.

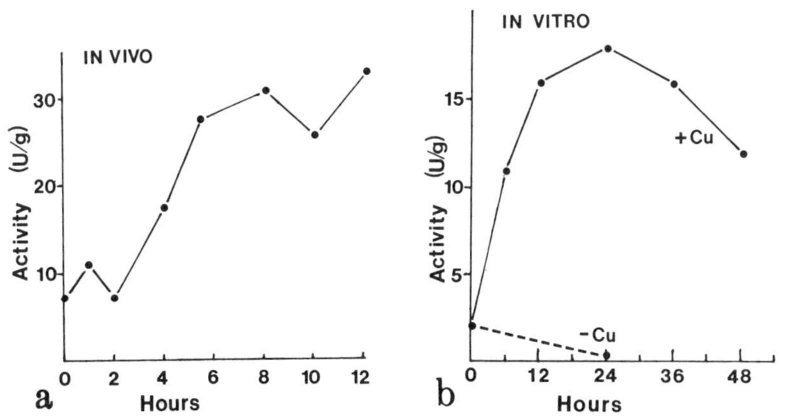

Figure 3. Activation of chick aortic CuZn-SOD. a. Aortic CuZn-SOD activity in deficient chicks following a single i.p. copper injection, 1 mg Cu/kg body weight. b. Activation of CuZn-SOD in copper-deficient aortic tissue cultured in Waymouth medium.

The activation achieved a maximum in approximately 24 hrs and did not show a lag period. On the other hand, the activation of LO, as explored by Rayton and Harris had a 5 hr lag period and required 16-20 hrs to reach its maximum[29]. Maximum stimulation of CuZn-SOD activity was achieved with only 0.5 micrograms Cu/ml. In contrast LO requires about 4 micrograms Cu/ml to achieve optimum levels of activation[29]. Both $CuSO_4$ and $Cu(NO_3)_2$ effectively activated the aortic CuZn-SOD at 37° but not at 0°. Interestingly, Cu-EDTA was not an effective activator of CuZn-SOD under the same conditions. Zinc and manganese did not enhance the CuZn-SOD activation or activate the mitochondrial superoxide dismutase, even though, as shown in Figure 2, both forms of the enzyme were depressed by the deficiency. Finally, activation of CuZn-SOD _in vitro_ occurred in the presence of the antiobiotics cycloheximide, actinomycin D and cordycepin, whereas LO[29] activation was inhibited by cycloheximide but not actinomycin D.

RESULTS AND DISCUSSION

Newly hatched chicks appear capable of synthesizing active aortic CuZn-SOD through the first 2-3 days of growth. When dietary copper is restricted soon after hatching the synthesis of CuZn-SOD is inhibited but may continue with the production of a catalytically inactive form. Restoration of CuZn-SOD activity in aortic tissue of deficient chicks is readily accomplished by adding copper to the chicks diets or by i.p. injections of copper sulfate. Activation of CuZn-SOD in isolated intact aortic tissue can be accomplished with less than 0.5 micrograms/ml copper in the medium, or about one-tenth the amount of copper needed for LO activation. The absence of a lag period in the CuZn-SOD activation _in vitro_, as compared to LO, suggests that the enzyme activaton is rapid and direct and may not require _de novo_ protein synthesis. This and other observations seem to support the existence of a preformed pool of apoenzyme. Additionally, the insensitivity of the activation to the antimetabolites cordycepin, actinomycin D, and cycloheximide further suggests that copper is reacting directly with preformed apoenzyme in the cytosol.

The characterization of CuZn-SOD in aorta is important because it provides another tool with which to study aortic copper metabolism. The intracellular response to copper depletion and repletion have heretofore been observed in tissues by measuring the cytochrome c oxidase and lysyl oxidase acitivity. Cytochrome C oxidase, a vitally important copper-dependent mitochondrial enzyme, has a long half-life and hence is probably not as responsive to rapid changes in copper availability. On the other hand, CuZn-SOD responds rapidly to dietary copper fluctuations. Lysyl oxidase also responds readily to sudden changes in copper availibilty but is not easily quanitated and is found only in connective tissues. Copper-zinc superoxide dismutase is quanitated easily and is found almost ubiquitously in animal cells. Because SOD shows sensitivity to copper status, it can be used as a target enzyme to measure copper uptake and incorporation into protein. For instance, the putative role of ceruloplasmin in the activation of copper enzymes can be further investigated. The circumstantial links between ceruloplasmin and the activation of cytochrome C oxidase[30] and LO[31] can be more critically evaluated when the activation of SOD is also studied, leading to a better understanding of how copper is distributed intracellularly.

Understanding the mechanisms used to incorporate copper into tissue enzymes is critical to understanding how copper and agents altering copper metabolism affect the nutrition and pathophysiology of animal systems. An important step in establishing the mechanisms is to identify the method or methods used to transport copper through the serum to the tissues and into cells. Because the aortic CuZn-SOD activity is dependent on the cellular availability of copper, the measurements of the activity can be used to detect changes in intracellular copper levels and test the efficacy of prospective copper transporting complexes to deliver copper to the cell.

REFERENCES

1. Schroeder, H.A. Clin. N. Amer. 1974, 58, 381-396.

2. Wester, P.O. Atherosclerosis 1971, 13, 395-412.

3. Netsky, M.G.; Harrison, W.W.; Brown, M.; Benson, C. Amer. J. Clin. Pathol. 1969, 51, 358-365.

4. Khandekar, J.D.; Sepaha, G.C.; Mukerji, D.P. New Engl. J. Med. 1970, 283, 319.

5. Owen, C.A., Jr. In: Physiological Aspects of Copper; Noyes Publications; Park Ridge NJ, 1982; pp 56-65.

6. Coulson, W.F.; Carns, W.F. Amer. J. Pathol. 1963, 43, 945-954.

7. Coulson, W.F.; Weissman, N.; Carns, W.F. Lab. Invest. 1965, 14, 303-309.

8. Coulson, W.F.; Carns, W.F. Amer. J. Pathol. 1963, 50, 861-868.

9. Chou, W.S.; Savage, J.E.; O'Dell, B.L. Proc. Soc. Exp. Biol. Med. 1968, 128, 948-954.

10. Weissman, N.; Shields, G.S.; Carns, W.F. J. Biol. Chem. 1963, 238, 3115-3118.

11. Weissman, N.; Reay, D. T.; Coulson, W.F.; Carns, W.F. Lab. Invest. 1965, 14, 372--380.

12. Waisman, J.; Cancilla, P.A.; Coulson, W.F.; Weissman, N.; Carns, W.F. Lab. Invest. 1969, 21, 548-554.

13. Smith, D. W.; Carns, W.F. Biochem. Biophys. Res. Comm. 1968, 31, 309-315.

14. Starcher, B.C.; Hill, C. H.; Matrone, G. J. Nutrition 1964, 82, 318-322.

15. Miller, E. J.; Martin, G. R.; Mecca, C. E.; Piez, K. A. J. Biol. Chem. 1965, 240, 3623-3627.

16. Miller, E. J.; Fullmer, H. M. J. Exp. Med. 1966, 123, 1097-1108.

17. O'Dell, B.L.; Bird, D.W.; Savage, J.E. J. Nutrition 1966, 88, 9-14.

18. Kim, C. S.; Hill, C. H. Biochem. Biophys. Res. Comm. 1966, 24, 395-450.

19. Coulson, W.F.; Lenker, A. Biochim. Biophys. Acta 1968, 158, 117-123.

20. Hill, C. H.; Starcher, B. C.; Kim, C. Fed. Proc. 1967, 26, 129-133.

21. Hill, C. H.; Nutr. Rev. 1969, 27, 99-100.

22. Owen, C.A., Jr. In: Biochemical Aspects of Copper; Noyes Publications; Park Ridge NJ, 1983; pp 11-12.

24. Siegel, R.C.; Martin, G.R. J. Biol. Chem. 1970, 245, 1653-1658.

25. Harris, E.D. Proc. Nat. Acad. Sci. (USA) 1976, 73, 371-374.

26. Balthrop, J.E.; Dameron, C.T.; Harris, E.D. Biochem. J. 1982, 204, 541-548.

27. Marceau, N.; Aspin, N. Biochim. Biophys. Acta 1973, 293, pp 338-350.

28. Marceau, N.; Aspin, N. Biochim. Biophys. Acta 1973, 293, pp 351-358.

29. Rayton, J. K.; Harris, E. D. J. Biol. Chem. 1979, 254, pp 621-626.

30. Frieden, E.; Hsieh, S.H. Biochem. Biophys. Res. Com. 1975, 67, pp 1326-1331

31. DiSilvestro, R.A.; Harris, E.D. J. Nutr. 1981, 111, pp 1964-1968.

DISCUSSION

LOVEN: We've done some studies with Cu-deficient rats and found an increase in Mn SOD activity in the kidney and I believe in the brain and no change in liver and skeletal muscle. We speculated that possibly there was a metabolic regulatory effect. How active are mitochondria in the aorta and what percentage of the SOD is Mn SOD?

DAMERON: About 40% of the activity in a normal control animal is the Mn form. The Mn form decreases with age in a control chick and it's very difficult to follow for that reason.

LOVEN: Did cytochrome oxidase (Cyt Ox) activity decrease as well?

DAMERON: I haven't assayed Cyt Ox activity in the aorta yet because these chick aortas are small, about 20 mg apiece. We have some preliminary evidence that it decreases in heart in Cu deficiency.

LOVEN: We also found that immunoreactive Cu-Zn SOD appears to decrease in Cu deficiency. The exact significance of this really isn't clear and I don't know how this ties in with your evidence for an apoenzyme.

DAMERON: I have not as yet been able to develop an antibody system sensitive enough to accurately quantitate the small control levels of this enzyme let alone apoenzyme levels. Initially I expected an increase in Mn SOD in a compensatory mechanism so I was quite surprised to see both of them.

DISILVESTRO: If you look at RBCs, which I've always been taught don't synthesize proteins, in animal refeeding studies, the return of SOD occurs well before the synthesis of all new cells. We also have preliminary evidence that RBC SOD is reactivated by incubation at 37°C in medium containing Cu salts or ceruloplasmin. However, red cells are likely to be totally different compared to aorta cells.

LOVEN: Measurement of enzymatic SOD activity requires care in avoiding artifactual interferences associated with free Cu and sulfhydryl groups. Were control experiments done to evaluate their presence?

DAMERON: I checked all assays with cyanide. On selected groups of aorta I used ethanol precipatation. I'm working with such small sample sizes that diluting it further is prohibited. I haven't used the pyrogallol or xanthine oxidase assays, particularily in the in vitro systems, because they did not give reliable results.

GREENAWAY: Did you try adding vitamin B_6 as well as Cu to lysyl oxidase (LO) preparations to determine whether or not it affected the lag period?

DAMERON: I have not tried that. Prof. Harris may have.

HARRIS: In trying to stimulate activity we did add pyridoxal and, as I recall, we didn't get any more activity than Cu alone produced. This concern regarding pyridoxal as a cofactor for LO is a very real one. It appears that this is no longer accepted dogma. I wonder if Prof. Greenaway was aware of that when he posed his question.

GREENAWAY: I am. We find that there is no pyridoxal phosphate in calf aorta LO. It has, however, been reported that pyridoxal phosphate does affect the reformation of LO in chicken aorta. I'm not sure that this necessarily means that that is the cofactor. It could still affect reincorporation of Cu.

HARRIS: I would agree with that.

SORENSON: I'm not aware of any data that unequivocly eliminates pyridoxal as a cofactor for LO or even the soluable amine oxidases, although this is a real warm question. Experimental work in this area needs to address the question as to whether or not pyridoxal or pyridoxyl alcohol either facilitates or has no effect on reestablishment of LO activity once Cu has been removed from it in the Cu-deficient animal. The reactivation by Cu may occur when the pyridoxal-containing apoprotein is recombined with Cu to fully activate the enzyme. Do you suppose that is reasonable?

HARRIS: It's reasonable but not likely. These animals are receiving an enriched semi-purified diet that contains all of the B_6 that they will ever need and possibly more. I think what you are suggesting implies two rate limiting components in the synthesis of LO, one being Cu incorporation and the other being pyridoxal incorporation. Three days after putting a Cu-deficient animal back onto a control diet there was a complete return of LO activity. This observation would suggest that pyridoxal is not a limiting factor in restoring enzyme activity.

SORENSON: Suppose the apoenzyme includes, in addition to the protein, pyridoxal. When LO activity is reduced by just providing a Cu-deficient diet, the enzyme has everything it needs to be activated when Cu is made available. Then reestablishment of LO activity can occur with the availability of Cu and additional pyridoxal is not required.

HARRIS: I think that is a fairly logical argument and I don't know how to get around it. If pyridoxal is actually part of the Cu binding site on the protein, which is not known with certainty, then your argument has more weight. We do not know anything about where Cu is inserted into this protein and by what mechanism, or what the donor ligands are, so we are on soft ground.

FIELDS: We have analyzed vitamin B_6 in plasma from our Cu-deficient animals and although these animals were supplemented with vitamin B_6 the level of vitamin B_6 in plasma is tremendously decreased.

HERNANDEZ: Is there a form of Cu-Zn SOD in normal animals that is inactive and can this form be activated by giving extra Cu?

DAMERON: I haven't tried that. We feed high amounts of Cu. I'm not sure how much we could increase it.

LOVEN: A group in Poland has reported induction of Cu-Zn SOD by giving high dietary Cu.

KEEN: If a control diet contains 8 to 12 ppm Cu, increasing it to a 1000 ppm, a toxic level, produces no change in Cu-Zn SOD activity. We were surprised by this result. We thought we would get to a certain point where we would start generating some free radical damage, perhaps even see an increase in Cu-Zn SOD, but under normal circumstances that does not occur.

I thought a different question was being asked. What would happen if the same study was done with animals at their peak Cu-Zn SOD activity and then they were given a Cu-deficient diet on day 10 or 12?

DAMERON: I have not done that experiment.

SORENSON: As I understand the published literature, following some insult, irritation, or stress, the amount of Cu-Zn SOD increases in the affected tissue as a physiologically mediated response, provided the organism has adequate stores of Cu.

Biochemical Aspects of Copper Complexes

INFLUENCE OF DIET AND INFLAMMATION ON SERUM COPPER ENZYMES

ROBERT A. DISILVESTRO

DEPARTMENT OF FOODS AND NUTRITION
PURDUE UNIVERSITY
WEST LAFAYETTE, IN 47907

Copper is believed to function primarily as a catalyst in copper metalloenzymes. A number of these enzymes perform functions which in principle should protect against tissue injury caused by stress such as inflammation. Three types of observations support the notion that copper enzymes work against stress induced tissue damage. First, dietary copper restriction reduces activity levels of several copper metalloenzymes while increasing susceptibility to injury (1). For instance, copper deficient rats show especially high sensitivity to CCl_4 hepatoxicity (2) and adjuvant arthritis stimulated footpaw swelling (3). A second set of observations implying a protective role for copper enzymes is the linking of many stress states with elevated serum levels of the copper protein ceruloplasmin (4). These elevations have been hypothesized to represent an attempt to minimize stress-induced tissue damage. Consistant with this hypothesis is the report that adjuvant arthritic inflammation is reduced by retinoic acid, an agent which enhances the inflammation induced rise in serum ceruloplasmin (5). Ceruloplasmin has been proposed to possess a number of enzymatic activities as well as to transport copper to nonhepatic tissue metalloenzymes (4). A third line of evidence for copper protective actions is the finding of beneficial pharmacological effects associated with certain low molecular weight copper complexes (6). These effects could result, at least in part, through activation of copper metalloenzymes (6).

At present, the full range of changes in copper enzyme levels during stress has not been defined. Furthermore, the influence of diet in modifying the response of copper enzymes to stress also requires further study. This article focuses primarily on inflammation and serum enzymes. Inflammation in experimental animals represents one of commonly used stress models for studying changes in copper metalloenzymes. Copper enzyme measurements in serum hold special importance in studying stress and copper metabolism since serum would be the most easily obtained sample for human studies. At least five copper containing enzymes can be measured in serum: ceruloplasmin, lysyl oxidase, extracellular (EC) superoxide dismutase (SOD), Cu-Zn SOD and dopamine-β-hydroxylase. Thus far, ceruloplasmin, a liver secreted serum protein (4), has been the best studied of these enzymes.

Dietary makeup has been shown to influence the rise in serum ceruloplasmin concentrations during inflammation. Marginal and severe restrictions of dietary copper in growing rats reduce, though to differing degrees, the rise in ceruloplasmin caused by adjuvant arthritis (3) and turpentine produced inflammation (Fig. 1). As mentioned previously, oral dosing with retinoic acid enhances the rise in serum ceruloplasmin during adjuvant arthritis (5). The converse situation, vitamin A deficiency, has not been studied in inflamed rats. However, vitamin A deficiency does reduce ceruloplasmin levels to some extent in noninflamed rats in a manner reversed by vitamin A repletion (7). Ascorbic acid injections raise ceruloplasmin levels in nonstressed chicks (8) and guinea pigs (9). Orally given ascorbic acid had a similar effect in a patient suffering from Menke's Syndrome, a genetic impairment of copper utilization (10). On the other hand, moderate ascorbic acid deficiency does not restrict the elevation in ceruloplasmin levels during turpentine induced inflammation (9).

Lysyl oxidase (11) and extracellular SOD (12) are found primarily in extracellular tissue matrices but small amounts could accumulate in serum. Lysyl oxidase initiates the crosslinking of collagen and elastin, a critical step in connective tissue repair (13). EC SOD is

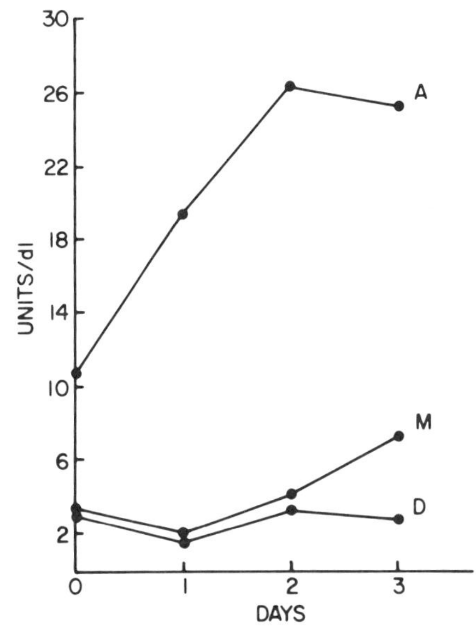

FIG. 1. Influence of diet and inflammation on rat serum ceruloplasmin levels. Rats were fed 6 ppm copper (Adequate-A), 1.5 ppm (Marginal-M) or 0.2 ppm (Deficient-D) for 6 weeks and then injected with turpentine (Day 0). Cp units are defined as # A_{550} in the p-phenylendeiamine oxidase assay of Rice (14).

proposed to dismutate superoxide radicals which could be potentially harmful to extracellular tissue (12). Lysyl oxidase activity is increased in the plasma by local inflammation of rat liver by CCl_4 and of human skin by burns (11, 15). Plasma lysyl oxidase is depressed in noninflamed pigs by copper deficiency (16). The combined effects of dietary variations and inflammation on serum lysyl oxidase contents has not been studied. Information gathering on serum lysyl oxidase has been hampered by the cumbersome and highly variable nature of the assay.

The influence of diet and inflammation on serum EC SOD activity levels has been difficult to evaluate. The low SOD activity levels found in serum do not allow much sample dilution. In contrast, assays on tissue Cu-Zn SOD activities use large sample dilution which eliminates the effects of many agents which can interfere with SOD enzyme assays. In addition, ceruloplasmin and sometimes Cu-Zn SOD can contribute along with EC SOD to overall serum SOD activity. Cu-Zn SOD is generally present in serum only at very low levels, but concentrations can rise drastically following kidney malfunction or red blood cell lysis (12, 17). Efforts are currently underway to determine the relative contributions of each copper protein to overall serum SOD activity under various circumstances. Recent studies found total rat serum SOD activity to be depressed by intake of a copper deficient diet and to a lesser extent by marginal intake (TABLE 1). Surprisingly, turpentine produced inflammation (third day after injection) raised SOD activity in rats from all 3 dietary groups to nearly the same value, one a little above that for noninflamed copper adequate rats. The explanation for this observation remains unknown at this time. In all these SOD studies, enzyme activity was determined using a recently published assay based on superoxide production by xanthine oxidase and colorimetric detection of the oxygen radical with sulfanilic acid and N-1-napthylethylene diamine (18). These serum studies will be repeated using other, more established SOD assays.

TABLE 1. Effects of Copper Intake on Serum SOD Activities

Diet	SOD Activity units[a]/dl $\pm$ SD
Adequate (6.0 ppm Cu)	1100 $\pm$ 63
Marginal (1.5 ppm Cu)	392 $\pm$ 123[b]
Deficient (0.2 ppm Cu)	90 $\pm$ 70[b]

Rats initially weighing 50-70g were fed diets for 6 weeks.

a. A unit is defined as the enzyme activity causing 50% inhibition in the assay used (18).

b. p<.001 (Student t test) compared to adequate.

Efforts will also be made to eliminate or take into account any interferences to the assay from various serum components.

Dopamine-β-hydroxylase (DBH) can be measured in serum though it functions primarily in the adrenal gland and in sympathetic nerve terminals in norepinephrine synthesis (19) . Sympathetic nerves release DBH into extracellular fluids during catecholamine secretion (19). The effects of inflammation on serum DBH have not been investigated but serum DBH activity levels are elevated by a wide spectrum of other stresses including exercise and childbirth (19). Recent unpublished studies in our laboratory found a highly variable increase in plasma DBH in lactating women, 1 month postpartum. Values ranged from close to normal up to a 4 fold increase over controls matched for age and sex. The influence of diet on serum DBH has not been examined though some studies have been carried out on the tissue enzymes. Rat adrenals and mouse brain actually show higher than normal DBH activity during copper deficiency (20, 21). These elevated readings could result from increased synthesis of DBH apoprotein _in vivo_ combined with copper activation _in vitro_ (21, 22). Copper ion is generally added to the assay mixture. Even when omitted the small amount of copper present in the assay ingredients could be sufficient for DBH activation (21). Reactivation may not be such a problem in assaying serum which contains many copper chelators such as albumin. This possibility requires investigation.

In summary, serum copper enzyme measurements hold great potential for studying the response of copper enzyme levels to the combined influence of diet and stress such as inflammation. However, information in this area is still largely lacking, partly due to technical difficulties.

REFERENCES

1. Mason, K.E. J. Nutr. 1979, 109, 1979.

2. Lawrence, R.A.; Jenkinson, S.G. Fed. Proc. 1983, 42, 818.

3. Kishore, V.; Latman, N.; Roberts, D.W.; Barnett, J.B.; Sorenson, J.R.J. Agents Actions 1984, 14, 274.

4. Cousins, R.J. Physiol. Rev. 1985, 65, 238.

5. Cousins, R.J.; Swerdel, M.R. Proc. Soc. Exp. Biol. Med. 1985, 179, 168.

6. Sorenson, J.R.J. Comprehensive Therapy 1985, 11, 49.

7. Cousins, R.J. This volume.

8. DiSilvestro, R.A.; Harris, E.D. J. Nutr. 1981, 111, 1964.

9. DiSilvestro, R.A. Nutrition Res. 1986, in press.

10. Ueki, Y.; Narazaki, O.; Toshio, T. Brain Develop. 1985, 7, 519.

11. Siegel, R.C.; Chen, K.H.; Greenspan, J.S.; Aguiar, J.M. Proc. Nat. Acad. Sci. 1978, 75, 2945.

12. Marklund, S.; Holme, E.; Hellner, L. Clin. Chim. Acta 1982, 126, 41.

13. Siegel, R.C. Int. Rev. Connect. Tiss. 1979, 8, 73.

14. Rice, E.W., Anal.Biochem. 1962, 3, 452.

15. Saleem, Y.; Sharada, K.; Muralidhar, K.; Habibullah, C.M. IRCS Med. Sci. 1985, 13, 62.

16. Ono, K.; Steele, N.; Richards, M.; Darcey, S.; Fields, M.; Schoifield, D.; Smith, J.; Reiser, S. Fed. Proc. 1986, 45, 357.

17. Bettger, W.J.; Fish, T.J.; O'Dell, B.L. Proc. Soc. Exp. Biol. Med. 1978, 158, 279.

18. Oyanagui, Y. Anal. Biochem. 1984, 142, 290.

19. Rush, R.A.; Geffen, L.B. Crit. Rev. Clin. Lab. Sci. 1980, 12, 241.

20. Hesketh, J.E. Gen. Pharmacy 1981, 12, 445.

21. Prohaska, J.R.; Smith, T.L. J. Nutr. 1982, 112, 1706.

22. Reis, D.J.; Joh, R.H.; Ross, R.A. J. Pharmacol. Exp. Ther. 1975, 193, 775.

DISCUSSION

CHIDAMBARAM: Have you observed any lysis when RBCs are incubated with "free" Cu?

DISILVESTRO: I haven't quanitatively measured lysis. I used reasonable levels of Cu, under 1 ug/ml. Those who want to study lysis of red cells will use much larger amounts of Cu. CHIDAMBARAM: I remember this study from Prof. Frieden's lab where they indeed found ceruloplasmin inhibited lysis. This led to the idea that there are ceruloplasmin receptors on cell membranes.

DISILVESTRO: Right. I haven't quanitatively assessed lysis but after centrifuging there is not a lot of red color. When I measure SOD in these red cells I first lyse them with water and then precipatate the SOD with ethanol-chloroform so there is no chance that ceruloplasmin will contribute to the SOD activity I measure. Some of these studies were also done with ceruloplasmin which will not lyse red cells.

DAMERON: Have you actually measured the half-life of these RBCs or are you just extrapolating from the literature? Could it be that deficiency actually decreases the last band of RBCs?

DISILVESTRO: The half-life is from the literature. I know that Zn deficiency produces an increased susceptibility to hemolysis. I don't think Cu deficiency does but it might and it could be influencing cell division rates or something like that.

WILLIAMS: Have you considered the possibility that the sudden change you see in red cell SOD might be related to something other than a change in SOD? Anemia associated with a decrease in reticulocyte count of the order of 30, 40, or 50% is a possibility.

DISILVESTRO: Right. I intend to examine that.

SORENSON: Our experience has been that red cells from Cu-deficient rats are very fragile and lyse upon withdrawal of blood using a syringe and needle.

DISILVESTRO: The only study I remember is one done by Boyd O'Dell to examine in vitro susceptability to hemolysis. Cells from Cu-deficient animals actually were less susceptable.

SMITH: We really don't need 120 days to show changes in RBCs. New cells are made everyday and younger cells may be less fragile. Also, protein doesn't have to be newly synthysized. It just needs the metal and that may have been provided in the vitamin and mineral supplement used in the human study. The point that you made concerning our need for sensitive parameters was important. Biochemical measurements in blood or plasma are the best although we can get biopsies. I don't think we have a Cu bullet yet, one single one, and I think we would like to have some functional test to evaluate it or them as they are developed.

COPPER AND CONNECTIVE TISSUE: THE MECHANISM OF LYSYL OXIDASE.

Frederick T. Greenaway and Coleen Young O'Gara.

Department of Chemistry, Clark University,

Worcester, MA 01610, U.S.A.

It has long been known that there is a relationship between copper-deficiency and solubilization of connective tissue. Although copper deficiency in humans is not common, cases have been reported in populations that subsist largely on a grain diet, and in such populations there has been noted a high incidence of musculo-skeletal weaknesses attributed to solubilization of collagen and elastin, the proteins of connective tissue. These are proteins with very organized structures. They are very strong and quite insoluble because of extensive cross-linking of the protein chains. As the degree of cross-linking decreases, the proteins become increasingly soluble. Several diseases and some inherited disorders are also characterized by solubilization of connective tissue and it is significant that these are also associated with abnormal copper distributions. Several recent reviews of the physiological aspects of copper are available[1].

On the basis of these observations and of dietary studies with animals, it was postulated that a copper-dependent enzyme was involved in the cross-linking process and indeed the copper-containing enzyme now known as lysyl oxidase (EC.1.4.3.13) was first isolated in an impure form by Pinnell and Martin[2] in 1968. This catalyzes the oxidation of lysyl ε-amino groups in collagen and elastin, the first step in the formation of cross-links in these essential components of connective tissue[3]. Lysyl oxidases have been isolated from several sources but do not appear to differ substantially from one another. All show activity with both collagen and elastin substrates. The current state of knowledge about lysyl oxidase has been recently reviewed[4].

The enzyme is, because of its mode of action, an amine oxidase, and since it contains copper it is reasonable to classify it with the other copper-containing amine oxidases. Like the other copper-containing amine oxidases[5], lysyl oxidase requires O_2 for activity, and ammonia and H_2O_2 are produced. There do appear to be several differences between lysyl oxidase and the other copper-containing amine oxidases of which bovine plasma amine oxidase, BPAO, [EC 1.4.3.6] is representative. Lysyl oxidase is a much smaller protein and appears to be active as a monomer, whereas BPAO acts as a dimer, and there are differences in substrate specificity and inhibition patterns. It is not known how similar are the organic cofactor or the copper in the two enzymes.

ENZYME ISOLATION

Since 1968, many advances have been made in techniques for purifying lysyl oxidase and monitoring its activity in vitro[2,3,6-7]. Most methods use 4 or 6M urea to extract the enzyme from tissue and various forms of chromatography to purify the enzyme. Because high concentrations of urea inhibit enzyme activity[8], urea is normally removed by dialysis, although a peroxidase-coupled fluorimetric assay for H_2O_2 produced by the enzyme is optimized in the presence of 1.2 M urea[9].

Lysyl oxidase is obtainable from any young growing source of connective tissue. We use the method of Kagan[6] modified by the use of Cibacron-Blue Sepharose instead of DEAE-cellulose[7], with calf aorta as the source of connective tissue. The solutions of enzyme we obtain have a concentration ranging from 20 μg/ml to 95 μg/ml (0.6 - 3.0 μM) and contain no contaminating protein, as shown by SDS-polyacrylamide gel electrophoresis of enzyme concentrated to 400 μg/ml which produces a single band corresponding to a molecular weight of 32,000. Although concentration of the enzyme five-fold by dehydration with "Aquacide" or Sephadex is possible, these concentrations are much lower than are typically obtained for other amine oxidases and many of the usual spectroscopic techniques can not be used.

We routinely test activity by a fluorimetric peroxidase-coupled assay[9,10], but since this assays for hydrogen peroxide and not specifically lysyl oxidase, we periodically perform the usual specific radioactive assay for the enzyme[2,11,12] as a check. We find an extremely good correlation between the two methods which attests to the lack of contaminating amine oxidases in our samples.

PROPERTIES

Lysyl oxidase is a smaller protein than the other amine oxidases, with a molecular weight of about 32,000. Kagan has isolated four isozymes of very similar size[8] but the significance of the different forms is not known. Atomic absorption studies have shown that copper is present in purified lysyl oxidase but normally only to the extent of 0.7 mole per mole of enzyme[11,13]. One EPR study of lysyl oxidase has been reported[14] but the EPR spectrum probably is of another amine oxidase because inhibitor studies showed marked differences from the results of other workers. Several studies have shown that copper is necessary for activity[4] and that copper can be removed by dialysis against 2,2'-bipyridyl with concomitant loss of activity[11,13]. Dialysis against copper has been reported to restore activity. The stoichiometry of the copper requirement, the nature of the ligand environment of the copper, and its location with respect to the active site are not known.

Although it is clear from inhibition studies[4,8,10,11] that there is an aldehyde cofactor in lysyl oxidase, the identity of this is still a matter of dispute. On the basis of vitamin B6 deficiency, which is associated with solubilization of connective tissue, it has been suggested that the cofactor is pyridoxal phosphate (PLP). A number of in vivo studies have supported this postulate[16-20] but such experiments have not established whether the PLP directly affects the lysyl oxidase or whether it acts in some indirect manner[21].

Attempts have been made to identify pyridoxal derivatives in lysyl oxidase preparations directly by their spectroscopic features. Bird and Levene[22] found that the fluorescence spectra of aldehyde-inhibited derivatives of the cofactor in chick embryo aortic lysyl oxidase were very similar to those of authentic PLP derivatives but we and other workers[4] have been unable to reproduce these results with bovine aortic lysyl oxidase.

Spectroscopic differences from systems known to contain PLP, such as the aminotransferases, suggest that the organic cofactor of lysyl oxidase might not be a pyridoxal derivative. The cofactor of BPAO was originally thought to be PLP[23] but is now known to be pyrroloquinoline quinone (PQQ)[24]. Williamson et al.[25] have carried out sensitive antibody experiments and report that PLP is not present in lysyl oxidase. A dinitrophenylhyrazine derivative of the cofactor has been isolated in small quantities and tentatively identified as the hydrazone of PQQ[25]. At the present time the weight of evidence indicates that PLP is not the cofactor at least in bovine aortic lysyl oxidase, although it is possible that the chick and bovine enzymes may have different aldehyde cofactors.

One of the most well known and most effective inhibitors of lysyl oxidase is β-aminopropionitrile (BAPN)[4] which irreversibly inhibits at a level of less than 0.1 mole per mole of enzyme[13]. This result along with our EPR results (see below) suggests that perhaps purified enzyme is up to 90% inactive. BAPN is thought to act at the aldehyde site by irreversibly binding to the protein near the aldehyde cofactor[13] but this has not been definitely established. Both carbonyl reagents and copper chelators inhibit activity[4].

Kinetic studies by Kagan have shown that lysyl oxidase operates through a double-displacement or "ping-pong" mechanism[4]. Several mechanisms of action have been suggested for copper-containing amine oxidases which appear to be appropriate to apply to lysyl oxidase at least in broad outline[4,5].

The common features of these are an aldehyde cofactor (until recently usually considered to be PLP) bound by an aldimine linkage to a lysine side chain of the protein, a copper(II) ion bound directly to the aldehyde, interaction of dioxygen with the copper(II) usually as a ligand bound in the first coordination sphere of the copper after displacement of a water molecule, attack by a substrate amine on the aldimine releasing the protein-bound lysine with the formation of another aldimine, a series of electron displacements which end with a two-electron reduction of dioxygen forming hydrogen peroxide possibly via radical intermediates[26,27], and formation of the cofactor as the aldehyde with release of ammonia. The most likely role of the copper is in the binding of O_2 and mediation of the reduction to H_2O_2 although there are no experimental data on this point.

There are several points of contention about such mechanisms, most importantly that there is little direct experimental evidence to support even the fundamental structural features of the reacting species. Substantial doubts as to the identity of the aldehyde cofactor, the nature and function of the copper in the enzyme, and the interactions between the two make these mechanistic schemes largely speculation. Furthermore, dioxygen does not commonly bind to monomeric Cu(II) species[28] and there is no evidence that it does so in the amine oxidases.

EPR STUDIES OF COPPER.

EPR is a technique that has been extensively utilized in the study of copper-containing enzymes as it is non-destructive, quite sensitive, and specific for paramagnetic species[29]. For Cu(II) species, analysis of the EPR parameters, particularly $g_{\parallel}$ and $A_{\parallel}(Cu)$ for monomeric species, and lineshapes and $\Delta m_s = 2$ signals for interacting species, can give information about the nature of the metal environment[30], about the nature of the copper ligands and their geometry, and also about indirect changes at the copper site as a result of protein conformational changes. For interacting paramagnetic species it is sometimes possible to obtain internuclear distances. By studying the effects of substrates and inhibitors, information may be obtained about reaction mechanisms. Because it is a quantitative technique, it is also able to indirectly monitor reduction to Cu(I) which commonly occurs in oxidative enzymes although so far no such evidence has been reported for the copper-containing amine oxidases.

We have quantitated the amount of EPR-detectable Cu(II) by integration of spectra and by comparison with standards, and have also determined the amount of total Cu in the EPR samples by atomic absorption spectroscopy. Our first experiments which followed the reported procedure of dialysis against buffer[9,15] gave the same results for the two methods within experimental error. However, a substantial portion of the copper is adventitiously bound and we now dialyze enzyme samples against EDTA which does not affect activity, but which removes much of the EPR-detectable copper. Thus published reports of copper:protein molar ratios which are typically about 0.7[9,15] are undoubtedly too high since they include adventitiously bound copper. Our preliminary results indicate a molar ratio of active copper:protein between 0.1 and 0.3 with some variation from preparation to preparation. The result suggests that pure enzyme is in fact at most 25% active and that much of the enzyme is copper-deficient.

The copper adventitiously bound to the enzyme in 6 M urea solutions is characterized by the EPR parameters: $g_{\parallel} = 2.28$ and $A_{\parallel}(Cu) = 0.017\ cm^{-1}$. This is typical for Cu(II) ions in a tetragonally-elongated octahedral site surrounded by at least two uncharged nitrogen atoms[30]. Enzyme dialyzed against EDTA has a much weaker EPR signal with no detectable "parallel" features. The observed signal has the features expected for interacting paramagnetic species separated by several Angstroms[29]. We have so far been unable to detect any signals near g=4.

We have also heat-denatured the enzyme in the presence of p-chloromercuribenzoate (to prevent reduction by sulfhydryl groups) and have quantitated the EPR-detectable Cu(II), which should represent the total Cu(II) in the enzyme. Upon heat-denaturation, the EPR signal changes to that typically found for other denatured copper proteins, with $g_{\parallel} = 2.236$ and $A_{\parallel}(Cu) = 0.0181\ cm^{-1}$, and increases substantially in intensity. The difference between the EPR-parameters of "adventitious" and denatured Cu(II) is significant but probably merely reflects the greater flexibility of the denatured protein which enables the copper to bind to more of the preferred nitrogen ligands of the protein.

The increase of EPR intensity observed on denaturation could be explained by the presence in the resting enzyme of (i) antiferromagnetically-coupled Cu(II) ions, (ii) Cu(I) ions, or (iii) an organic radical interacting with a Cu(II) ion. If any of these cases are true, then lysyl oxidase is not like the other copper-containing amine oxidases. Because we have no absolute measure of the number of active enzyme molecules it may be that the enzyme was partly trapped in an inactive mode which had one of these species. PQQ is known to form stable radicals under such conditions[31,32].

The EPR spectrum is not affected by the urea concentration between 0 and 6 M, nor by BAPN. These results indicate that the modes of inactivation by urea and BAPN do not involve copper.

SPIN LABEL STUDIES

One of the most important questions about which there is no experimental information at the present time, is where the copper is positioned relative to the aldehyde moiety in the copper-containing amine oxidases. Although many mechanisms have been proposed, most of which assume direct binding of Cu(II) to the aldehyde moiety no experiments have been carried out to test this hypothesis. We have carried out EPR studies of the enzyme which has been spin-labelled by the use of a spin-labelled inhibitor.

After aerobic incubation of the label with the enzyme, and reduction of the aldimine C=N bond, the excess free label is removed by extensive dialysis against buffer, and EPR spectra are recorded at room temperature and at 77K. Room temperature EPR spectra yield information about the mobility of the spin label whereas low temperature spectra are more useful in investigating interactions between the label and neighbouring paramagnetic species, such as Cu(II)[33].

We have performed this experiment with commercially-available BPAO which has a PQQ cofactor[4,24] using spin label I, and with lysyl oxidase using spin label II. The spin labels attach irreversibly to the enzyme causing loss of activity, and the amount of spin label that binds is the same as the amount of ^{14}C-phenylhydrazine that binds to the enzymes. We therefore believe that all of the active aldehyde sites have been labeled. For BPAO the room temperature EPR signals show that 90% of the labels are strongly immobilized and that 10% are quite mobile. The 10% probably represents enzyme that has denatured after labeling. In lysyl oxidase the label is quite mobile.

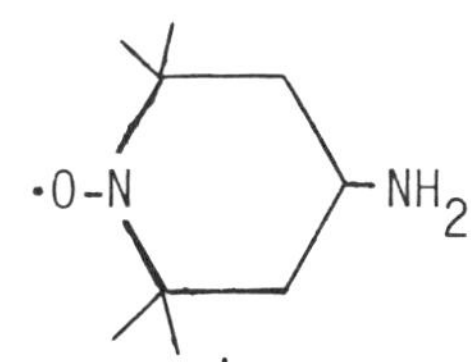
I

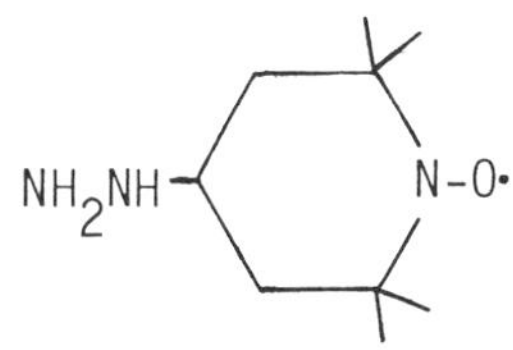
II

By titrating solutions of the labeled enzymes with ferricyanide which results in paramagnetic line broadening of spin labels within 10A of the iron(III)[34], we have also shown that the labeled site in BPAO is not accessible to ferricyanide whereas the labeled site in lysyl oxidase is quite accessible. We conclude that the spin label in lysyl oxidase is probably near the surface of the protein and accessible to solvent, whereas for BPAO the label is probably in the interior of the protein and not accessible to solvent. The nitroxide is not close (>10A) to the Cu(II) in either enzyme since fairly normal nitroxide spectra[35] are observed. This result is confirmed when we remove the copper by dialysis against diethyldithiocarbamate in the case of BPAO or 2,2'-bipyridyl in the case of lysyl oxidase. The EPR spectra do not change in either case.

DISCUSSION

Our EPR studies have been directed at answering some very basic questions about the copper and aldehyde moieties in lysyl oxidase. We find that the total amount of strongly

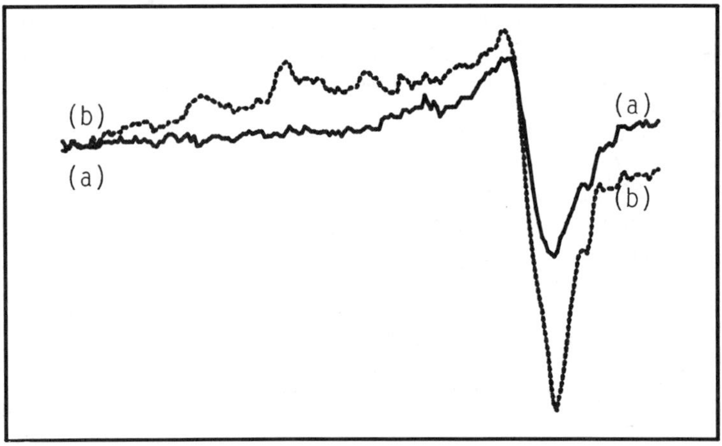

Copper EPR spectra of lysyl oxidase samples at 110K. (a) native enzyme, (b) enzyme after denaturation.

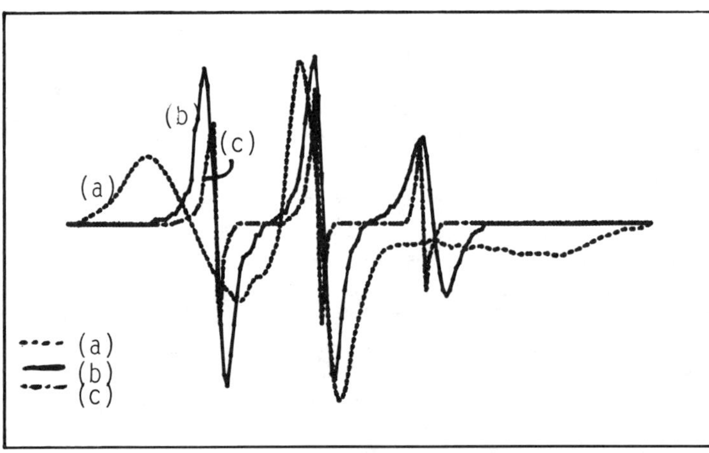

EPR spectra of (a) BPAO showing the immobilized label, (b) BPAO showing the freely moving label thought to be of denatured enzyme, and (c) lysyl oxidase showing the freely moving label. Spectral intensities are not to scale. All spectra were obtained at room temperature.

bound copper in the enzyme is always much less that one mole per mole of protein. The amount of spin label or ^{14}C-phenylhydrazine that binds is similar to the amount of copper present in the enzyme and it seems probable that only a small proportion of the isolated enzyme has both copper and an active aldehyde. The enzyme isolation procedure appears to cause loss of relatively large amounts of copper. Our data do not show whether the aldehyde cofactor has also been lost or whether it has been merely irreversibly inhibited. We are thus unable to address the question of the stoichiometry of the copper and aldehyde requirements of the enzyme although they appear to be the same.

The question of the oxidation state of the copper during enzyme turnover is yet to be answered and will require stop-flow experiments. We observe significant amounts of EPR-undetectable copper in the resting enzyme but have yet to correlate these findings with activity measurements. At the present time we are not able to say much about the environment of the copper since no clear "parallel" structure was observed in the EPR spectra. Urea inhibits the enzyme reversibly and BAPN causes irreversible inhibition, but neither cause any change in the copper(II) EPR spectra of the enzyme and it appears that neither inhibit at the copper site.

The spin-labeled aldehyde site appears to be fully accessible to the solvent and the label has a high degree of mobility, in contrast to the aldehyde site in BPAO. These results suggest that the aldehyde is near the protein surface in lysyl oxidase. The cofactor is not close to the copper and is certainly not directly bound to it.

Our results suggest that some aspects of the mechanism suggested above can not be valid. In particular, since the copper is not near the aldehyde, no direct interaction between the two can be occurring. Thus if there is electron transfer between the two, this must be across a considerable distance. Since no previous examples have been reported where O_2 binds to monomeric copper(II) species, we believe that this is not probable in lysyl oxidase either. We do see an increase in copper(II) upon denaturation which suggests that Cu(I), or possibly coupled Cu(II), species are present. Both of these are known to bind O_2. This would indicate that the first step of the mechanism was transfer of an electron to the copper, followed by oxygen binding rather than the reverse.

ACKNOWLEDGEMENTS

We wish to thank Peter Spacciapoli and Dr. Herbert M. Kagan for assistance with the enzyme isolation. This work was supported by a grant from the Kroc Foundation.

REFERENCES.

1. (a) "Copper and Lymphomas", ed. Hrgovcic, M. J; Shullenberger, C. C. CRC Press, Boca Raton, FL, 1984.
   (b) "Inflammatory Diseases and Copper", ed. Sorenson, J. R. J. Humana Press, Clifton, NJ, 198 .
2. Pinnell, S. R; Martin, G. R. Proc. Natl Acad. Sci. U.S.A. 1968, 61, 708 .
3. Siegel, R. C; Martin, G. R. J. Biol. Chem. 1970, 245, 1653.
4. (a) Kagan, H. M. Intl Rev. Conn. Tissue Res. 1986; in press. (b) Buffoni, F. in "Structure and Functions of Amine Oxidases", ed. Mondovi, B. CRC Press, Boca Raton 1985; p 77.
5. (a) Knowles, P. F; Yadav, K. D. S. in "Copper Proteins and Copper Enzymes", ed. Lontie, R. CRC Press, Boca Raton, 1984, 2, 103. (b) "Structure and Functions of Amine Oxidases", ed. Mondovi, B. CRC Press, Boca Raton, 1985. (c) Hamilton, G. A. in "Copper Proteins", ed. Spiro, T. G. Wiley-Intescience, New York 1981; p 193.
6. Kagan, H. M; and Sullivan, K. A. Methods Enzymol. 1982, 82, 637.
7. Williams, M. A; Kagan, H. M. Anal. Biochem. 1985, 149, 430.
8. Kagan, H. M; Sullivan, K. A; Olsson, T. A; Cronlund, A. L. Biochem. J. 1979, 177, 203.
9. Trackman, P. C; Zoski, C. G; Kagan, H. M. Anal. Biochem. 1981, 113, 336.
10. Trackman, P. C; Kagan, H. M. J. Biol. Chem. 1979, 254, 7831.
11. Harris, E. D; Gonnerman, W. A; Savage, J. E; O'Dell, B. L. Biochim. Biophys. Acta 1974, 341, 332.
12. Melet, J; Vianden, G. D. N. E; Bachra, B. N. Analyt. Biochem. 1977, 77, 141.
13. Tang, S.-S; Trackman, P. C; Kagan, H. M. J. Biol. Chem. 1983, 258, 4331.
14. Shieh, J. J; Yasunobu, K. T. Adv. Exp. Med. Biol. 1976, 74, 447.
15. Siegel, R. C; Pinnell, S. R; Martin, G. R. Biochem.

1970, 9, 4486.
16. Starcher, B. C. Proc. Soc. Exp. Biol. Med. 1969, 132, 379.
17. Rucker, R. B; O'Dell, B. L. Fed. Proc. 1970, 29, 2387.
18. Murray, J. C; Levene, C. I. Biochem. J. 1977, 167, 463.
19. Bird, T. A; Levene, C. I. Biochem. J. 1983, 210, 633.
20. Carrington, M. J; Bird, T. A; Levene, C. I. Biochem. J. 1984, 221, 837.
21. Myers, B. A; Dubick, M. A; Reynolds, R. D; Rucker, R. B. Biochem. J. 1985, 229, 153.
22. Bird, T. A; Levene, C. I. Biochem. Biophys. Res. Commun. 1982, 108, 1172.
23. Yamada, H; Yasunobu, K. T. J. Biol. Chem. 1963, 238, 2669.
24. Lobenstein-Verbeek, C. L; Jongejan, J. A; Frank, J; Duine, J. A. FEBS Letters 1984, 170, 305.
25. Williamson, P. R; Kittler, J. M; Thanassi, J. W; and Kagan, H. M. Biochem. J. 1986, 235, 597.
26. Kagan, H. M; Soucy, D. M; Zoski, C. G; Resnick, R. J; Tang, S.-S. Arch. Biochem. Biophys. 1983, 221, 158.
27. Dooley, D. M; Coolbaugh, T. S. Biochem. Biophys. Res. Commun. 1980, 96, 823.
28. "Copper Coordination Chemistry. Biochemical and Inorganic Perspectives", ed. Karlin, K. D; Zubieta, J. Adenine Press, Guilderland, NY, 1983.
29. Boas, J. F. in "Copper Proteins and Copper Enzymes", ed. Lontie, R. CRC Press, Boca Raton, 1984, 1, 5.
30. Peisach, J; Blumberg, W. E. Arch. Biochem. Biophys. 1974, 165, 691.
31. De Beer, R, Duine, J. A; Frank, Jr, J; Large, P. J. Biochim. Biophys. Acta 1980, 622, 370.
32. J. A. Duine, J. Frank, Jr, R. De Beer, Arch. Biochem. Biophys. 233, 708 (1984).
33. Eaton, S. S; Eaton, G. K. Coord. Chem. Rev. 1978, 26, 207.
34. Timofeev, V. P; Polyanovskii, O. L; Vol'kenshtein, M. V; Preobrazhenskaya, O. V; Likhtenshtein, G. I; Kokhanov, Yu. V. Mol. Biol. 1972, 6, 302.
35. There are many references to the theory and applications of spin-label EPR. See, for example: "Spin Labelling. Theory and Applications", ed. Berliner, L. J. Academic Press, New York, NY, 1976, vol. 1; and 1979, vol. 2.

DISCUSSION

HARRIS: In your collaboration with Prof. Kagan you use bovine enzyme and you find a relatively high concentration of adventitious Cu. I agree that if we are going to do relevant experiments we must use clean buffer and so forth. Why, though, have you correlated lysyl oxidase (LO) activity with Cu? We have always had a difficult time trying to stoichiometrically correlate enzyme activity with Cu content. That suggests to me that your correlation of activity values might be erroneous. Biochemically, we would like to speak of activity as maximum activity.

GREENAWAY: There is a real problem with measuring LO activity in a radioactive assay which does not relate to moles of Cu, as we chemists like to do. We can do another assay, a fluorometric-coupled assay, which does give good correlations between Cu and activity as measured by the radioactive assay. We think what we measure in the fluorometric assay is a true reflection of LO activity. This assay directly measures how many moles of hydrogen peroxide are produced. Earlier we obtained enzyme which lost its activity very rapidly and we could not understand why. Fortunately, we have managed to obtain enzyme which retains its activity. What we cannot do is satisfactorily correlate Cu and aldehyde levels with enzyme activity because anything we do with the Cu or the aldehyde completely destroys activity. We have tried to measure Cu concentrations by EPR and measure aldehyde concentrations by titrating with ^{14}C labeled phenylhydrazine. When we measure enzyme activity and compare these three parameters, there is no good correlation. Cu may come out of some enzyme sites, the aldehyde is inhibited at some other enzyme sites, and we have only a little bit of truly active enzyme left which contains both cofactors.

HARRIS: You might have discovered something that we have long suspected and that is that we have a fairly good pro-enzyme, a Cu free form of the enzyme, in tissue. Is your thinking along that line?

GREENAWAY: Yes, that's exactly it. That's why I am so disappointed that we can't put the Cu back because if we could put it back then we would have active enzyme. There's certainly no other protein in our preparation.

SORENSON: How, mechanistically, does oxygen function with Cu to convert the amino form of the cofactor to the aldehyde form?

GREENAWAY: Oxygen probably binds at the reduced Cu site. As our studies have shown, that's not near the amine group. There has to be some way for electrons to cross several angstroms. This is not unknown for enzymes. Oxidation occurs at the amine with electrons being passed on to oxygen through Cu. This is speculation. We have no detailed mechanistic information.

COPPER - A MODULATOR OF PEPTIDE RELEASE IN THE BRAIN

Ayalla Barnea

Departments of OB/GYN and Physiology,
Cecil H. and Ida Green Center for
Reproductive Biology Sciences
5323 Harry Hines Blvd., Dallas, TX 75235

A major conceptual understanding of brain function is the discovery that neurons are specialized secretory cells which secrete biologically active substances, *e.g.*, biogenic amines, small peptides and polypeptides. Although the chemical nature of the secretory substances varies from one cell-type to the other, there are several features of the secretory process that are common to all secretory cells: a) the secretory substance(s) is sequestered within secretory granules; b) secretion occurs by means of exocytosis; and c) the trigger for exocytosis is of extracellular nature, operating at the level of the plasma membrane of the cell. In exocytosis, a granule approaches the plasma membrane, merges with it, and after rupture of the fused granule-plasma membranes, voids its content into the extracellular space (1). In this presentation, I would like to develop the concept that copper can serve as a neuromodulator of peptide release in the brain, using as an example the release of the decapeptide - luteinizing hormone releasing hormone (LHRH). This peptide is stored in and secreted from hypothalamic neurons, the axonal terminals of which are highly concentrated in the region of the median eminence. The latter region of the rat brain was used in our *in vitro* studies demonstrating a dual role of copper in LHRH release: copper stimulating the release of LHRH and copper amplifying prostaglandin E_2 (PGE_2) stimulation of LHRH release.

SECRETORY GRANULES - A SOURCE FOR EXTRACELLULAR COPPER

In order for copper to serve as a neuromodulator, the following pre-requisite should be met. Copper should be present in high concentrations in brain tissue, particularly in the extracellular space of the LHRH neuron. In support of this pre-requisite are the findings of high concentrations of copper, 1-2 orders of magnitude greater than in blood, in hypothalamic tissues, in isolated synaptosomes and in synaptic vesicles (2-4). Since a number of intra-granule soluble enzymes are copper dependent [e.g. dopamine beta-hydroxylase, alpha-amidating enzyme] and since the soluble contents of secretory granules are co-released during exocytosis (5-6), it is conceivable that copper is released from hypothalamic neurons which creates a condition of high extracellular levels of copper. Although this copper release hypothesis has not yet been validated, the co-release of Zn and insulin from pancreatic islets (7) and the release of newly taken-up ^{65}Zn from brain tissue (8) are supportive of this hypothesis.

COPPER - A SECRETAGOGUE OF LHRH

In a series of studies, we have established that copper is a potent secretagogue of LHRH from LHRH axonal terminals and that one site of copper action is the LHRH secretory granule itself (9-16). The characteristics of the copper-triggered release process are summarized in the following model for copper action on the LHRH axonal terminals (Fig. 1).

Step I - Copper uptake. The observations that complexed copper (CuHis), but not ionic copper, is the active form of the metal (12), that a limited number of copper interactive sites are involved (14), and that LHRH release is inhibited by ouabain and Li^+ (16) are consistent with the proposition that CuHis is taken-up by the LHRH axonal terminals by co-transport with Na^+; a co-transport that is driven by the electrochemical Na^+ gradient generated by Na^+/K^+ ATPase. This uptake process results in high concentrations of chelated copper in the plasma membrane and in its vicinity, the active zone.

A MODEL FOR COPPER ACTION ON LHRH NEURONS

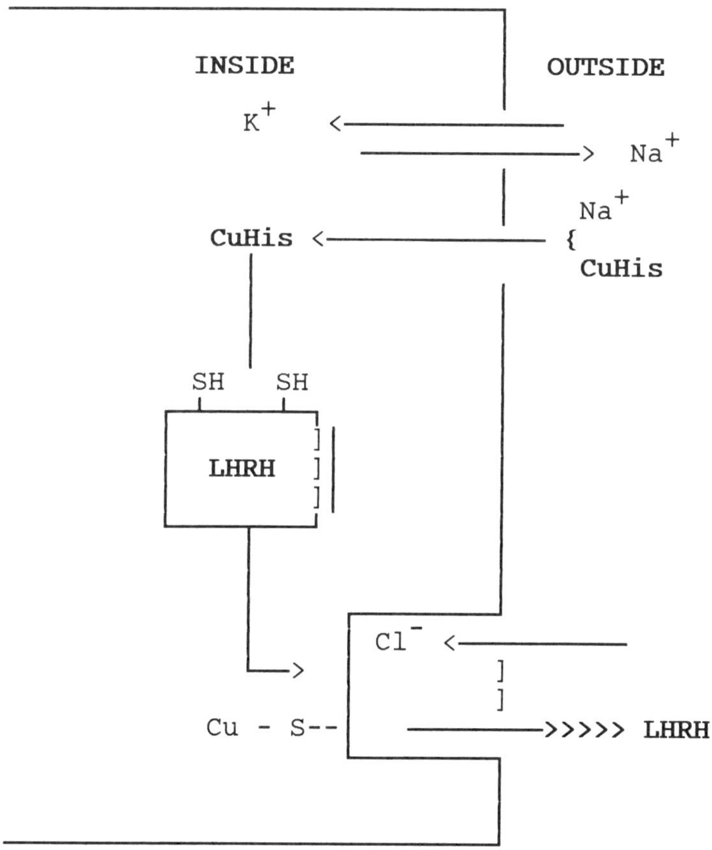

Figure 1

Step II - Interaction with the LHRH secretory granules. We have demonstrated that CuHis stimulates the release of LHRH from isolated secretory granules (Kmof about 5 µM CuHis) (11) and that an oxidative interaction with thiol groups is involved (10). Thus, we envision that the newly taken-up copper interacts with thiol groups of LHRH granules that are in close proximity to or in the process of fusion with the plasma membrane. The interaction of copper with the LHRH granules has two consequences. One, changes in granule membrane permeability and two, facilitation of fusion of the granules with the plasma membrane as well as fusion of one granule to another. It has been shown that Cu^{2+} binds to synaptic vesicles with high affinity (17) and that Cu^{2+} enhances the attachment of synaptic vesicle membranes to plasma membranes (18). Recently, we have demonstrated that transport of monovalent anions is essential for copper-stimulation of LHRH release and that the transporter has a size-selectivity for monovalent anions (15). The exact role of this anion transporter and its location has yet to be elucidated. Based on the postulated role for chloride transport in the secretory process in other secretory cells (19), an attractive proposition is that the anion transporter is located on the membrane of the LHRH granule itself and that the fusion of the granule membrane with the plasma membrane exposes the transporter to the high extracellular Cl^-, which allows Cl^- influx into the granule. There are still many unresolved questions as to the exact role of chloride transport in LHRH release. Is a unidirectional flux of chloride involved? If so, is peptide release a consequence of increased osmotic pressure, i.e., chemiosmolysis, increased salt concentrations, or changes in pH? What is the subcellular location of the transporter? It should be emphasized that this release process does not require calcium in the extracellular space (13). Whether this process is totally independent of an increase in intracellular free calcium or whether copper causes mobilization of calcium from the endoplasmic reticulum, as do other secretagogues (20, 21), has yet to be elucidated. In support of the latter possibility is the finding that copper leads to calcium release from isolated endoplasmic reticulum vesicles (22).

It has been shown that cupric acetate, administered

to female rabbits, induces ovulation and that the hypothalamus is the major site of copper action (23). Most interesting is the demonstration that systemically administered cupric acetate leads to an increase in the concentration of LHRH in the hypophysial portal blood (24) as well as in the extracellular space of the LHRH neuron, the CSF (25). These observations in the intact animal taken together with our results generated in the in vitro studies lend support to the view that copper plays an important role in the regulation of LHRH release.

COPPER - A MODULATOR OF PROSTAGLANDIN E_2 ACTION

It is well established that PGE_2 is a secretagogue of LHRH under in vivo (26) as well as under in vitro conditions (27, 28). The current view is that the calcium-cAMP pathway, an increase in cAMP formation coupled to influx of extracellular calcium, is the major intracellular pathway mediating this action of PGE_2 (29, 30). Since disulfide bonds appear to be important for the binding of PGE_2 to its receptor (31) and for PGE_2 action in other cellular systems (32), we hypothesized that copper will enhance the stimulatory action of PGE_2 on the LHRH neuronal terminals. We have tested this hypothesis and found that a short (5 min) pre-incubation of median eminence explants with copper markedly amplifies PGE_2 stimulation of LHRH release (33, 34). The fact that extracellular calcium is also required for copper-amplified PGE_2 stimulation of LHRH release and that the Ca^{2+} is required for the action of PGE_2 and not for copper action (35) is consistent with the involvement of the Ca-cAMP pathway in this amplified release process. The important question is: Does copper modulate the PGE_2 receptor or a post receptor component? If it is a post receptor component, what is its identity (Fig. 2)?

The observations that complexed copper (CuHis), but not ionic copper, is the active form of the metal(II) (36), that a limited number of copper interactive sites are involved (34), and that copper amplification of PGE_2 action is inhibited by ouabain and by Li^+ (unpublished observation), are suggestive that CuHis uptake is required and that it occurs via co-transport with Na^+ the driving force of which is the $[Na^+]$ gradient generated by

COPPER AMPLIFICATION OF PGE_2 ACTION

POTENTIAL SITES (??) FOR COPPER ACTION

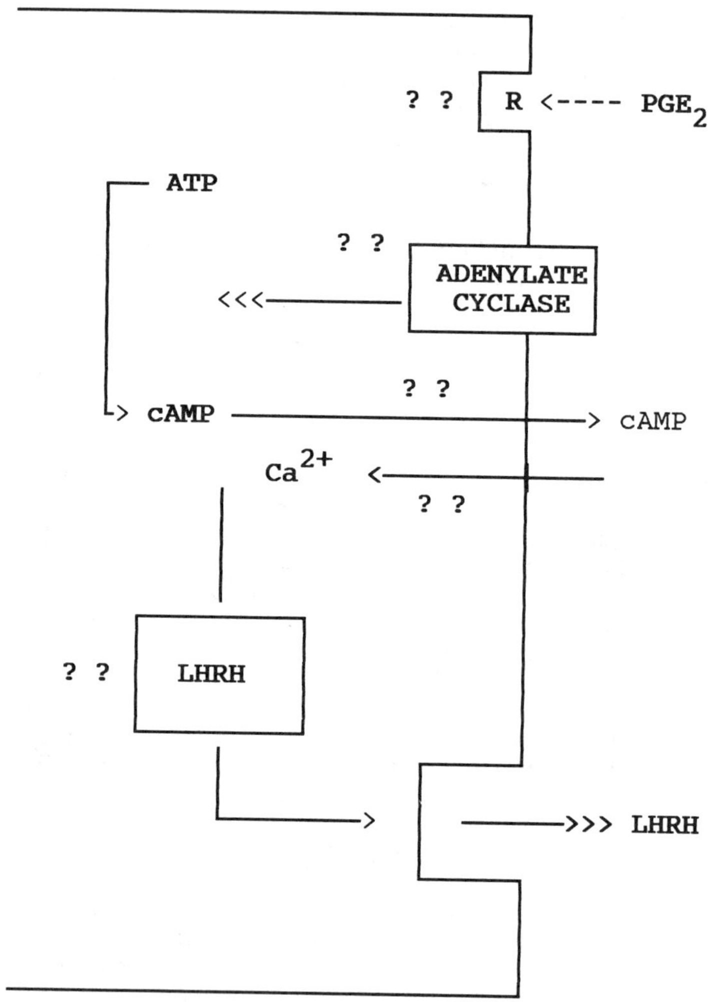

Figure 2

Na^+/K^+ ATPase. Several lines of evidence support the identification of the copper-interactive site as a PGE_2 post-receptor component, most likely the adenylate cyclase system. One, the ED_{50} for PGE_2 stimulation of LHRH release is not altered by copper (34). Two, the time courses of the PGE_2-stimulated release and of the copper-amplified, PGE_2-stimulated release of LHRH are different (33, 34). Three, forskolin, which directly activates the catalytic subunit of adenylate cyclase (37), stimulates LHRH release and copper amplifies this stimulatory action of forskolin (38). Four, the copper-amplified PGE_2 stimulation of LHRH release and the copper-amplified forskolin stimulation of LHRH release are not additive, indicative of a common mechanism of action. Five, copper does not amplify 8-Br-cAMP stimulation of LHRH release. Six, copper does not amplify LHRH release operating by a calcium pathway that does not involve cAMP, e.g., K^+-induced depolarization (13, 38).

Although these results are consistent with copper modulating the functional state of the adenylate cyclase system, one should consider other possible sites of copper action. It is well known that the newly formed cAMP does not accumulate to a large extent within the cell and that a large portion of the cAMP is secreted. Recently, it has been shown that taken up PGA_1 rapidly forms an adduct with glutathione, and it has been postulated that this PGA_1 -glutathione complex inhibits cAMP efflux (39, 40). One possibility is that the intracellular copper-complex inhibits cAMP efflux, and hence, amplifying PGE_2 action.

TERMINATION OF COPPER ACTION

Since copper is found in relatively high concentrations in axonal terminals in the brain (4) and since copper uptake is involved in copper action, an important question is: Why copper does not deplete the intracellular content of LHRH? A reasonable explanation is that intracellular copper is inactivated by glutathione and copper binding proteins, both are found in high concentrations in the brain (41, 42). In support of this explanation is the observation that the stimulatory effect of CuHis on LHRH release from the median eminence tissue is transient [rate of LHRH release declines exponentially 15 min after transfer of the tissue to copper-free medium (12, 14)] and that thiol reducing agents (glutathione, dithiothreitol) inhibit copper action on isolated secretory granules (10, 11). It is interesting that the time course of the copper amplification of PGE_2 action is similar to that of copper stimulation of LHRH release and that these two effects of copper are reversible. Thus, it appears that the presence of copper in the vicinity of the plasma membrane, the active zone, is essential for the manifestations of copper action, and once copper is removed from this location it is no longer active.

METAL SPECIFICITY OF COPPER ACTION

Both copper-related release processes, copper stimulation of LHRH release and copper-amplified PGE_2 stimulation of LHRH have a high degree of metal(II) specificity. Of several metal(II)-His complexes tested (Ni, Fe, Zn, Cd, Mn, Ba), Ni^{2+} was the only metal that stimulated LHRH release, albeit to a lesser extent than Cu^{2+} (13). Moreover, none of these metals amplified PGE_2 stimulation of LHRH release (36).

QUESTIONS FOR THE FUTURE

1) Is copper a secretagogue of other peptides, polypeptides, or biogenic amines? While CuHis was found to stimulate the release of the peptide alpha-melanotropin from isolated hypothalamic granules (43), it is not yet clear whether copper exerts an effect on other peptidergic/aminergic neuronal terminals.

2) Does copper modulate the adenylate cyclase system in other cellular systems?

3) Could copper amplification of PGE_2 action occur at another intracellular site, e.g., inhibition of efflux of newly formed cAMP?

4) Does copper modulate the function of other neuromodulators?

ACKNOWLEDGEMENTS

These studies were supported by NIH Research grants AM25692 and HD11149.

REFERENCES

1. Douglas, WW. Ciba Found Symp 1978, 54, 61.
2. Colburn, RW: Maas JW. Nature 1965, 208, 37.
3. Merriam, GR: Nunnelley, LL: Trish, JWV: Naftolin, F. Brain Res 1979, 171, 503.
4. Rajan, KS: Colburn, RW: Davis, JM. Life Sci 1976, 18, 423.
5. Levine, M: Asher, A: Pollard, H: Zinder, O. J Biol Chem 1983, 258, 13111.
6. Mains, RE: Glembotski, CC: Eipper, BA. Endocrinology 1984, 114, 1522.
7. Grodsky, GM: Schmid-Formby, F. Endocrinology 1985, 117, 704.
8. Howell, GA: Welch, MG: Frederickson, CJ. Nature 1984, 308, 736.
9. Burrows, GH: Barnea, A. Endocrinology 1982, 110, 1456.
10. Rice, GE: Barnea, A. J Neurochem 1983, 41, 1672.
11. Barnea, A: Cho, G. Endocrinology 1984, 115, 936.
12. Barnea, A: Colombani-Vidal, M. Proc Natl Acad Sci USA 1984, 81, 7656.
13. Colombani-Vidal, M: Barnea, A. Neuroendocrinology 1986, In Press.
14. Colombani-Vidal, M: Barnea A. Neuroendocrinology 1985, 41, 454.
15. Colombani-Vidal, M: Barnea, A. Neuroendocrinology 1986, In Press.
16. Colombani-Vidal, M: Barnea, A. Neuroendocrinology 1986, In Press.
17. Hoss, W: Okumura, K: Formaniak, M: Tanaka, R. Arch Biochem Biophys 1980, 203, 647.
18. Hoss, W: Formaniak, M. Neurochem Res 1980, 5, 795.
19. Pollard, HB: Pazoles, CJ: Creutz, CE: Scott, JH: Zinder, O: Hotchkiss, A. J Biol Chem 1984, 259, 1114.
20. Albert, PR: Tashjian, AH Jr. J Biol Chem 1984, 259, 15350.
21. Gershengorn, MC: Thaw C. Endocrinology 1985, 116, 591.
22. Abramson, JJ: Trimm, JL: Weden, L: Salama, G. Proc Natl Acad Sci USA 1983, 80, 1526.
23. Hiroi, M: Sugita, S: Suzuki, M. Endocrinology 1965, 77, 963.

24. Tsou, RC: Dailey, RA: McLanahan, CS: Parent, AD: Tindall, GT: Neill, JD. Endocrinology 1977, 101, 534.
25. Pau, KYF: Spies, HG. Neuroendocrinology 1986, 43, 197.
26. Ojeda, SR: Wheaton, JE: McCann, SM. Neuroendocrinology 1975, 17, 283.
27. Ramirez, VD: Gallardo, E: Hartter, D. J Endocrinol Invest 1980, 3, 29.
28. Heaulme, M: Dray, F. Neuroendocrinology 1984, 39, 403.
29. Ojeda, SR: Negro-Vilar, A. Endocrinology 1985, 116, 1763.
30. Ojeda, SR: Urbanski, HF: Katz, KH: Costa, ME. Endocrinology 1985, 117, 1175.
31. Malet, C: Scherrer, H: Saavedra, JM: Dray, F. Brain Res 1982, 236, 227.
32. Johnson, M: Jessup, R: Ramwell, PW. Prostaglandins 1974, 5, 125.
33. Barnea, A: Cho, G: Colombani-Vidal, M. Endocrinology 1985, 117, 415.
34. Barnea, A: Cho, G: Colombani-Vidal, M. Endocrinology 1986, 119, 1254.
35. Barnea, A: Cho, G: Colombani-Vidal, M. Endocrinology 1986, 119, 1262.
36. Barnea, A: Cho, G: Colombani-Vidal, M. Brain Res 1986, In Press.
37. Seamon, KB: Padgett, W: Daly, JW. Proc Natl Acad Sci USA 1981, 78, 3363.
38. Barnea, A: Cho, G. Proc Natl Acad Sci USA 1986, In Press.
39. Heasley, LE: Azari, J: Brunton, LL. J Cyclic Nucleo Prot Phosphor Res 1985, 10, 3.
40. Heasley, LE: Watson, MJ: Brunton, LL. J Biol Chem 1985, 260, 11520.
41. Kosower, NS; Kosower, EM. In: Free Radicals in Biology; Academic Press, New York, 1976; p 55.
42. Nalbandyan, RM. Neurochem Res 1983, 8, 1211.
43. Rice, GE: Barnea, A. Brain Res Bull 1983, 11, 527.

DISCUSSION

RISK: I gathered that the naturally occurring complex, whether it's Cu-histidine or some other molecule, would have to be small, no larger than a 1000 Daltons or so, to pass through the small opening of the Na-K transport channel.

BARNEA: Actually I don't know if an amino acid is required. It could be something else. It could be something that is more lipophilic, that doesn't need Na-K ATPase for passage. It's very possible that the reason why we need co-transport is that this is required for histidine transportation and histidine is essentially a carrier for Cu. It may be fortuitous that the co-transporter can also accept Cu-histidine complex. If we had another complex or if we could solubilize the Cu(II)(histidinate)$_2$ complex that Prof. Sorenson sent us or Cu(II)(3,5-Dips)$_2$, we may not need the co-transport system. It's possible that what we need is some way for extracellar Cu to enter the neuron and when it's in an amino acid complex the best way to move it into the neuron is via the co-transport mechanism.

CROOK: If I understood correctly, the majority of what you are doing is in vitro. We made ^{64}Cu complexes of D and L valine and tryptophan because we were interested in studying them in vivo. Interestingly enough, when these were injected intravenously into rats their tissue distributions were exactly the same which implies that the Cu, unfortunately, had come apart from the amino acids we were using. This is then a word of caution. What takes place in vitro, although fascinating, may not necessary hold in vivo. Also, when you showed LHRH release amplification effects of PGE$_2$ by Cu you didn't show data for LHRH release with Cu alone. What does the curve for LHRH release look like when only Cu is added?

BARNEA: What I showed were data obtained with a single concentration of Cu and different concentrations of PGE$_2$. Actually, the basal release is higher and when we start adding PGE$_2$ superimposed on the effect of Cu we see that the time course of the Cu amplified PGE$_2$-stimulated release of LHRH is very similar to Cu by itself, although, it is always much greater than Cu by itself.

CROOK: One thing you might consider in regard to your schema is that perhaps PGE_2 is now undergoing metabolism and that may explain some differences you see between Cu and PGE_2 and just Cu alone. One final point. Your issue concerning Na-K transport system and Cu co-transport is novel. It would be interesting to know if this co-transport system in these synaptosomes is inhibited by Furosemide.

BHATHENA: Consistent with your results, it appears that Cu may be involved in hormone secretion from pancreas as well.

BARNEA: In response to Dr. Crook's comment, I'd like to point out that in the fifties it was shown that $Cu(II)_2$-(acetate)$_4$ administered systemically to rabbits induces ovulation and there are at least two papers showing that systemically administered Cu leads to an increase in concentration of LHRH in the hypophysial portal blood and a recent paper published by Harold Spiers showed that systemically administered $Cu(II)_2$(acetate)$_4$ increases the concentration of LHRH in the extracellular space. They did push-pull cannulation to infuse the brain and collect the CSF. There was a marked increase in LHRH release. So if we can't figure out the exact characteristics and properties of the system in vitro, it does operate in vivo. Cu can do the job in vivo. How it does it and whether it's the same mechanism as we find in vitro may be another question.

BHATHENA: Do you have any evidence as to whether Cu may stimulate LHRH synthesis which then may be responsible for increased release?

BARNEA: I'm sure there is no synthesis in our system because the tissue we incubate in vitro contains only axonal terminals of LHRH neurons. Cell bodies containing the synthesizing machinery are in another region in the brain and are not in our incubation medium.

PICKART: The Cu binding tripeptide from human serum, glycylhistyllysine, induces axonal and dendrite development in cultured brain tissue at 5 ng/ml. Could you speculate as to why a Cu complex would do this?

BARNEA: I don't know. Actually we found your publications concerning this tripeptide very exciting and we started to study it but, unfortunately, it was not effective in stimulating LHRH release. Ceruloplasmin and Cu-albumin were also ineffective.

PICKART: There is also a report, sent to me by John Sorenson, showing that Cu(II)(3,5-dips)$_2$ also induces axonal outgrowth in a neuroblastoma cell line.

Copper-Dependent α-Amidation in the Synthesis of Neuroendocrine Peptide Hormones

Thomas R. Gibson and Christopher C. Glembotski

Mol. Biology Institute and Dept. of Biology

San Diego State University

San Diego, CA 92182

COPPER AND THE NEUROENDOCRINE SYSTEM

The absolute requirement for copper by many enzymes and biological processes has been well established. The low activity *in vitro* of many copper-requiring enzymes has been shown to be responsible for symptoms associated with copper deficiency. Among these are altered mental status due to changes in norepinephrine levels (dopamine β-hydroxylase)[1], loss of hair pigmentation (tyrosinase)[2], and reduced bone and artery competence (lysyl oxidase)[3]. In addition, the symptoms of copper deficiency include other manifestations of neurological dysfunction.[4]

Experimental evidence using copper-deficient rats has also suggested that the altered biological activity of neuroendocrine peptides may be responsible for some of these symptoms. In one study, while the ability of the pituitary under these conditions to respond to exogenously administered thyrotropin releasing hormone (TRH) was normal, the levels of immunoactive plasma thyrotropin and total thyroxine were both depressed.[5] These results could be explained by the lack of bioactive TRH produced by the hypothalamus under these condition of copper deficiency. In addition, atrophy of pancreatic acinar cells results from dietary copper restriction, while the remainder of the pancreas appears normal.[6] It could be that the lack of bioactive cholecystokinin (CCK) normally produced by the intestine, which is known to stimulate the growth of pancreatic acinar cells, might be involved in this copper-deficiency induced atrophy.

NEUROENDOCRINE PEPTIDE HORMONE PROCESSING

The bioactivity of peptide hormones is determined by both amino acid sequence and post-translational modifications. Neuroendocrine peptide hormones are synthesized as large protein precursors, containing an NH_2-terminal signal sequence that directs the elongating peptide into the lumen of the RER and is then co-translationally cleaved, forming the pro-hormone (Fig. 1). This protein is then transported to the Golgi apparatus, where early post-translational processing occurs while being packaged into secretory granules. Within secretory granules, many post-translational modifications which are critical to the bioactivity of the peptide hormone take place, such as NH_2-terminal α-N-acetylation and COOH-terminal α-amidation.[7]

Our laboratory has been working for several years on the post-translational modifications of pro-ACTH/endorphin, the pro-hormone that gives rise to adrenocorticotropic hormone (ACTH), α-melanocyte stimulating hormone (αMSH), and β-endorphin. The post-translational processing of pro-ACTH/endorphin provides an interesting model system, since it has been shown that this pro-hormone is cleaved in a tissue-specific manner - different peptides with different hormone activities are produced from pro-ACTH/endorphin in the anterior and intermediate lobes of the pituitary. While the anterior lobe produces primarily ACTH and β-lipotropin, the intermediate lobe processes these peptides further to αMSH and inactive β-endorphin fragments, respectively.[8] Thus, pro-ACTH/endorphin is an excellent system for studying neuroendocrine peptide hormone biosynthesis and the enzymes which catalyze these modifications.

Since it appears that the biological activities of several neuroendocrine peptide hormones may be altered

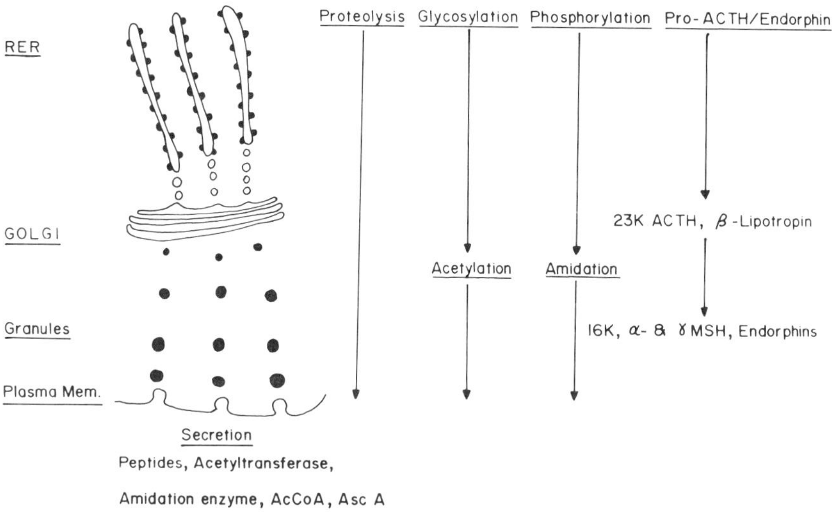

Figure 1. Schematic of Neuroendocrine Peptide Processing

under conditions of copper deficiency, the various neuroendocrine peptide processing events should be examined for their effects on biological activity. Limited proteolytic cleavages are important in creating or destroying a given biological activity. For example, the cleavage of ACTH in the intermediate pituitary causes the loss of the steroidogenic activity of ACTH, while generating the sequence for αMSH (des, mono, and di-acetyl-ACTH(1-13)NH_2), which possesses completely different hormonal actions.[9] However, the arginyl esterases that have been postulated to catalyze these cleavages have not been shown to be copper-dependent,[10] and thus may not play a role in the copper-deficient activity modification of these peptides. NH_2-terminal α-N-acetylation is also an important modulator of the bioactivity of αMSH, for example,[11] but this is a relatively rare post-translational modification, and thus cannot be responsible for these effects. For similar reasons, phosphorylation and glycosylation are probably not responsible as well.

However, COOH-terminal α-amidation of neuroendocrine peptides is a common modification, and all peptide hormones described above as possible contributors to the symptoms of copper deficiency are indeed amidated. An examination of peptide hormone precursor sequences reveals that amidated peptides possess glycine residues in their precursor sequence immediately COOH-terminal to the amino acid residue that will be amidated in the final hormone. In addition, the amidated forms of these peptide hormones are almost universally required for maximal biological activity. Thus, if the amidation of neuroendocrine peptide hormones was copper sensitive, copper deficiency could drastically modify the effective levels of many bioactive hormones.

AMIDATION OF α-MELANOCYTE STIMULATING HORMONE

The examination of the α-amidation of αMSH was begun by placing enzymatically-dissociated rat intermediate pituitary cells into serum-free culture, labeling the cultures for various times with [^{3}H]tryptophan, and extracting the cells in an acidic media. The labeled αMSH-related material was then purified by HPLC. When the cells were labeled immediately after plating, the major labeled αMSH-related peptide produced by the cultures was acetyl-ACTH(1-13)NH_2, which is amidated αMSH found in the intact intermediate pituitary.[12] However, when the cells were incubated in culture for 20 hours before labeling, the major labeled product was unexpectedly acetyl-ACTH(1-14), the glycine extended form of αMSH.

Thus it appeared that the primary intermediate pituitary cultures were losing the ability to form amidated αMSH, and that perhaps the immediate precursor to amidated αMSH is the glycine extended hormone. All other post-translational processing events examined in these cells were shown to be proceeding normally.

Since cell culture is a relatively complex experimental system, it was decided that the enzyme(s) responsible for peptidyl α-amidation should be examined <u>in vitro</u> to assess the need for various cofactors. Amidation ability in secretory granule preparations of various tissues was determined utilizing a modification of an assay developed earlier.[13] Briefly, an iodinated synthetic peptide, D-Tyr-Val-Gly, was used as a substrate, and an ion-exchange column was used to separate the labeled substrate from the labeled product, D-Tyr-Val-NH_2. With

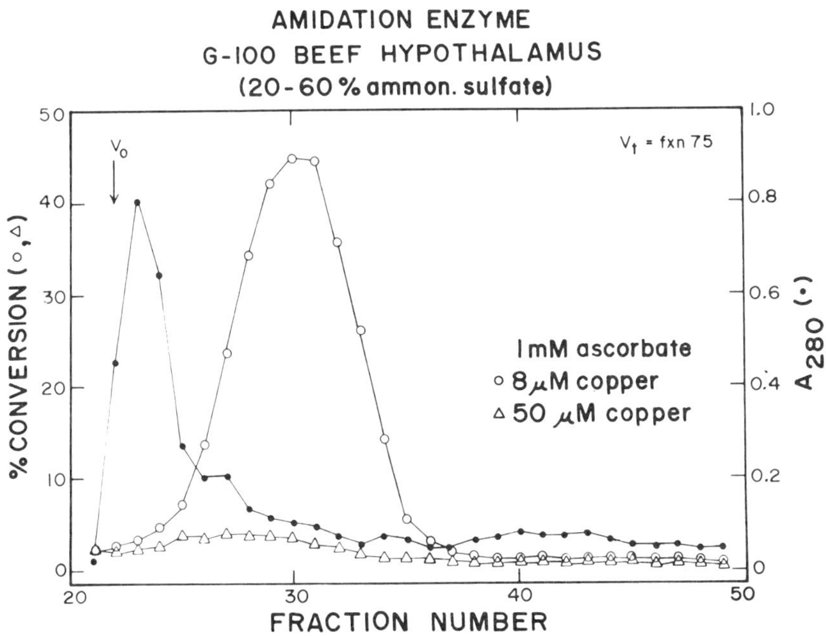

Figure 2. Sephadex G-100 Elution Profile of Beef Hypothalamic Secretory Granules - Amidation activity of the resulting fractions was assayed as described in the text at 2 copper concentrations.

this assay, it was determined that copper, among all cations tested, had the ability to stimulate amidation enzyme activity.[14] When extracts of pituitary secretory granules or hypothalamic synaptosomes were purified by G-100 Sephadex gel filtration chromatography (see Fig. 2 for example), the copper concentration required for maximal activity in all tissues tested was about 3 μM (Fig. 3).[15,16] The difference between the copper concentration maxima was apparently due to copper chelators in the crude extracts that were removed by the G-100 treatment. The amidation activity was inhibited by EDTA and o-phenanthroline with micromolar potency, while DDC inhibited the activity with 10-fold greater potency.[14]

In addition to the copper requirement, other factors required for maximal activity were also determined.

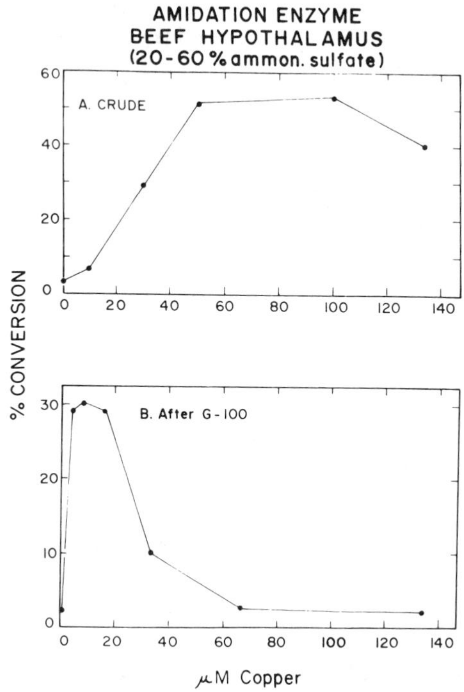

Figure 3. Copper Response of Crude and G-100 Purified Amidation Activity.

Reduced ascorbate stimulated the amidation enzyme by over 10-fold, while the activity was abolished when the assays were performed in an argon atmosphere, indicating that molecular oxygen was also required.[14] Because of the similarity to dopamine β-hydroxylase (a monooxygenase), the amidation enzyme was named peptidyl α-amidating monooxygenase, or PAM. Thus, it has been proposed that the same or very similar enzymes are found in most neuroendocrine tissues, and is/are responsible for the amidation of virtually all amidated peptide hormones.

When the rat intermediate pituitary cultures were re-examined utilizing the knowlege gained from the in vitro studies described above, it was found that, while copper and oxygen were already present in the culture medium, the addition of ascorbate to the medium resulted in a rapid reestablishment of amidation ability.[17] This concentration of ascorbate was also found to allow maximal transport of ascorbate into AtT-20 anterior pituitary-derived tumor cells,[18] thus suggesting that all phases of the amidation process require about the same concentration of ascorbate for maximal activity. When DDC was added to the cultures in preliminary experiments, the ability to amidate αMSH was lost, indicating that copper is required by these cultures for amidation as well. Also, treatment of rats with disulfiram (Antibuse) or DDC resulted in the appearance of glycine-extended αMSH in the pituitaries of these animals, demonstrating an in vivo copper requirement.[19]

AMIDATION OF THYROTROPIN RELEASING HORMONE

To extend the determination of the mechanism of amidation from the endocrine system to the central nervous system, studies of the biosynthesis of thyrotropin releasing hormone (TRH) in the rat hypothalamus were undertaken. The structure of TRH is atypical of most neuroendocrine peptides. It consists of only 3 amino acids, pyroGlu-His-Pro-NH_2, and thus very little is known about the mechanism of biosynthesis and post-translational processing of this unusual peptide hormone. In this study, it was postulated that the precursor to TRH, like αMSH, possessed a COOH-terminal glycine residue as the signal for eventual amidation. Since the culturing of the intermediate pituitary cells yielded important information on the immediate precursor to αMSH due to the lack of ascorbate in the medium, it was decided to culture fetal rat hypothalamic neurons in the presence and absence of ascorbate, and to determine if the levels of TRH, and later TRH-Gly, changed as did αMSH and αMSH-Gly in the experiments described above.

The hypothalamic cells were cultured for 1 to 7 days, extracted, and the TRH content was determined by RIA. The antibody used in this assay only binds to amidated TRH, and thus cannot detect TRH-Gly. While the TRH content of the cultures without ascorbate remained constant or even dropped slightly between 3 and 7 days, the cultures containing 250 μM ascorbate increased in TRH content during the same period (Fig. 4).[20] Thus, the level of amidated (and thus bioactive) TRH increased only when ascorbate was

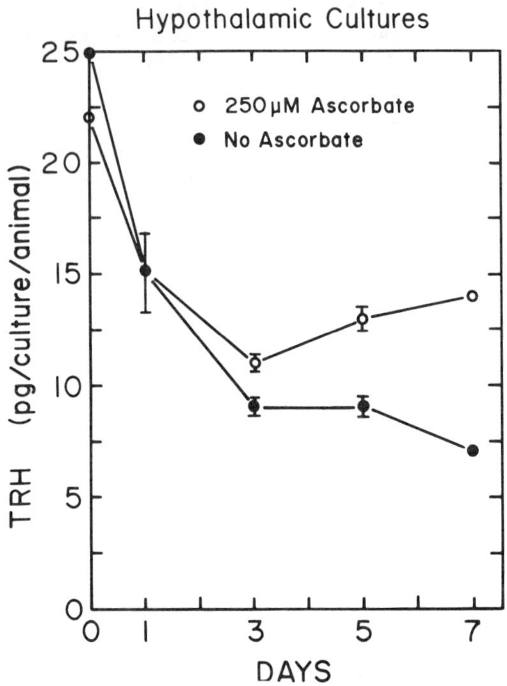

Figure 4. Time Course of TRH Content in of Rat Neonatal Hypothalamic Cultures. (From reference 20.)

included in the medium, which indicated that there may be a similar cofactor requirement for amidation in the CNS as was previously shown for the pituitary. This dependence was quantitated by a dose response experiment, and it was shown that the the half-maximal response occurred at 20 μM ascorbate, which was similar to that observed for the pituitary amidation system.[15 16]

At this point, an antiserum was produced against the synthetic glycine-extended form of TRH, or TRH-Gly. When the ratio of TRH to TRH-Gly was examined in the presence and absence of ascorbate, it was shown that after 7 days the cultures with ascorbate contained 100 pg of TRH and 106 pg of TRH-Gly per hypothalamus. This result was almost identical to the concentrations of these two peptides found in the fetal rat hypothalamus,[20] which indicated that the culture system was yielding similar values as the intact tissue. However, in cultures not containing ascorbate, the TRH concentration was 3-fold lower than in ascorbate-containing cultures, while the TRH-Gly levels were 3-fold higher. These results strongly suggest that TRH-Gly is the direct precursor to TRH. Due to the similarity of the CNS and pituitary amidation systems, it is likely that the amidation of TRH in the CNS is also copper dependent. Thus, copper may be important for the generation of bioactive neuropeptides as well.

The importance of copper in the post-translational processing of neuroendocrine peptide hormones is thus well established. However, much work remains to correlate the known symptoms of copper deficiency with the altered bioactivity of amidated hormones, and to evaluate the possible regulatory importance of varying copper levels in these tissues. In addition, the actual role of copper in the amidation process is not well understood, and the elucidation of this mechanism will provide important information on this modification that is so critical to the biological activity of these peptide hormones.

ACKNOWLEDGEMENTS

We would like to thank Paul Shields and Gary Wildey for their helpful discussions during the preparation of this presentation and manuscript. Supported by NIH grant NS-20396 to C.C.G. C.C.G. is an Established Investigator of the American Heart Assn., and T.R.G. is a Special Investigator of the Southeastern Pennsylvania Chapter of the AHA.

REFERENCES

1. Feller, DJ, O'Dell, BL. J. Neurochem. 1980, 34, 1259.
2. Danks, DM, Camakaris, J, Herd, S, Mann, JR, Phillips, M. In: Biological Aspects of Metals and Metal-Related Diseases, Raven Press, New York, p 133.
3. Opsahl, W, Zeronian, H, Ellison, M, Lewis, D, Rucker, RB, and Riggins, RS. J. Nutr. 1982, 112, 708.
4. Hunt, DM. Ciba Found. Symp. (N.S.) 1980, 79, 247.
5. Allen, DK, Hassel, CA, and Lei, KY. J. Nutr. 1982, 112, 2043.
6. Fell, BF, King, TP, Davies, NT. Histochem. J. 1982, 14, 665.
7. Mains, RE, Eipper, BA, Glembotski, CC, Dores, RM. Trends Neurosci. 1983, 6, 229.
8. Mains, RE, Eipper, BA. J. Biol. Chem. 1981, 256, 5683.
9. Mains, RE, Eipper, BA. Ann. New York Acad. Sci. 1980, 343, 94.
10. Loh, YP, Brownstein, MJ, Gainer, H. Ann. Rev. Neurosci. 1984, 7, 189.
11. O'Donohue, TL, Handlemann, GE, Chaconas, T, Miller, RL, Jacobowitz, DM. Peptides 1981, 2, 333.
12. Glembotski, CC, Eipper, BA, Mains, RE. J. Biol. Chem. 1983, 258, 7299.
13. Bradbury, AF, Finnie, MDA, Smyth, DG. Nature 1982, 298, 686.
14. Eipper, BA, Mains, RE, Glembotski, CC. Proc. Natl. Sci. USA 1983, 80, 5144.
15. Mains, RE, Glembotski, CC, Eipper, BA. Endocrinology 1984, 114, 1522.
16. Eipper, BA, Mains, RE, Glembotski, CC. In: Biogenetics of Neurohormonal Peptides, Academic Press, London, 1985, p 187.
17. Glembotski, CC. J. Biol. Chem. 1984, 259, 13041.
18. Shields, PP, Gibson, TR, Glembotski, CC. Endocrinology 1986, 118, 1452.
19. Mains, RE, Park, L, Eipper, BA. Endocrinology 1986, 118 (Supplement), 227.
20. Glembotski, CC, Manaker, S, Winokur, A, Gibson, TR. J. Neurosci. 1986, 6, 1796.

DISCUSSION

HARRIS: How strong is the case for this being an enzyme? Basically most Cu-dependent enzymes generally resist DDC removal of Cu. On another point, there have been occasions where Cu in combination with ascorbate has led to production of artifactural catalytic reactions that appear to be enzymatic in nature but basically they are not. I was waiting to see a slide in which you showed that Cu and ascorbate plus peptide substrate generated the same amidated product, to document that this is an enzymatic reaction.

GIBSON: Yes, those experiments have been done and indeed you do require enzymatic activity. If you boil the enzyme and add Cu and ascorbate, enzyme activity is not observed. In addition, the enzyme has now been virtually purified to give a single band on a gel column. From measurements of enzymatic activity it's clear that indeed it is an enzyme and cloning is in progress.

SORENSON: (Comment directed to Prof. Harris) DDC is a well known dopamine β hydroxylase (DβH) inhibitor and it inhibits by removing Cu by a chelation mechanism. DDC is also a well known inhibitor of Cu-Zn SOD and alcohol dehydrogenase, which is a Zn-dependent enzyme. DDC will complex with metalloelements in protein matrices and remove those metalloelements or else prevent enzyme catalysis by blocking substrate interaction at the active site.

GIBSON: Yes, there are actually many parallels between DβH and the amidating enzyme.

SORENSON: I have a comment pertaining to DβH and the amidating enzyme. At one time it was thought that DβH required vitamin C to provide the two reducing equivalents that this mixed-function oxidase requires. However, Boyadzhyan (Biochem. 1985, 50, 75) recently suggested that Extremely Acidic Copper Containing Protein (EACCP) which has two forms similar to DβH, one is cytosolic and the other is membrane bound, co-exists with DβH and supplies the two reducing equivalents required by DβH rather than vitamin C based upon the observation that EACCP is much more effective than vitamin C in catalyzing dopamine β hydroxylation in adrenal medullary cells. This leaves open the possibility that in the pituitary and in the hypothalamus, perhaps, there is some EACCP that provides the needed reducing equivalents. Has that been looked at?

GIBSON: Absolutely not. I wish it had. We did quite a few studies on the reducing agent itself and ascorbate was the best of all of those that we tested. There were a few that were partially effective. Tetrahydrobiopterin and catecholamines are actually a very good source of reducing equivalents but not as good as ascorbate. That's similar to the DβH story. You are right, it just has not been examined. The problem is that you can obtain DβH in large amounts with relative ease from adrenals compared to the amount in the pituitary secretory granules and for that reason alone those studies haven't been done. The hypothalamus would be expected to have an even lower concentration than the pituitary.

SORENSON: Might it be an NADPH-dependent reductase that's providing these reducing equivalents?

GIBSON: NADPH doesn't have any effect in crude extracts. It doesn't support the reaction.

BHATHENA: Now that opiate peptides and neuro peptides have been found outside of the brain for example in pancreas, adrenals, and heart, do you think that amidating enzymes are in those tissues?

GIBSON: Yes , there have been reports of amidating enzyme in non-neuronal tissues including pancreas and the GI tract. TRH is also found in the pancreas.

BHATHENA: In the nerve terminals or in granules?

GIBSON: I don't know if they have gotten that close. I think it's just from tissue.

BHATHENA: You mentioned something about atrial naturetic peptide. Can you expand on that a little more?

GIBSON: It's not an amidated peptide. We are working on it because it's a very intriguing biosynthetic peptide project. It's stored in a precurser form yet it seems to be released in the finally processed form, unlike the usual stored and released forms of endocrine peptides.

MELETHIL: On one of your slides you showed reversal of amidation as the concentration of Cu was increased from 2 uM to 10 uM and I was wondering why.

GIBSON: Actually there is a lot of voodoo involved in some of these reactions. The problem is associated with the addition of Cu and ascorbate at the same time. They are mutually incompatible. Cu tends to catalyze the oxidation of ascorbate.

HARRIS: Was the indicated Cu concentration added or was it already there?

GIBSON: It was added to the optimal concentration.

DISCUSSION

ELLIOTT (Graham): Prof. Barnea, other groups have shown Cu-induced expression of a high specificity prostaglandin receptors. Have you ever looked at specific prostaglandin E_2 receptors in your system? Could that account for some of these effects that you find?

BARNEA: I found only one paper that dealt with Cu-mediated specific binding. Membranes prepared from mammary glands in the presence of dithiothreitol lost their high affinity binding for PGE_2 which was restored with Cu. These data led us to question as to whether or not Cu might be similarily involved in our system. In a desperate attempt to support a grant application, I tried to do some PGE receptor binding studies using brain membranes and I found that 80% to 90% of binding was non-specific binding. The answer to your question is that there is very little known about a role for Cu in PGE receptor binding, especially in the brain. So it's very hard to prove or disprove the hypothesis that Cu indeed modifies PGE receptors in any way, either the number of binding sites or affinity.

HARRIS: As we close our session on the basic aspects of Cu transport and incorporation into protein structures we leave you with some very interesting thoughts. We heard from several of our speakers this morning that Cu does not easily find its way into proteins. Nature is able to do things that we are not able to do or reproduce in the test tube. We do not understand all of those factors that seem to be involved. We heard from Prof. Greenaway that the enzyme our laboratory has been studying, lysyl oxidase, contains Cu(I) which means that if Cu is going to be incorporated into the apoprotein it's very likely that there would have to be a reduction step somewhere along the pathway of incorporation. I raise this issue because as we start thinking of some of the pharmacological actions of these Cu complexes we are confronted with the multivalent nature of Cu and we have to design our complexes to accomodate that property as well as solubility properties that allow the complex to penetrate into the cell and reach the target. Perhaps some of these issues will be addressed during the remainder of the week. We are very interested in knowing the chemistry and fate of these Cu complexes as well as their biological function.

GREENAWAY: In general, Cu(I) complexes are very very uncommon under aerobic conditions. I don't think it's feasible to make Cu(I) salicylate or something like that and even if one could, it's not possible to get it into the body before it is oxidized to Cu(II) salicylate. It's a very difficult problem you're suggesting. What I am suggesting is that no one really has looked at solving this type of problem. I think inorganic chemists probably have a wide open field here. They need to look at more of this sort of thing.

HARRIS: I recall some work that was done with a stable Cu(I) complex, in acetonitrile, which was used in reactivating cytochrome oxidase. We were quite excited about those results until they turned out to be artifactual, unfortunately. Basically, we did have a stable Cu(I) complex.

GREENAWAY: But even those complexes readily oxidize under aerobic conditions to Cu(II).

CROOK: The conversion of Cu(II) to Cu(I) may very well be the situation under pathological conditions where it's no longer aerobic but anerobic, in pus pockets or in tumors.

HARRIS: The amount of reduced glutathione is quite high in cells. As Prof. Barnea has pointed out, that would generally be conducive to keeping Cu in the Cu(I) oxidation state.

DISILVESTRO: I have a question for Prof. Greenaway. Catechin plus Cu is an outstanding pseudo lysyl oxidase (LO). I would like to know if the structure of Catechin is such that it could stabilize Cu in the Cu(I) state. Maybe that's why it's able to increase LO activity in vitro.

GREENAWAY: When I saw the structure of that molecule I thought it looked a bit like PQQ. If those hydroxyl groups were oxidized to either acids or aldehydes then you might have some sort of model that could bond with Cu. I don't think that unaltered Catechin would have such effects. Phenols don't bond that strongly with Cu.

HARRIS: As we follow the literature on Catechin we see that this agent does indeed mimic some of the properties of

LO such as decreasing collagen solubility in a lathyritic animal, suggesting that there is some type of cross-linking that occurs as a result of using this compound. In Prof. DiSilvestro's studies, he actually showed that Cu complexes did activate LO in vivo. It is not known as to whether or not Catechin can act as a cofactor for LO.

GIBSON: I'd like to point out that there are subcompartments, for example secretory granules, that concentrate ascorbate to very high levels and there could very well be reducing environments within cells that could support the existance of Cu(I) complexes.

RISK: I know that mycrobial catechol sulfates are used as artificial chelators. Mammals secrete catechols as sulfates. Is there any evidence that they may complex with Cu and serve as transport molecules?

SORENSON: Rajan, Colburn, and Davis (Metal Ions In Biological Systems, volume 6, 292, 1976) have made Cu-ATP-catecholamine complexes which they suggest as possible catecholamine storage forms. These complexes are made under anerobic conditions to avoid catecholamine oxidation. This work has relevance to anerobic storage and utilization situations. The problem with Catechin as I look at it is that it has a catechol group which is easily oxidized. There may be some opening of the flavan and catechol rings, which are generally not stable when exposed to metallo-elements like Fe or Cu in air, that yield aldehydic compounds capable of forming Schiff bases that might facilitate appropriate oxidations or cross-linking.

BARNEA: Actually, there is a lot of Cu associated with catecholamines and ATP in chromatin granules and the catecholamines are not oxidized. In the early sixties it was hypothesized that a Cu-ATP-catecholamine complex existed in secretory granules and released as such a complex. In spite of the presence of Cu and catecholamines, which are so readily oxidized by Cu, catecholamines are not oxidized in the granule. Perhaps intragranule pH, 6.0, and reduction potential account for the existance of this Cu-ATP-catecholamine complex.

SORENSON: Those may be Cu(I) complexes. I don't know that anybody has ever characterized those. Rajan is trying to account for the existence of Cu-ATP-catecholamine complex in secretory granules. We've made very few Cu(I) complexes because I'm sure Cu is easily transformed in vivo to whatever oxidation state is needed.

Copper Complexes in Essential Metalloelement Nutriture and Disease States

INDICES FOR ASSESSING COPPER NUTRITURE IN ANIMALS AND HUMANS

J. Cecil Smith, Jr.

USDA, ARS, Beltsville Human Nutrition
Research Center, Vitamin and Mineral
Nutrition Laboratory
Beltsville, MD 20705

The importance of copper in animal nutrition was first demonstrated by feeding a diet (milk) low in copper (1). Copper deficiency also developed in human infants fed diets similar to those employed for inducing copper deficiency states in experimental animals (2,3,4). Although the importance of adequate copper content of animal diets has long been recognized, only recently has the importance of adequate dietary copper for humans received renewed interest. Indeed, the recommendation of a "safe and adequate" daily dietary intake of 2-3 mg by the Food and Nutrition Board of the U.S. National Academy of Sciences in 1980 (5) has resulted in increased attention to methods for assessing copper status in both animals and humans.

DIETARY INTAKE

Since the major source of copper is dietary, measurement of the copper intake from food is one of the prime indices for assessing copper status. In regard to humans, the daily copper intake can be calculated using nutrient data bases, although at present there is a need for more complete and accurate data concerning the copper content of individual foods. Another problem using this method concerns estimating the contribution of water and other beverages because they can vary greatly in their copper content. This technique is often used for large dietary surveys or for institutions, e.g. hospitals.

Where a more precise method for measuring copper intake is required the "duplicate meal" or "plate" has been employed. This method involves a collection of duplicate portions of all foods and beverages consumed. The major advantage of this technique is that it allows accurate analytic measurements of the actual intake of copper including any exogenous contamination during the preparation of the food. The disadvantages are that it is time-consuming, expensive and not practical for many settings including assessment of large population groups.

Measurement of the daily copper ingested from self-selected meals by free-living indivuduals has indicated intakes well below the lower recommended limit. Indeed, a recent one-year study of such individuals reported average intakes of 1.2 mg for 28 adults living in the Washington, DC area (6). The health consequences of habitually consuming less than the recommended levels are unknown. However, a link between ischemic heart disease and inadequate copper nutriture has been hypothesized (7,8).

For assessment of animal feeds, analytic measurement is rather simple with present-day instrumentation. Likewise, sample collection of animal diets is obviously less complicated than for humans.

PLASMA/SERUM CONCENTRATION

The most frequently employed parameter for assessing copper nutriture has been serum or plasma concentration (9). However, there are complicating factors which must be considered. For example, Lahey et al. (10) reported that serum copper concentrations were significantly higher in females than in males. In addition, higher copper concentrations have been noted in serum than in plasma (11) apparently as a result of specific anticoagulates such as citrate causing fluid shifts from the blood cells (12). A diurnal variation in plasma copper also has been demonstrated (13). Several diseases or conditions may alter the blood copper concentration. Diuretics (14) and regular strenuous exercise (15) are among other factors reported to affect serum copper concentrations. Hypercupremia has been associated with pregnancy, oral contraceptive agents (16), infectious diseases, hematologic disorders, neoplasms, neurological, cardiovascular, and rheumatic diseases (2). In contrast, hypocupremia ($<$ 80 ug Cu/100 ml) has been associated with Wilson's Disease, kwashiorkor, marasmus, tropical sprue, macrocytic anemia, malabsorption due to small bowel disease, hyperchromic and hypochromic anemia, and nephrosis (17,18).

"Normal" values for serum or plasma copper concentrations may vary widely among laboratories because the accuracy of the methods used has not always been validated. Numerous methods for plasma or serum copper concentrations have been published; however, they are only recently becoming standarized (19). Likewise, recently a certified standard reference material (bovine serum) (20) has become available through the National Bureau of Standards and can aid in monitoring accuracy of methodologies.

Although an abnormal concentration of copper in plasma or serum must be interpreted with caution, demonstrable copper deficiency in man and animal is accompanied by a low copper concentration. In addition, this technique remains the most readily available, quickest and simplest.

CERULOPLASMIN

Another frequently used parameter is the measurement of plasma ceruloplasmin oxidase activity. A major portion of the copper present in mammalian plasma is associated with this alpha-globulin protein. The fraction of total plasma copper associated with ceruloplasmin has recently been reported to approximate 60% (21) although earlier reports indicated more than 90% (22). At any rate, the serum ceruloplasmin oxidase activity is frequently used for assessing copper nutriture in a clinical setting. If ceruloplasmin oxidase activity is extremely sensitive to copper status in humans, it may be helpful in detecting marginal status. However, results from recent animal studies indicate that ceruloplasmin oxidase activity is not correlated with liver copper concentrations when the animals are severely deficient because the ceruloplasmin activity decreased to non-detectable levels prior to development of an overt deficiency (23). Although, the oxidase activity may become non-detectable soon after animals are fed a copper deficient diet, it may be that the plasma concentration of ceruloplasmin is more closely related to copper status than is the oxidase activity. Thus the use of determining ceruloplasmin concentration as well as oxidase activity may be helpful.

Liver is considered the prime storage organ for copper and thus liver copper concentration is usually the parameter of choice for assessing copper status in animals. Obviously, this tissue is not available for routine assessment of copper nutriture in humans.

SUPEROXIDE DISMUTASE ACTIVITY (SOD)

Another promising parameter for copper assessment is the SOD activity of the erythrocytes. This copper metalloenzyme catalyzes the conversion of the superoxide radical to hydrogen peroxide and oxygen. Recent animal studies have indicated a strong association between copper status (as determined by liver copper concentration) and liver superoxide dismutase activity (24). Insufficient data from human studies involving SOD activity of erythrocytes are available to establish its value but a preliminary report indicates its potential (25). At present, a deterrent to the use of SOD for assessing status is the difficulty in methodology.

HAIR

Copper concentration in hair has also been used as an index of copper nutriture (8). However, a report concerning two children with overt copper deficiency indicated normal hair copper levels and (26) concluded that the "considerable variation in the copper content of hair is probably independent of body stores and may be more closely related to hair production needs . . . . .". It appears that hair copper concentrations may not accurately reflect the copper status. Thus, as stated by Hambidge (27) "interpretation of analytical data on hair copper requires great caution".

URINE

Urinary copper excretion via the urine is another potential parameter. However, estimates of normal 24 hour urinary excretion indicate a relatively low concentration with wide variations among laboratories. The apparent discrepancies probably results from methodologic differences and possible contamination. The low concentration of urinary copper precludes its analysis using conventional flame atomic absorption spectroscopy. More sensitive methodologies such as AAS graphite furnace (flameless) techniques must be developed for urinary copper.

BALANCE STUDIES

The metabolic balance technique has been routinely used to assess nutrient requirements, retention and/or bioavailability. However, it cannot be considered a method for assessment and is not practical. Obviously, negative balance extended over a long duration would result in a compromised status. The use of stable isotopes (copper) coupled with the balance technique may provide more definitive data (28). Problems associated with metabolic balance studies involving minerals have been discussed by others (29,30).

SUPPLEMENTATION OR REPLETION

Although not helpful for situations where an expedited assessment is required, the most definitive index for assessing overt copper deficiency is a significant physiological clinical response to copper supplementation under controlled conditions (25). Usually the supplementation is oral, although copper is routinely administered intravenously to subjects receiving total parenteral nutrition. The design of supplementation studies may vary but may include depletion-repletion or cross-over schemes. In animal studies, supplementation is usually accomplished by adding a copper compound to a deficient diet or deionized (copper-free) drinking water. The response indices usually include specific biochemical measurements and/or functional assessments such as electrocardiograms, glucose tolerance tests, or tissue pathology.

ELECTROCARDIOGRAM (ECG) AND OTHER NON-SPECIFIC INDICES

Recent data have indicated abnormal cardiac function in experimental copper deficient animals as measured by ECG. It has been suggested that decreased copper status may affect the electrophysical aspects of the intraventricular conducting system of the heart. As a result, abnormal ECGs and premature beats may occur (25, 31). Abnormal ECGs are non-specific and alone can not definitively assess copper status.

Several other non-specific parameters have been reported to be abnormal in humans or animals exhibiting copper depletion; these include decreased hematocrit, increased serum, cholesterol and triglycerides (31), impaired glucose tolerence (32), altered blood pressure (33), and tissue pathology (gross and micro) including hypertrophy of the heart (33,34).

POTENTIAL INDICES

Lysyl oxidase Recent experiments have added evidence that the activity of the copper-dependent metalloenzyme lysyl oxidase is decreased in the tissue of copper-deficient animals (35). Specifically, there was a significant decrease in the enzyme's activity of both aortic tissue and plasma in swine fed a copper deficient diet compared to supplemented controls. The accuracy and sensitivity of the enzyme to reflect copper status has yet to be determined. In addition, and of utmost importance, is the problem that the method is difficult, time-consuming, and presently not available for routine analysis. If validated as an accurate and sensitive method, automation of the technique would increase its application, especially if plasma can be used as the biological sample.

Plasma Enkephalins A recent report from our Beltsville Human Nutrition Research Center has indicated that plasma enkephalin concentrations decreased in male subjects fed a low copper (1.0 mg/day) diet for 11 weeks (36). After repletion with copper (3 mg/day), the plasma enkephalins returned to pretest concentrations. Although this is the first evidence that plasma enkephalins may be useful in assessing copper status in humans (males), and must be considered preliminary, these findings suggest a promising "new" parameter.

Cerebrospinal Fluid (CSF) Studies have measured the copper concentration of CSF and this biologic fluid has potential as an index for assessing copper status. A normal value for the copper concentration of CSF in humans was reported several years ago (37). However, it is doubtful if sampling CSF fluid of humans can be justified for copper assessment alone. Perhaps of greater importance, it must first be established that CSF accurately reflects copper status.

SUMMARY

Dietary copper content, plasma/serum copper concentration, and ceruloplasmin activity have been most frequently used for assessing copper status of animals and humans. Other parameters have included both biochemical and functional indices. At present, the most definitive method for establishing that a true copper deficiency exists is controlled supplementation studies. Several impractical or inaccurate parameters were discussed and included hair and urinary copper concentrations and metabolic balance studies. Likewise, non-specific indices which may be altered by an impaired copper status but which should not be used alone for copper assessment were discussed.

FUTURE NEEDS

Although several indices are presently available for measuring overt copper deficiency, the challenge is to develop techniques sufficiently sensitive, accurate, and specific to detect marginal copper status. These methods are needed in order to definitively establish if the typical U.S. dietary intake, which is less than recommended, is compromising copper nutriture in our country. Hopefully, the contributions made at this Symposium will help answer this important question which may have long-reaching health-related effects.

REFERENCES

1) Hart, EB, Steenbock, H, Waddell, J, and Elvehjem, CA, J. Biol. Chem. 1928, 77, 797.
2) Mason, KE, J. Nutr. 1979, 109, 1979.
3) Graham, GG and Cordano, A, Johns Hopkins Med. J. 1969, 124, 139.
4) Naveh, Y, Hazani, A and Berant, M, Pediatrics 1981, 68, 397.
5) National Academy of Sciences, National Research Council, Food and Nutrition Board: "Recommended Dietary Allowances" 9th Ed., Washington, DC 1980.
6) Patterson, KY, Holbrook, JT, Bodner, JE, Kelsey, JL, Smith, Jr JC and Veillon, C, Am. J. Clin. Nutr. 1984, 40, 1397.
7) Klevay, LM, Am. J. Clin. Nutr 1970, 23, 1194.
8) Klevay, LM, Ann. NY Acad. Sci. 1980, 355, 140.
9) Solomons, NW, J. Am. College Nutr. 1985, 4, 105.
10) Lahey, ME, Gubler, CJ, Cartwright, GE and Wintrobe, MM, J. Clin. Invest. 1953, 34, 322.
11) Rosenthal, RW and Blackburn, A, Clin. Chem. 1974, 20, 1233.
12) Smith, JC, Holbrook, JT and Danford, DE, J. Am. Coll. Nutr. 1985, 4, 627.
13) Henkin, RI, In: Mertz, W, Cornatzer, WE, eds, Newer Trace ELements in Nutrition. New York, NY, Marcel Dekker 1971, p. 255.
14) Sempos, CT, Greger, JL, Johnson, NE, Smith, EL and Seyedabadi, FM, Nutr. Rep. Int. 1983, 27, 1013.
15) Lukaski, HC, Bolonchuk, WW, Klevay, LM, Miline, DB and Sandstead, HH, Am. J. Clin. Nutr. 1983, 37, 407.
16) Smith, Jr, JC and Brown, ED, In: Prasad, AS, Oberleas, D, eds. Trace Elements in Human Health and Disease, New York, NY, Academic Press 1976, p. 315.
17) Wintrobe, MM, Cartwright, GE and Gubler, CJ, J. Nutr. 1953, 50, 395.
18) Wintrobe, MM, Am. J. Clin. Nutr. 1958, 6, 75.
19) Osheim, DL, J. Assoc. Off. Anal. Chem. 1983, 66, 1140.
20) Veillon, C, Lewis, SA, Patterson, KY, Wolf, WR, Harnly, JM, Versieck, J, Vanballenberghe, L, Cornelis, R, and O'Haver, TC, Anal. Chem. 1985; 57, 2106.
21) Linder, M, J. Nutr., Growth and Cancer 1983, 1, 27.
22) Delves, HT, Clin. Chim. Acta. 1976, 71, 495.

23) Fields, M, Ferretti, RJ, Smith, Jr, JC and Reiser, S, J. Nutr. 1983, 113, 1335.
24) Fields, M, Ferretti, RJ, Judge, JM, Smith, JC and Reiser, S, Proc. Soc. Expt. Biol. Med. 1985, 178, 362.
25) Reiser, S, Smith, Jr, JC, Mertz, W, Holbrook, JT, Scholfield, DJ, Powell, AS, Canfield, WK and Canary, JJ, Am. J. Clin. Nutr. 1985, 42, 242.
26) Bradfield, RB, Cordano, A, Baertl, J and Graham, GG, Lancet 1980, 2, 343.
27) Hambidge, KM, Am. J. Clin. Nutr. 1973, 26, 1212.
28) Turnland, JR, Michel, MC, Keyes, WR, Schutz, Y, Margen, S. Am. J. Clin.. Nutr. 1982, 36, 587.
29) Hegsted, DM, J. Nutr. 1975, 106, 307.
30) Spencer, H, In: Zipkin, I, ed., Biological Mineralization, New York, NY, Wiley and Sons 1973, p. 689.
31) Klevay, LM, In: Rennert, OW, Chan, WY, eds., Metabolism of Trace Metals in Man Vol I, Boca Raton, FL CRC Press, 1984, p. 129.
32) Fields, M, Ferretti, RJ, Smith, Jr, JC and Reiser, S, J. Nutr. 1984, 114, 393.
33) Fields, M, Ferretti, RJ, Smith, Jr, JC and Reiser, S, Life Sciences 1984, 34, 763.
34) Steele, N, Richards, M, Darcey, S, Fields, M, Smith, J, Reiser, S, Fed. Proc. 1985, 357. Abst.
35) Ono, K, Steele, N, Richards, M, Darcey, S, Fields, M, Scholfield, D, Smith, J and Reiser, S, Fed. Proc. 1986, 45, 357 Abst.
36) Bhathena, SJ, Recant, L, Voyles, NR, Timmers, KI, Reiser, S, Smith, Jr, JC and Powell, AS, Am. J. Clin. Nutr. 1986, 43,42.
37) Kjellin, K, J. Neurochem. 1963, 10, 89.

DISCUSSION

<u>MELETHIL</u>: Why isn't whole blood analyzed as opposed to plasma or serum?

<u>SMITH</u>: It is generally understood that plasma or serum contain the more mobile forms of Cu and Zn. We do analyze other components of whole blood if we wish to obtain information concerning formed elements in whole blood.

<u>KEEN</u>: If you wish to assess the Cu status of a pregnant female, with regard to the developing fetus, her plasma is what the developing fetus is going to be using so plasma is the only relevent sample. If you have an individual receiving D-penicillamine for treatment of rheumatoid arthritis or Wilson's disease, it is much more practical to look at plasma rather than whole blood to evaluate effects of treatment. Red blood cells, platelets, and white cells would cause unwanted interferences.

<u>MELETHIL</u>: With regard to yesterday's discussions, is SOD available in a pure form, and if so, with the recent advancements in liquid chromatographic separations of proteins, do you know if any attempt has been made to measure SOD by liquid chromatography?

<u>SMITH</u>: Yes, pure bovine liver derived Cu-Zn SOD is commercially available. I don't know whether or not it has been measured by anyone using liquid chromatography.

<u>SORENSON</u>: SOD is usually determined by measurement of enzyme activity in the presence of enzymatically produced O_2^- which is detected with cytochrome C or nitroblue tetrazolium. These are reduced by O_2^- that's formed in the system. This turns out to be a very easy spectrophotometric method. If both Cu-Zn SOD and Mn SOD are present, the Cu-Zn SOD is inhibited with cyanide and then the difference between total SOD activity and the Mn SOD activity gives a measure of the Cu-Zn SOD activity. There are also immuno-methods for measuring SODs. It has turned out that immunologically reactive protein may or may not have enzymatic activity which is a problem. Since Cu-Zn SOD is water soluble and insoluble in an ethanol and chloroform mixture while Mn SOD is soluble in ethanol and chloroform, chromatographic determinations are possible.

Chromatographic separations coupled with biochemical and immunochemical measurements of activity, which are both precise and accurate, would be an important contribution.

<u>MELETHIL</u>: A lot of reports of protein separations by liquid chromatography claiming retention of activity are being published. I think there is a lot of interest now in protein separation using chromatography and, because of many problems associated with SOD determinations, it may be worthwhile looking into.

<u>BALL</u>: I'd like to point out that in addition to loss of Cu in urine there is about a 10-fold greater loss of Cu in sweat but people don't seem to analyze sweat. This is true for Mg also. I just wondered what effect fructose might have on the amount of Mg lost in sweat.

<u>SMITH</u>: Frankly, I am leery of data on sweat because of the difficulties associated with obtaining sweat samples. We have not done sweat analyses. As a result, we have no information concerning the effect of fructose on Mg loss in sweat.

<u>BALL</u>: You indicated that ceruloplasmin levels didn't change much for those individuals in your study. It's a comparable problem with plasma Mg concentration which doesn't change in disease states and therefore people don't use it as a diagnostic criterian. However, red cell SOD falls with Cu deficiency. I noticed that SOD values for these 20 patients were all at the low end of normal, and I wonder if this might be because these patients tended to be more insulin dependent.

<u>SMITH</u>: I don't want to say that ceruloplasmin measurements aren't sensitive. I also don't know that we've got good firm data on what's normal for red cell SOD in this group of individuals but these levels may indicate some basis for the possible indication of some insulin impairment.

<u>BALL</u>: Since you gave these individuals demineralized water, the amount of Mg in their foods becomes important. For 11 weeks these patients got no Mg from water and this can be a significant percentage of the Mg that people get. A deficiency of Mg certainly plays hell with the heart.

EGNER: What is the take home message for Jane Doe who is a homemaker with three growing children? Should she give them vitamins with mineral supplementation or should she give them a lot of water, a lot of juice, and hope they have Cu tubing in the house? What really is the take home message, say for dietitians?

SMITH: We would like to think that we can do this with the typical American diet. We, in general, are against, I think, supplementing certainly taking Cu alone or Zn alone. Eating a balanced diet would be the best way to proceed. On the other hand, if I were pregnant or lactating I probably would want a balanced intake as far as supplementation and I think you are all aware that Fe is always supplemented. I don't have the final answer and indeed this is the whole dilemma. Whenever the Department of Agriculture suggests that we eat a balanced diet, but we know that a balanced diet doesn't provide 15 mg of Zn or 2 to 3 mg of Cu, we are at a loss to know exactly what we should do. I don't have the answer to this question but it's the one question that we are hoping to answer. Farmers can fertilize with Cu and Zn but that's not economical and that may not be recommended by others. In some countries it's being done. In Finland for instance, they are putting Se in the fertilizer. I haven't answered your question. It's an excellent question and one that's a challenge to all of us especially those young people coming up with lots of energy. Please help us.

DIETARY SIMPLE SUGARS INCREASE THE SEVERITY OF COPPER DEFICIENCY IN EXPERIMENTAL ANIMALS.

Meira Fields

USDA, ARS, Beltsville Human Nutrition
Research Center Carbohydrate and Vitamin
and Mineral Nutrition Laboratories
Beltsville, MD 20705

INTRODUCTION

It is well established that diets consumed by subjects living in industrialized societies including the USA contain relatively high levels of simple sugars (1) but marginal amounts of copper (2). The feeding of simple sugars such as sucrose and fructose as compared to starch, to humans and experimental animals, produces undesirable changes in blood risk factor metabolites associated with heart disease (3). Low copper intake has a similar effect on these indices (4,5). Does sucrose feeding in combination with copper deficiency have an additive effect that is greater than the effect of either of these constituents alone? The purpose of our preliminary studies was to determine if diet composition can influence the severity of copper deficiency. Specifically, can a certain dietary component such as carbohydrate magnify or depress the deficiency.

This manuscript describes observations on the interaction between dietary carbohydrate and copper deficiency on metabolic and biochemical indices in experimental animals, and provides evidence that feeding fructose containing diets deficient in copper magnifies the symptoms associated with copper deficiency.

Copper-carbohydrate interaction in the rat

A. Effects on heart related abnormalities.

In order to determine the consequeces of an interaction between dietary sucrose or starch and copper on the severity of copper deficiency, weanling male Sprague-Dawley rats were randomly divided into 4 dietary groups based on the type of dietary carbohydrate and copper levels. Group 1 - sucrose, copper deficient, Group 2 - sucrose, copper supplemented, Group 3 - cornstarch, copper deficient, Group 4 - cornstarch, copper supplemented. The composition of the diets and the levels of dietary copper were described earlier (6).

Although the starch and sucrose, copper deficient diets contained the same concentration of copper, severe copper deficiency developed only in the rats fed sucrose, which resulted in 60% mortality of the animals during the 8 week study. Post mortem examination of the animals revealed a large quantity of clotted blood in the pleural cavity. Hearts from these animals were signficantly hypertrophied. Hearts from rats fed the sucrose diet deficient in copper had histopathologic features that included necrosis, thrombi formation, pericarditis and lack of mature collagen in the areas of fibrosblastic proliferation and repair. Hearts from rats fed the starch diet deficient in copper were normal in appearance and showed no abnormal histopathologic charges (6)

The reasons for the more severe copper deficiency when sucrose is fed as compared to starch are not fully understood. However, many of the metabolic effects observed after feeding sucrose to rats have been attributed to the fructose moiety of sucrose. Increases in the levels of hepatic lipogenic enzymes, liver lipid and blood triglycerides have been observed after feeding rats diets containing sucrose or fructose as compared to glucose or glucose polymers (7). Rats fed sucrose as compared to starch have been reported to have significantly higher insulin and glucose responses after a glycemic stress (8-11), and reduced insulin sensitivity, (12,13). These parameters are similarly affected when rats are fed fructose as compared to starch (14,15).

In order to determine whether the more severe symptoms of copper deficiency found when rats are fed sucrose as compared to starch are due to a specific molecule, (fructose versus glucose), or to a molecular size (simple versus complex carbohydrate), metabolic and biochemical indices were measured in rats fed diets supplemented or deficient in copper and containing either starch, sucrose, fructose or glucose (16,17).

The most dramatic interaction between copper status and the type of dietary carbohydrate was on rat mortality. Fatal heart lesions in copper deficient rats fed fructose or sucrose were more numerous than in copper deficient rats fed starch or glucose. Copper deficient rats fed sucrose or fructose as compared with starch or glucose developed anemia . The high levels of blood cholesterol which has been observed both after sucrose intake and during copper deficiency were further increased by fructose feeding (16,18) However, no correlation was found between the levels of blood cholesterol and the extent of cardiac damage.

B. Effect on insulin binding, lipogenesis, glucose oxidation and blood glucose and insulin.

Copper has been shown to stimulate insulin binding, deoxyglucose transport and lipogenesis by adipose tissues and adipocytes of copper adequate rats (6,19). The metabolism of glucose by the isolated adipocytes and adipose tissue such as lipogenesis and glucose oxidation was reduced by sucrose feeding when compared to starch feeding and was further reduced by copper defiency (6). Insulin binding was also impaired in copper deficient rats fed sucrose compared to those fed starch (6). The decrease in insulin binding was due to a decreased number of receptors with no change in binding affinity (6). Copper deficiency produced significant and large increases in the glucose response to the glucose load only in rats fed sucrose, fructose and glucose, but not in those fed starch. The increased blood glucose was associated with decreased insulin levels (16,20).

C. Effect on Lipid peroxidation of rat tissues.

The role of dietary copper in maintaining tissue integrity, through the effect of this metal on activity of the superoxide dismutase (SOD) enzyme is well recognized. However, it remained to be established if dietary carbohydrates affect the activity of the copper enzyme SOD and if other antioxidant systems are also dietary dependent. Regardless of the nature of dietary carbohydrate, copper deficient rats exhibited decreased activity of SOD. However, the decreased activity was more pronounced in rats fed fructose (21). In addition, feeding diets containing fructose decreased the activity of the seleno-enzyme glutathione peroxidase (GSH-Px) by 25 and 50% in the copper supplemented and copper deficient rats respectively, compared to enzyme activities in rats fed similar diets containing either starch or glucose. The decreased SOD and GSH-Px activities in rats fed the fructose diet deficient in copper was associated with increased tissue peroxidation and decreased hepatic ATP (21).

D. Effect on liver functions.

As the metabolism of fructose and copper occurs mainly in the liver, it would be expected that indices of liver function would be impaired by copper deficiency in rats fed a fructose diet deficient in copper. Increased liver weight (17) decreased hepatic ATP, (16,21) decreased plasma albumin (18) increased blood ammonia (17) decreased activities of the copper metalloenzyme SOD and the selenoenzyme GSH-Px (21) could be attributed to impaired liver function. The reduced activities of SOD and GSH-Px have been shown to accentuate the increased lipid peroxidation of tissue membranes (21). Thus, it is expected that in copper deficiency the activities of several hepatic enzymes of regulatory importance in the metabolism of fructose could also be affected by the severity of the deficiency. However, except for the activity of phosphenolpyruvate carboxykinase which was unaffected either by copper deficiency or by the type of dietary carbohydrate, the hepatic activities of glucose 6'-phosphate dehydrogenase, malic enzyme, L-glycerophosphate dehydrogenase and fructose 1,6-diphosphatase were unaffected by copper deficiency but were affected by the type of the carbohydrate in the diet (22,23). Thus, the deleterious effects of a fructose diet deficient in copper on biochemcial and physiological indices could not be due to an immediate metabolite of fructose. However, the involvement of a subsequent metabolite of fructose in the mechanism of copper utilization/or requirement cannot be excluded.

E. Is the effect of the interaction reversible?

It was important to determine whether the severity of the metabolic and cellular lesions caused by copper deficiency and fructose feeding could be reversed by changing the type of dietary carbohydrate from fructose to either starch or glucose. During the 5th week of dietary treatments, the carbohydrate of the copper deficient fructose diet was replaced by either starch or glucose and the remaining rats continued to be fed the copper deficient fructose diet. The study was terminated 5 weeks later. The results of that study showed that fructose has no adverse effect that is permanent and cannot be reversed. The interaction between dietary fructose and copper deficiency is specific for fructose. Once fructose is removed from the diet, the deleterious effects cease to exist. The enhanced severity induced by feeding fructose was reversible and the impaired metabolic and cellular indices were ameliorated by altering only the type of dietary carbohydrate. Although, both starch and glucose were protective against the severity of copper deficiency, starch was more effective (17,21).

F. Absorption, tissue distribution and excretion of copper.

Carbohydrates, especially fructose, have been reported to form chelates with minerals in the intestine, which may alter the absorption of the inorganic element (24). In addition, alterations could occur within the intestine when fructose is the sole dietary source of carbohydrate which could affect copper bioavailability.

As copper transport has been reported to occur also from the stomach, it was possible that copper absorption is inhibited by fructose already from that tissue. Under anesthesia, stomachs of copper deficient and supplemented rats fed fructose or starch were ligated prior to the oral administration of ^{64}Cu. ^{64}Cu was not absorbed from the stomach regardless of the type of dietary treatment, copper status or whether copper was administered either in diet or in water (25). When ^{64}Cu was administered with diets containing fructose or starch directly into the isolated duodenal loops in situ, the absorption of ^{64}Cu in the presence of fructose was delayed when compared to starch. When ^{67}Cu was administered in a diet containing either fructose or starch directly into the stomach, although not always significant, the gastrointestinal (GI) contents of rats fed the copper-deficient fructose diet exhibited higher radioactivity than those fed the copper-deficient starch diet (26). However, this apparent retention of ^{67}Cu in GI contents was accompanied by decreased whole-body radioactivity but only at 8 h. post load. At 24, 48 and 96 h. post load the absorption of ^{67}Cu was similar for all copper deficient rats regardless of the nature of dietary carbohydrate. The cumulative excretion of ^{67}Cu via feces over the 96-h period of collection was similar for both groups of copper-deficient rats, regardless of whether the dietary carbohydrate was fructose or starch.

In order to bypass the gastrointestinal tract as a factor in the metabolism of ^{67}Cu, tissue distribution and excretion of ^{67}Cu was studied following its administration via an intraperitoneal injection. Regardless of dietary carbohydrate, copper deficient rats retained similar levels of radioactivity in various tissues and excreted similar amounts of ^{67}Cu in feces and urine (27). However, since rats supplemented with copper and fed fructose retained more ^{67}Cu in the tissues and excreted less in the feces than the rats fed starch may indicate that effects of fructose on copper metabolism may also occur postabsorptively. The requirements for copper may be greater when fructose in fed.

G. Uptake of radiolabeled copper by the liver in vivo.

It is established that the liver plays a major role in fructose and copper metabolism. Thus, an interaction may occur in the liver between dietary fructose or its metabolites and copper which can affect copper utilization. Although it has been shown that copper absorption was delayed in copper deficient rats fed fructose when compared to those fed starch, a defective postabsorptive mechanism involving the liver cannot be ruled out. The next experiment was designed to study the uptake of copper by the liver and extrahepatic tissues when injected directly into the liver via the portal vein in plasma in combination with either fructose or glucose (28). Regardless of copper status, rats fed diets containing fructose retained more radioactivity in extrahepatic tissues than rats fed starch. In addition, there was an increased uptake of ^{67}Cu by the liver and several extrahepatic tissues when fructose as compared to glucose was injected in combination with the isotope (28). These data strongly suggest that copper requirements or utilization are greater when fructose is the main dietary carbohydrate. The results may also in part explain the reason for the increased severity of copper deficiency in rats fed fructose.

<u>Copper-Carbohydrate interaction in the growing pig.</u> In order to determine whether the same type of interaction between copper status and dietary carbohydrate found in rats also occurs in other animal model, studies were also conducted in the growing pig. Swine were chosen as the animal model for this project since their circulatory system is

morphologically similar to that of humans. Weanling pigs, 21 days of age, were fed diets containing 20% of calories as glucose or fructose, low (1.5 ppm Cu) or supplemented with copper (40 ppm) for 10 weeks.

In agreement with data from rat studies, copper deficient pig exhbited anemia, decreased ceruloplasmin and Cu SOD activities. In addition, the serum and hepatic copper concentration were low (29). Dietary copper deficiency reduced lysyl oxidase activity in both plasma and aortic tissues. The feeding of fructose in conjunction with low copper intake resulted in a greater reduction of lysyl oxidase activity, in the aorta, implying that connective tissue metabolism and cardiac muscle integrity could be adversely affected (30). In addition, heart hypertrophy developed only in copper deficient pigs fed fructose. The results indicate that the pig, too, is a sensitive animal model, particularly with reference to cardiac hypertrophy, for the study of copper-carbohydrate interactions as relevant to the human population (29).

Conclusion

The results from the rat studies have prompted investigators at the USDA, Human Nutrition Research Center in Beltsville, Maryland, to conduct a human study to determine whether the same type of interaction between copper status and dietary carbohydrate found in rats occurs in humans. During the course of feeding the diets for 11 weeks, four of the subjects exhibited heart related abnormalities and were removed from the study (31). Thus, the results from experimental animals taken together with those of the human study (31) suggest that dietary copper deficiency produces a significant risk. It is apparent that further studies are required to determine the mechanism that is responsible for the more pronounced severity of copper deficiency when dietary carbohydrates are fed.

References

1. Woteki, C.E., Welsh, S.O., Raper, W., Manstom, R.M. In: Reiser S., ed. Metabolic effects of utilizable carbohydrates. New York, NY, Mancel Dekker 1982, p. 1.
2. Klevay, L.M. Ann. NY Acad. Sci. 1980, 355, 140.
3. Reiser, S. Nutr. & Health 1985, 3, 203.
4. Allan, K.G.D., Klevay, L.M. Atherosclerosis 1978, 29, 81.
5. Klevay, L.M. Biol. Trace. Elem. Res. 1983, 5, 245.
6. Fields, M., Ferretti, R.J., Smith, J.C., and Reiser, S. J. Nutr. 1983, 113, 1335.
7. Reiser, S. In: Rechcigl Jr. M., ed. CRC Handbook Series in Nutrition and Food. Section E: Nutritional Disorders. West Palm Beach, FL: CRC Press, 1978, p. 409.
8. Cohen, A.M., Teitelbaum, A., Briller, S., Yanko, L., Rosemann, E., Shafrir, E. In: Sepple, H., McNutt, K., eds. Sugars in Nutrition. New York, NY: Academic Press, 1974, p. 483.
9. Pfeiffer, E.F. In: Greten, H., Levine, R., Pfeiffer, E.F., Renold, A.E., eds. Lipid metabolism, obesity and diabetes mellitus; impact upon athersclerosis. Stuttgart, Germany: Georg Thieme Publishers 1974, p. 143.
10. Laube, H., Wojcikowski, C., Schatz, H., Pfeiffer, E.F. Horm. Met. Res. 1978, 10, 192.
11. Hallfirsch, J., Lazar, F., Jorgensen, C. Reiser, S. Am. J. Clin. Nutr. 1979, 32, 787.
12. Vrana, A. Kazdova, L. Life Science. 1970, 9, 257.
13. Reiser, S., Hallfirsch, J. J. Nutr. 1977, 107, 147.
14. Cohen, A.M., Tetelbaum, A., Rosenman, E. Metabolism. 1977, 25, 17.
15. Blakely, S.R., Hallfrisch, J., Reiser, S., Prather, E.S. J. Nutr. 1981, 111, 307.
16. Reiser, S., Ferretti, R.J., Fields, M., Smith, J.C., Role of dietary fructose in the enhancement of mortality and biochemical changes associated with copper deficiency in rats. Am. J. Clin. Nutr. 1983, 38, 214.
17. Fields, M., Ferretti, R.J., Reiser, S., Smith, J.C. Proc. Soc. Expt. Biol & Med. 1984, 175, 530.

18. Fields, M., Ferretti, R.J., Smith, J.C., Reiser, S. Am. J. Clin. Nutr. 1984, 39, 289.
19. Fields, M., Reiser,S., Smith J.C. Nutr. Rep. Int. 1983, 28, 165.
20. Fields, M., Ferretti, R.J., Smith, J.C., Reiser, S. J. Nutr. 1984, 114, 393.
21. Fields, M., Ferretti, R.J., Smith, J.C., Reiser, S. Biol. Trace. Elem. Res. 1984, 6, 379.
22. Fields, M., Michaelis, O.E.IV, Hallfrisch, J., Reiser, S., Smith, J.C.. Nutr. Rep. Int. 1983, 28, 123.
23. Fields, M., Ferretti, R.J., Judge, J.M., Smith, J.C., Reiser, S. Proc. Soc. Exp. Biol. & Med. 1985, 178, 362.
24. Davis, P.S., Deller, D.J. Aust. Ann. Med. 1967, 16, 70.
25. Fields, M., Craft, N., Lewis, C., Holbrook, J., Rose, A., Reiser, S. Smith, J.C. J. Nutr. 1986, 116, in press.
26. Fields, M., Holbrook, J., Scholfield, D., Smith, J.C., Reiser, S.
J. Nutr. 1986, 116, 625.
27. Holbrook, J., Fields, M., Smith, J.C., Reiser, S. J. Nutr. 1986, 116, 831.
28. Fields, M., Lewis, C.G., Rose, A., Smith, J.C., Reiser, S. Biol. Trace. Elem. Res. 1986, 10,(3) in press.
29. Steele, N., Richards, M., Darcey, S., Fields, M., Smith, J.C., Reiser, S. Fed. Proc. 1986, 45, 357 (abstact).
30. Ono, K., Steele, N., Richards, M., Darcey, S., Fields, M., Scholfield, D., Smith, J.C., Reiser, S. Fed. Proc. 1986, 45, 357. (abstract)
31. Reiser, S., Smith, J.C., Mertz, W., Holbrook, J.T. Scholfield, D., Powel, A.S., Canfield, W.K., Canary, J.J. Am. J. Clin. Nutr. 1985, 42, 242.

DISCUSSION

RISK: An ancient test for reducing carbohydrates is the Fehling Test in which Cu(II) is reduced to Cu(I). I wonder if there could be a chemical mechanism such as this occurring here?

FIELDS: I am sure that some of the Cu might be reduced but these data don't seem to support that mechanism.

ELLIOTT (Graham): Could you reverse these effects of fructose if you feed glucose or starch at the same time? What sort of concentrations of these would protect against the effect of fructose? Could you also place these fructose concentrations in a context of an ordinary diet? What would be dangerous for example?

FIELDS: In the first study we used sucrose, which is 50% fructose. If fructose is 31% of the diet the effect is still seen. In the pig study and the human study the diet contained only 20% fructose and the effect was still seen. If the diet contains starch, there is protection according to literature reports.

CROOK: Were these 4 patients in your 10 week study who had heart problems symptomatic? Did they have abnormal chest x-rays or EKGs or did they actually have myocardial infarction or some other heart abnormality?

FIELDS: One patient had severe heart attacks and 2 or 3 had arrhythmias. All carried monitors and this is how it was detected early.

CROOK: Were they Cu deficient according to plasma analysis?

FIELDS: No. They were not deficient. Serum Cu was normal. Ceruloplasmin was still within the normal range. The most sensitive index was the erythrocyte SOD which was reduced, but still within the normal range. When the study was stopped and they were repleted with Cu, the only measure that changed was the erythrocyte SOD, it went up.

CROOK: Was the Cu content of these pig hearts markedly reduced in association with their increase in weight?

FIELDS: Yes, Cu was decreased in the heart of rats and pigs.

CROOK: We have just obtained some very preliminary results with ^{64}Cu and scanning dog hearts with Positron Emittion Tomography to describe kinetics as far as uptake and disappearance. We do have an approved IND for human studies which we are in the process of amending to include cardiovascular disease. That might be of interest to you.

SORENSON: Might ligation of the duodenum affect results you obtained in your study of gastric absorption?

FIELDS: This is true. Tying the duodenum produces a huge secretion volume and the stomach is tremendously distended. In spite of this we hoped we would get some absorption but it was not absorbed. Obviously this is not at all physiological but this was an approach to study gastric absorption in the absence of absorption from the small intestine.

SORENSON: Right, but the increased gastric secretion may have impeded absorption.

FIELDS: The study which does say that Cu is absorbed from the stomach was done exactly the same way. It was tied and they found that Cu was absorbed from the stomach. I think the reason why that study appeared to show it was absorbed is because the radio-labeled Cu was injected into the stomach and the radioactivity could have leaked out from the site of injection. We gave an oral dose so I don't think there would have been any way that the meal could have escaped from the tied stomach.

SORENSON: On another matter, does anybody have a rationale for why there's lysyl oxidase activity in plasma, or is something else really being measured?

FIELDS: I think you are right. This is really a tissue enzyme. It might be more correct to measure it in tissues other than blood.

SORENSON: All of your data are consistent with the fact that it is the reducing sugar content of the diet that accounts for inhibition of Cu absorption since Cu(II) is reduced to Cu(I) which, as I understand it and I ask others to comment, is less readily absorbed throughout the GI tract.

FIELDS: As I said before, these animals really did not have an impaired absorption, it was delayed. It is slower. If measurements are made up to eight hours, absorption is less. But afterwards, there are no differences in absorption. Now if fructose reduces Cu it should do that also in a supplemented animal. Fructose will not reduce Cu in a deficient animal and not reduce it in a supplemented animal. If it's the reducing property of the sugar it should reduce it under both circumstances. I have never seen reduced or impaired Cu absorption or impaired excretion or impairment in any of the Cu measurements that we make that show that fructose will impair anything in a supplemented animal. Impairment is observed when the animal is deficient, the animal is so sick after being fed a Cu deficient diet for 5 to 6 weeks that anything you measure is abnormal. When various indeces of Cu are measured during the first 3 weeks when they are not sick there are no differences. After 7 to 8 weeks they are very sick and anything you measure is abnormal.

HARRIS: The initial observation was made with sucrose fed animals, correct?

FIELDS: Right.

HARRIS: Some of these questions regarding reduction potential become problematical because we know sucrose is broken down to reducing sugars. It itself is not a reducing sugar. It is broken down as it actually penetrates through the brush border and so forth, so events that you are seeing are apparently going to be eminating from that source. In that regard I don't quite understand your rationale for running the stomach absorption test. What were you looking for there?

FIELDS: We were interested in studying absorption from the stomach since it was reported that Cu is absorbed from the stomach. Fructose could have been inhibiting Cu absorption from that organ. Therefore, we had to eliminate that and we did.

HARRIS: Again, I return to the fact that you were dealing

initially with a non-reducing sugar and still saw the effect.

FIELDS: We have established that fructose is the reason for the aggravated deficiency, so we no longer study the surose effect.

DISILVESTRO: With regard to lysyl oxidase (LO), a rationale for why it might be in serum is that it is an extracellular enzyme and some could find its way into serum. It's thought to be LO because of its sensitivity to BAPN. One way to really know if it's really LO is to develop an antibody assay, which hasn't been done yet.

HYPERTENSION INDUCED ALTERATIONS IN COPPER AND ZINC METABOLISM: A LINK TO VASCULAR DISEASE?

Carl L. Keen[1,2], Michael S. Clegg[1], Fay Ferrell[1], Glenn C. Hunter[3], and Michael A. Dubick[2,4]

Departments of Nutrition[1] and Internal Medicine[2], University of California, Davis; Departments of Surgery[3] and Medicine[4], Veterans Administration Medical Center, Martinez, California

Ischemic heart disease is the leading cause of death in the United States and other industrialized countries. Although the etiology of disease is without question multifactorial in nature, several potential risk factors have been agreed upon. These risk factors include hypercholesterolemia, hypertension, electrocardiogram abnormalities, cigarette smoking and diabetes.[1] During the last decade it has been proposed that Cu deficiency may also be a significant risk factor for vascular disease. This hypothesis has been based on the observations that severe Cu deficiency in experimental and domestic animals can result in connective tissue defects producing myocardial damage and aortic aneurysms, and that the dietary intake of Cu is often below that suggested by health agencies.[2,3] Furthermore, the risk of Cu deficiency is thought to be increased by the consumption of excess Zn in the diet, as these two elements can be biologically antagonistic.[3-5]

We have recently initiated a series of investigations to test the idea that some of the above risk factors may in some way be interrelated. Our initial studies have focused on the potential interrelationship between hypertension and Cu deficiency. This has been studied as a result of the dual observation that abnormal Cu and Zn metabolism has been reported in some hypertensive patients,[3,6] and one effect of Cu deficiency has been reported to be an induction of hypertension.[7]

Hypertension Induced Changes in Cu and Zn Metabolism

To investigate the effects of hypertension on Cu and Zn metabolism, we have made use of the Dahl salt-sensitive (S) rat. The Dahl S rat is thought to be a model for salt-sensitive, low renin hypertensive humans.[8] In our studies Dahl S rats are fed either an 0.4% or 8.0% NaCl diet. The diets, which have been specifically formulated for the Dahl rat by Brookhaven National Laboratory (Upton, NJ), contain 12 ug Cu and 55 ug Zn/g diet. In the experiment discussed below, the rats were fed their respective diets ad libitum for a period of 32 days. Systolic blood pressures were recorded using the tail cuff method. Additional details on these studies have been published.[9]

The Dahl S rats fed the high NaCl diet were characterized by a low final body weight (Figure 1), and a markedly elevated blood pressure (Figure 2), compared to the rats fed the 0.4% NaCl diet. When the animals were killed at day 32 it was observed that the high NaCl rats were also characterized by low hematocrits, splenomegaly and cardiomegaly. Plasma and soft tissues were collected and analyzed for Cu and Zn concentrations.

Plasma Zn concentrations were significantly lower in the high NaCl group than in controls, while plasma Cu concentrations were higher (Table 1). A possible explanation for the observed changes in plasma Zn and Cu may be that the hypertensive condition is resulting in a stress to the animal which results in an increase in glucocorticoid synthesis and release. Glucocorticoids can markedly increase the synthesis of the liver Zn-binding protein metallothionein, which can result in a redistribution of plasma Zn into the liver.[10] Glucocorticoids have also been reported to stimulate the synthesis and release of ceruloplasmin.[10] Since ceruloplasmin is a Cu-containing protein, an increase in its synthesis and release could explain the observed high plasma Cu concentrations. Epinephrine can also stimulate the synthesis of ceruloplasmin and metallothionein.[10] This observation is

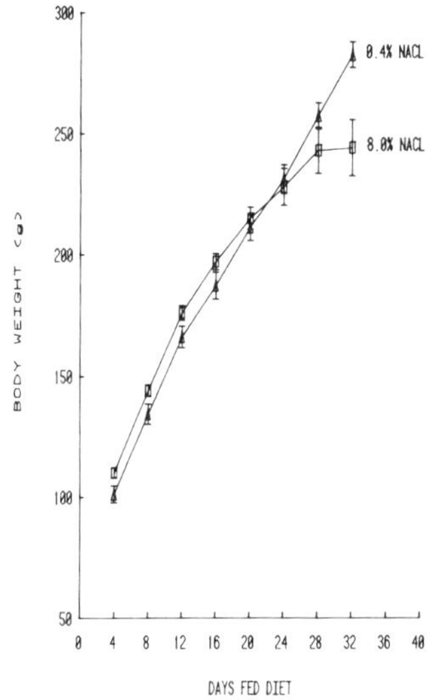

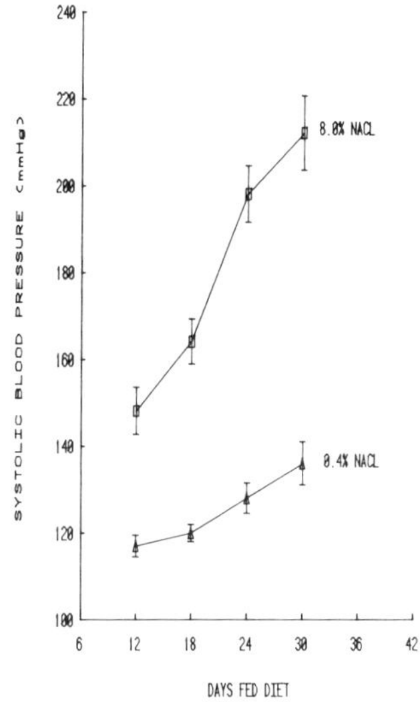

Figure 1. Body weights of Dahl S rats fed a 0.4% (Δ) or 8.0% (□) NaCl diet for 32 days.
Figure 2. Systolic blood pressure of Dahl S rats fed a 0.4% (Δ) or 8.0% (□) NaCl diet for 32 days.

Table 1. Plasma, liver, heart and kidney Cu and Zn concentrations in Dahl S rats fed 0.4% or 8.0% NaCl diets for 32 days.

Diet	Tissue			
	Plasma	Liver	Heart	Kidney
0.4% NaCl (n=31)				
Cu (ug/g)	0.89±0.02	4.4±0.2	5.1±0.2	6.9±0.2
Zn (ug/g)	1.53±0.08	26.4±0.8	14.7±0.5	18.7±0.4
8.0% NaCl (n=15)				
Cu (ug/g)	1.63±0.20*	3.6±0.1*	4.3±0.1*	3.5±0.3*
Zn (ug/g)	1.17±0.10*	31.0±1.5*	16.6±0.4*	18.4±0.7

*Statistically different from 0.4% values at $p < 0.05$. Values are shown on a wet weight basis.

particularly important as elevated plasma catecholamine levels are characteristic of many essential hypertensives.[11] The observation of high plasma Cu concentrations in the hypertensive Dahl S rat is consistent with the finding of high plasma Cu concentrations in human hypertensives,[6,12,13] and supports the hypothesis that there is a link between hypertension and altered Cu metabolism.

A question that can be asked about the above observations is: Are the changes in plasma Zn and Cu etiologic factors in the development of hypertension, or are they physiological responses to the disease? And, if they are responses to the disease, do the changes in plasma Zn and Cu play a role in the remission of the disease, or do they result in additional vascular damage?

The biological reason for the increase in ceruloplasmin synthesis and release during periods of tissue injury and stress are not clear. It has been proposed that ceruloplasmin may be an important anti-inflammatory agent due to its ability to scavenge superoxide radicals.[14] A second function which has been ascribed to this protein is the transfer of Cu to extrahepatic tissues via a receptor mediated process.[15] However, as is discussed below, we have obtained evidence suggesting that delivery of Cu to extrahepatic tissues may be impaired in hypertensive animals. There is some evidence that the reduction in plasma Zn may be a benefit to the hypertensive animal. Reeves and O'Dell[16] have reported that the serum activity of angiotensin-converting enzyme (ACE) in rats is dependent on serum Zn concentrations. The two roles of ACE in animals are to catalyze the formation of angiotensin II and to degrade bradykinin. Angiotensin II acts to increase blood pressure while bradykinin acts to reduce it; thus a reduction in ACE activity due to low plasma Zn concentrations should have a hypotensive effect.

Consistent with our idea that the lower plasma Zn concentrations in hypertensive rats were due in part to a redistribution of body pools of Zn, was the observation that liver and heart Zn concentrations were high in this group compared to controls (Table 1). In marked contrast to tissue Zn concentrations, Cu concentrations in liver, heart and kidney were significantly lower in hypertensive rats compared to controls (Table 1). The finding of low liver Cu concentrations in the hypertensive animals is consistent with the idea that the high circulating Cu concentrations are the result of liver Cu being mobilized in the form of ceruloplasmin. However, the observation of low heart and kidney Cu concentrations was not expected given the high concentrations of circulating Cu. Our interpretation of these data is that in the hypertensive animal there is a reduction in extrahepatic uptake of Cu, possibly due to alterations in the ceruloplasmin receptor protein. Regardless of the biochemical lesion underlying the lower tissue Cu concentrations in the hypertensive rats, the observation of the lower concentrations is very important, as extrahepatic tissue stores of Cu are thought to be minimal. Thus, a reduction in Cu concentration in these tissues may result in reductions in the activities of Cu-containing enzymes such as Cu,Zn-superoxide dismutase (Cu,ZnSOD), cytochrome C oxidase and lysyl oxidase. If pronounced reductions in the activities of these enzymes occurred, this could result in significant tissue damage independent of the hypertensive condition itself. Based on the above considerations, an important question that must be asked is whether we have any similar evidence of abnormal Cu metabolism in humans suffering from hypertensive disorders. Below, we discuss some recent data we have obtained that supports the hypothesis that a factor underlying vascular disease in humans is altered vascular wall Cu metabolism.

Abnormal Cu Metabolism in Human Aneurysmal and Occlusive Diseases

In our laboratories we have been able to obtain tissues from patients with two types of vascular disease to test the hypothesis that abnormal Cu metabolism may be one factor underlying vascular disease in humans. The tissues have been collected from patients undergoing surgery for either occlusive disease or aneurysmal disease of the abdominal (infrarenal) aorta. In both groups of patients examined to date, 35-38% were hypertensive. Additional details on the subjects have been published.[17] The collected specimens were stripped of adventitia and all surface thrombi were removed. The cleaned samples were immediately frozen and maintained at -70°C until analysis. Aortic tissue removed from the same region of the aorta was obtained at autopsy from patients who showed no signs of vascular disease or calcified plaque. These tissues served as controls.

As shown in Table 2, tissue Cu concentrations were 50% higher in patients undergoing corrective surgery compared with controls, while Zn concentrations were not markedly different. However, in aneurysmal patients with hypertension, Zn concentrations were significantly lower in comparison to both controls and patients with occlusive disease, while Cu concentrations were significantly higher than controls. Although these observations may argue against a role of Cu deficiency in the etiology of vascular disease, data on Cu,ZnSOD activity (Table 2) would indicate abnormal Cu distribution within aortic tissues. In contrast to Cu,ZnSOD, the activity of MnSOD was higher in the aorta samples from patients compared to controls (Table 2). This observation suggests that the lower activity determined in the patient aortic tissue is not merely a reflection of dead tissue. Thus the distribution of Cu within the aortic tissue is different between control and diseased tissue. Further studies on the localization of Cu and Zn in these tissues are needed. An additional observation we can make from the above study is that since low aorta Cu,ZnSOD activity was found in all of the patients examined, it is unlikely that the disturbance in Cu metabolism was due to a simple dietary Cu deficiency, but rather that it was a manifestation of a different physiological condition, such as hypertension.

Evidence of abnormal Cu metabolism in patients with vascular disease has also been presented by Tilson and co-workers, who have reported that patients with abdominal aortic aneurysms can be characterized by low liver and skin Cu concentrations.[18,19] In contrast to our results, Tilson did not find evidence of abnormal Cu metabolism in these tissues in patients with occlusive disease. This suggests that there may be some tissue specificity with regard to the changes in Cu metabolism. Taken together, the above observations support the idea that abnormal Cu metabolism may be an etiologic factor in vascular disease.

At this point we would like to suggest that one mechanism by which a local Cu deficiency could be affecting vascular wall integrity in patients with either aneurysmal or occlusive disease is through enhanced tissue lipid peroxidation. Increased rates of vascular wall lipid

Table 2. Cu and Zn concentrations, SOD activity and ascorbic acid levels in human aortic tissue.

Parameter	Controls (n=8)	Aneurysm Disease (n=26)	Occlusive Disease (n=14)	Aneurysmal Disease + Hypertension (n=10)
Cu (ug/g wet wt)	0.62±0.09	0.91±0.08*	0.93±0.13	0.97±0.09*
Zn (ug/g wet wt)	16.1±0.9	13.4±1.1	21.2±3.4	10.6±0.9*
Cu,ZnSOD (U/g wet wt)	156±20	24±5*	17±5*	17±5*
MnSOD (U/g wet wt)	61±8	103±9*	91±8*	125±14*
Ascorbic acid (mg/100 g wet wt	2.7±0.4	1.3±0.2*	0.9±0.1*	1.5±0.2*

*Significantly different from control values at $p < 0.05$.

peroxidation may be predicted to be a consequence of the lower Cu,ZnSOD activity in these vessels. In addition, since Fe is a powerful prooxidant,[20] the high Fe concentrations in the aorta samples from the patients may result in an increased production of superoxide anion in the vessel wall, further stressing the compromised Cu,ZnSOD system. Supporting this idea is the observation of high MnSOD activity in these tissues. We suggest that the increase in MnSOD activity in these tissues is in response to increased substrate (superoxide anion) pressure. Further evidence for an increased rate of vascular wall lipid peroxidation in these patients is given by our observation that aortic wall ascorbic acid concentrations are low in these subjects compared to controls (Table 2).

Finally, we have recently made the observation that serum uric acid concentrations are often abnormal in the patient population we are studying. As shown in Table 3, less than half of the patients tested had serum uric acid concentrations within the normal range. Hyperuricemia is commonly considered as a cardiovascular risk factor.[21] Klevay has reported that in rats dietary Cu deficiency can result in hyperuricemia. He has used this observation in support of his hypothesis that dietary Cu deficiency is a

Table 3. Serum uric acid levels (mg/100 ml) in humans.

Normal range	4.2-8.5
Occlusive Disease	3.9; 7.0; 10.5
Aneurysms	4.6; 4.7; 5.7; 6.5; 9.1; 9.7; 10.0; 10.2; 10.2

Underlined values are outside of the normal range.

major cardiovascular risk factor that may give rise to other "so called risk factors" such as hyperuricemia.[22] We would like to suggest that an alternative explanation for the high uric acid levels in patients with vascular disease, and in animals given Cu deficient diets, is that it represents a positive response of the body to increased rates of vascular wall lipid peroxidation. This suggestion is based on the recent reports that serum uric acid functions as an antioxidant through the complexation of free serum Fe.[23]. If this hypothesis is correct, then the observation of high serum uric acid levels in the aneurysmal patient are indicative of soft tissue free radical damage. Again, as stated above, one of the postulated roles of ceruloplasmin is that it functions to scavenge free radicals.[14] Thus, the high serum Cu concentrations reported in hypertensive patients may also be interpreted as a response of the body to increased soft tissue (vascular wall) lipid peroxidation.

Conclusions

To summarize the above results, we have shown that one effect of experimental hypertension is an alteration in the metabolism of Cu and Zn. This alteration is reflected in part by increases in plasma Cu concentrations, an observation which is similar to what has been reported for human hypertensives. The time sequence and magnitude of these changes needs further study. A second expression of the altered Cu metabolism in our experimental hypertensive rat is a reduction in soft tissue Cu concentrations. We suggest that this reduction in tissue Cu is due to an impairment of normal tissue Cu uptake, possibly resulting from a block in normal ceruloplasmin receptor function. We hypothesize that the hypertension-induced reductions in tissue Cu concentrations are functionally significant, and indeed may be one of the biochemical lesions underlying vascular damage in hypertension. Support for this hypothesis is provided by data collected from human patients with either aneurysmal or occlusive disease of the infrarenal aorta; these patients are characterized by low aortic tissue Cu,ZnSOD levels. Data from both experimental animal studies and human studies suggest that one biochemical lesion resulting from the disturbance in Cu metabolism is an increased rate of soft tissue lipid peroxidation.

References

1. Insull, W Jr. In: Coronary Risk Handbook, American Heart Association, New York, 1973; p 3.
2. Tinker, D; Rucker, RB. Physiol. Rev. 1985, 65, 607.
3. Klevay, LM. In: Metabolism of Trace Elements in Man, vol. 1. OW Rennert & W-Y Chang, eds.; CRC Press, Boca Raton, FL, 1984; p 129.
4. Hurley, LS; Lonnerdal, B; Keen, CL. Fed. Proc. 1983, 42, 1735.
5. Fisher, PWF; Giroux, A; L'Abbe, MR. Am. J. Clin. Nutr. 1984, 40, 743.
6. Sempos, CT; Greger, JL; Johnson, NE; Smith, EL; Seyedabadi, FM. Nutr. Rep. Inter. 1983, 27, 1013.
7. Klevay, LM. Fed. Proc. 1986, 45, 235.
8. Rapp, JP. Hypertension 1982, 4, 753.
9. Clegg, MS; Ferrell, F; Keen, CL. Hypertension 1987, in press.
10. Cousins, RJ. Physiol. Rev. 1985, 65, 238.
11. DeChamplain, J; Farley, L; Cousineau, D; Ameringen, MR. Circ. Res. 1976, 38, 109.
12. Sinha, SN; Gabrieli, ER. Am. J. Clin. Path. 1970, 54, 570.
13. Ahmed, T; Sackner, MA. Respiration 1985, 47, 243.
14. Goldstein, IM; Kaplan, HB; Edelson, HS; Weissman, G. J. Biol. Chem. 1979, 254, 4040.
15. Stevens, MD; DiSilvestro, RA; Harris, ED. Biochemistry 1984, 23, 261.
16. Reeves, PG; O'Dell, BL. J. Nutr. 1986, 116, 128.
17. Dubick, MA; Hunter, GC; Casey, SM; Keen, CL. Proc. Soc. Exp. Biol. Med. 1987, in press.
18. Tilson, MD. Arch. Surg. 1982, 115, 1173.
19. Tilson, MD; Davis, G. Surgery 1983, 94, 134.

20. Aust, SD; Sviingen, BA. In: Free Radicals in Biology, vol. 5. WA Pryor, ed.; New York, Academic Press, 1982; p 1.
21. Kannel, WB. Am. J. Cardiol. 1976, 37, 269.
22. Klevay, LM. Nutr. Rep. Inter. 1980, 22, 617.
23. Davies, KJA; Sevanian, A; Muakkassah-Kelly, SF; Hochstein, P. Biochem. J. 1986, 235, 747.

DISCUSSION

LOVEN: Uric acid is potentially generated by xanthine oxidase, after ATP is degraded to hypoxanthine, along with the production of O_2^-. It's been shown by McCord's group that this is a consequence of ischemia. Is this tissue in any way ischemic prior to surgery as a result of an aneurysm or something like that?

KEEN: With aneurysms you sometimes see some ischemia. That certainly is a problem. The surgeon who removed these tissues did not think that there was all that much ischemia. However, it certainly is possible that this is part of the problem. The idea that uric acid acts as an antioxidant is relatively new. The suggestion is that uric acid can actually scavenge hydroxyl radical on its way to be converted to allantoin. Hochstein at the University of Southern California goes one step further to suggest that uric acid may be one of the major low molecular weight ligands for Fe in serum and once it binds to Fe it effectively quenches Fe's oxidizing activity.

LOVEN: You mentioned that Mn SOD was elevated, which suggests that it's an inducible enzyme. Is there any information on the activity of cytochrome oxidase in these tissues?

KEEN: I really can't comment on a change in O_2^- production. We did not measure cytochrome oxidase in this study, we just didn't have enough tissue. I have been impressed by Joe Prohaska's observations that a copper-deficiency-induced 80% decrease in cytochrome oxidase produced absolutely no effect, that they could measure, on ATP production. Many people have argued that in vivo there may be marked reductions in cytochrome oxidase without much of an effect.

DISILVESTRO: You found virtually no change in aortic Cu and, still, big differences in SOD. The same is true for lysyl oxidase (LO). Aortic Cu doesn't change much but you can't measure LO. Regarding your observation that kidney Cu decreased where as plasma Cu increased, we measured induction of metallothioneine by interleukin-1 and found that kidney metallothioneine was way down which could account for the decrease in kidney Cu levels. Essentially the Cu in kidney metallothioneine was leaving. Maybe that's where ceruloplasmin gets its Cu, in the long run.

KEEN: From kidney Cu?

DISILVESTRO: Yes. When you give interleukin-1 injections to rats, even though you induce metallothioneine in the kidney, in terms of increasing messenger RNA, if you look at the amount of Cu bound to that metallothioneine it's down about 50% where the Zn is up. You measured total kidney Zn and didn't see much of a change but there's a lot of other Zn proteins in this tissue so that might obscure any change. In the case of Cu, there's not as much Cu in this tissue so you might see the impact of reducing Cu more on a tissue level than you would with Zn, where metallothioneine is only one of many Zn proteins. Kidney Cu associated with metallothioneine is decreased in interleukin-1 treated animals.

CROOK: I think the anemia you found in rats might require a different interpretation. It's well known in clinical situations that individuals with severe heart failure will appear to be anemic because they retain so much fluid, it's a dilutional anemia. If you're in right-sided heart failure, liver weight increases and there is marked splenomegaly. If you want to invoke hemolytic anemia as a reason for the anemia and splenomegaly, you have to have really marked gross hemolysis for it to be visible to the eye.

KEEN: We had a little bit of kidney staining. I think it was possibly hemolytic anemia. It was very similar to what I've seen in Cu toxicosis cases in terms of the amount of pigment staining.

CROOK: As I recall, rats have haptoglobin and zero haptoglobin levels would be good evidence for plasma hemoglobin. Uric acid is another of these floating parameters. I didn't see a sex break down on your occlusive and aortic aneurysms but there are differences for males versus females.

KEEN: These were males. There is an interesting phenomena that occurs in that the incidence of abdominal aneurysms in occlusive disease is much higher in males. The factor I've seen is as high as 7 or 8 to 1 for males versus females.

CROOK: But that is related to males at risk with their hypercholesterolemias and hypertension etc. Did you have occasion to do any electron microscopy on any of the tissue sections from your animal hearts and any observation concerning mitochondria?

KEEN: No, unfortunately, these tissues were taken and banked. We got our exciting results after that fact. We currently have another study going where we have titrated levels of NaCl and we are also beginning to manipulate amounts of Zn and Cu in the diet to see if we can perturb these phenomena. In these animals we will be taking tissues for histopathology.

CROOK: There was a very nice study done by the group at the Oak Ridge National Laboratories using Dahl-S rats and looking at differences in substrate utilization depending on whether they have big hearts or small hearts. That may also have some influence on what you have observed concerning the stage in development of cardiomegaly.

HERNANDEZ: Cu-Zn SOD is an enzyme that is induced in lung following chronic O_2^- production. Humans in your study have chronic problems. Do you think that the absence of SOD in their tissues was due to the lack of induction or is it due to fibrosis or new tissue that's formed with low levels of SOD?

KEEN: We are merely measuring activity. It is possible that we have inhibition of the Cu-Zn SOD which is there. We have not ruled that out. We have to get the antibody to test that. I don't think it's just a function of the fibrotic tissue per se because we are not getting Cu into that tissue. If it were a case of aneurysm severity then there would be graded levels of Cu-Zn SOD. We have compared different sizes of aneurysms in an attempt to correlate changes in Cu-Zn SOD with aneurysm size. We didn't find a good correlation. I think if we had, that would have strongly supported what you are suggesting. I think we're just simply not getting enough Cu into that tissue to synthesize the Cu-Zn SOD. I don't know why this might occur. We might have an unusual situation.

DAMERON: Have you measured any other parameters of O_2^- damage, like soluble collagen or malonyldialdehyde?

KEEN: That's what we hope to do in the next round. Under ideal circumstances I'd like to measure something like TBA reactants in vessel walls but we can't get the tissue soon enough to do that. We ideally have to come up with a good animal model but it's going to be very difficult, these are spontaneously hypertensive animals. In individuals, we may be studying events that took 30 or 40 years to develop. Remember many of these patients are 70 to 80 years old, so modeling will be difficult.

MANIFESTATIONS OF COPPER DEFICIENCY IN DAIRY CATTLE

Alfred W. Anderson

Veterinary Medicine
19383 Ave 300
Exeter, California 93221

Marginal copper deficiency in cattle is widespread and costly due to lack of recognition of the more subtle clinical symptoms. The result of this situation is decreased feed efficiency, decreased growth and production, increased disease incidence and quite possibly economic failure of the farm or ranch involved. The advent of intense management, improved genetics and greater production naturally increases mineral demands.

Environmental Factors:

The concept of deficient areas is a hindrance to sound animal husbandry practices.[1] Transportation of forages and feed grains can displace these feeds from their origin to areas not experienced with the particular problem associated with the feedstuffs involved.[2] Feed processing with the use of by products, heat and chemical treating, and lasting synthetic additives present a confusing mineral analysis. Fertilization or the lack of it with a resultant mining of the soils over the years may have changed the level of inorganic element uptake by plants.[3] Total mineral composition of forages and grains may be very different from when the land was originally put into production. Irrigation with water from streams, rivers, deep wells and shallow wells may effectively change plant uptake of inorganic elements. The effect is transporting soluble inorganic elements from one area to another with possible concentration of elements which could interact or interfere with mineral utilization by both plant and animal. Climate variation, both seasonal and within a particular season, affect plant mineral composition. The stage of harvest is equally important and can be used to control mineral composition. Environmental factors are important since the interrelationship of copper to the excesses of other minerals such as S, Mo, Fe, Ca, P, Zn, Cadmium, protein and other nutrients are vital to understanding methods which can be utilized to correct severe problems.[4,5]

Case Background:

This 100 cow dairy was located in the high desert area of Central Oregon and is not unlike much of the Western United States. The herd was being pastured during the summer and fed alfalfa hay the remainder of the year. A 14 per cent protein commercial mix was fed in the milking parlor; average feeding rate was 20 pounds daily per cow. The water source was both well and irrigation pond which were available at all times. Free-choice salt and minerals were readily available. The herd was milked in a 12-stall trigon parlor and housed in a free stall barn with open access to corrals.[6]

General Clinical Symptoms:

On the evening of September 28, 1983 the cows began to eat manure and dirt while being moved to the holding pen for their evening milking. This was to continue with the consumption of 50 pounds of free-choice minerals every 12 hours until edema was noted in the jaw and brisket areas of several cows. Supplementation was discontinued until specific laboratory analysis of blood samples could be completed.

Milk production began to plummet drastically from the 62 pound daily average. The incidence of foot problems began to increase over the past previous months. Heifer growth was poor and they appeared to be very unthrifty while being fed very well.

Specific Clinical Symptoms:

Anemia: Anemia was not evident in this herd but is a frequent symptom in cattle deficient of copper. Copper is necessary for both production and maintenance of red blood cells. Ceruloplasmin is involved in the release of Fe to the plasma from the reticuloendothelial system of the

liver. Production of ceruloplasmin is directly related to the copper levels in the cattle.[7]
Diarrhia: Diarrhia was very much in evidence in this herd in both young and adult animals. The incidence of diarrhia in adult cattle can involve entire herds and be rather severe. Apparently supplementation, either by the oral or injection route, can give a rather rapid response accompanied by alteration of the implicated diet.

The implications in calves and young stock is much more serious and difficult to treat successfully for yet unexplained reasons. The calf is dependent on the copper in the colostrum as its initial source. The level of copper drops significantly within 48 hours to a range of 0.15 0.37 ppm.[8] When Mo is involved the peculiar situation occurs where the Mo content in milk responds rapidly and directly to increase Mo intake by the cow.[5,9] Milk Cu is hemostatically controlled and will not respond to dietary changes in copper.[5,9] When this circumstance occurs diarrhia in the calf and young animals under one year of age becomes nearly impossible to treat. Prevention with high levels of copper supplementation then becomes the only satisfactory approach to this problem. The diarrhia involved with copper deficiency has been associated with a reduction in cytochrome oxidase in the small intestine resulting in partial villus atrophy.[5,10,11,12,13]
Bone Disorders: Spontaneous fractures of long bones and of the pelvis occasionally occured in this herd. Generally stiffness during walking was noted in both young and adult animals due to laminitis and small internal hemorrhages in hooves of affected animals. While on concrete or in the milking barn they would tend to shift weight from one leg to another. Swollen joints were in evidence in some cattle with no noticeable pain when palpated. The most important skeletal lesion was the high incidence of epipisitis in younger animals.[4,9] The skeletal lesions described are caused by a decrease in amine or lysyl oxidase.
Hair Depigmentation: Hair depigmentation is a very sensitive indicator of copper deficiency with very subtle signs observed in this Holstein herd.[5,7,12,14] The obvious and apparent changes occur with black hair becoming red tinged, with complete graying occuring on any part of the body, but most noticable in the head and neck areas. White hair frequently will have a yellow or soiled appearance to it. Calves at birth exhibit gray hair ventral to the ears

with the black hair of the face appearing brown or in combination with gray throughout. Failure of pigmentation occurs because of insufficient copper containing polyphenyl oxidase for conversion of tyrosine to melanin.[5]

Keratinization Defects: This herd experienced hoof and hair defects as generally described. Their hair exhibited growth and physical appearance changes which resulted in a dry, dull, faded, and coarse hair.[14] Hooves exhibited heel cracks, excessive heel growth, heel abscesses, elongated lateral toes, separation of wall from the sole, foot rot, and lastly horizontal stress bands on the hoof itself. The exact biochemical involvement of copper at this time apparently is unknown.[15]

Infertility: Infertility was prominent in this herd and symptoms included anestrus, pyometra, repeat breeders, second trimester abortions and embryonic deaths.[5,12] Infertility of females and sterility of bulls has been demonstrated for some time.[16] It is possible that copper is required for the epithelial lining of the uterus and in luteolytic activity of the estrus cycle.

Cardiovascular Disease: Acute heart failure and abomasal ulcers were not noted in this herd. Cardiovascular disorders such as acute heart failure have been described.[5,17] The lesion is usually a slow progressive degeneration of myocardium with the replacement of fibrotic tissue. Abomasal ulcers due to impairment of the integrity of the microvascular system of the mucosa has been linked to copper deficient cattle.[13] Small internal hemorrhages in hooves were observed in this herd. Apparently decreased cytochrome activity with disruption of collagen formation could be implicated.

Neonatal Ataxia: Neonatal ataxia was a minor factor in this herd. Neonatal ataxia is a nerve condition in young calves caused by myelin aplasia and is irreversible.[5] The symptoms observed in this calf herd were weakness, incoordination, inability to walk, inability to suckle which would respond to copper injections. Occasionally a calf several weeks old not injected at birth would develop diarrhia, respond to treatment for diarrhia, but become progressively weaker, convulse and die. Apparently both cytochrome oxidase and amine oxidase are implicated in this disorder.

Immunosuppression: Respiratory infections were common in adult and young animals. Routine antibiotic treatments did little to return these animals to health. Most usually recovered on their own while some eventually succumbed. Laboratory isolations of infectious agents were inconclusive with no significant isolates being the usual report. The cytochrome activity of circulating leukocytes is decreased in copper deficient cattle.[2,18] Intense vaccination programs were disappointing until copper supplementation was started.

Subclinical mastitis with extremely high somatic cell counts were common in this herd. Milk cultured revealed both Strep. and Staph. infections involved which is rather common. Since bovine mastitis is considered primarily a management problem which is controlled with various sanitary procedures, separation techniques and culling at the time of the crises. Consideration was not given to copper deficiency at the time and in reflection should have been considered. It is quite possible decreased cytochrome oxidase activity could have compromised the epithelial lining of the udder with concurrent susceptibility of the mammary system. The decreased cytochrome oxidase activity of the leukocytes called forth to eliminate the bacterial invaders in the udder could cause repeated waves of debilitated leukocytes unable to complete their task. This could create high leukocyte counts which are used to determine the quality of milk by processors and government regulatory agencies.

Lab Analysis:

Blood samples in October 1983 indicated copper levels of 0.51 ppm and selenium levels of 0.1544 ppm. Comparative samples in August 1985, revealed a copper level 0.95 ppm and selenium levels of 0.208 ppm. The selenium finding was interesting since all selenium supplementation had been removed from this herd for a year. Lambs were used in a controlled situation on the dairy with similar responses in regard to selenium.[19,20,21]

Water samples revealed low levels of molybdenum in irrigation water.

Alfalfa hay samples indicated molybdenum levels ranging from 0.8 to 19.6 ppm. A series of pasture clippings taken throughout the summer during the grazing season revealed a tightening of the copper: molybdenum ratio from 4:1 to 1.2:1 toward the fall.

Commercial feed samples revealed copper levels anywhere from one-sixth to one-half of the level stated on the tags.

Mineral Management:

Mineral management is not an exact science but must begin with a biochemical analysis of the animals involved with a complete analysis of all material ingested. A complete analysis of all material ingested is required because of the interrelationships of other elements to copper with potential of interference created by excesses of various elements.

Animal Evaluation: The most common methods used to determine copper deficiency in cattle has been to determine copper levels of liver and blood. Liver is the tissue of choice because of accuracy, but blood is used most often because of the comparative ease required to obtain samples.

Feed Analysis: A complete feed analysis is required to determine elements which are in excess, thereby creating interference with the absorption of copper. Absolute values are important, but most important is the ratio to other interfering minerals. Water must be taken into account and emphasized since a dairy cow will consume 100 to 300 pounds of water per day.[22] Small quantities of soluble elements in water consumed by a cow producing 100 pounds of milk can become significant.

Prevention and Control:

Once the mineral status of the animal, forage, feed and water are known, then control can begin. Control can be obtained with selective purchases of feed and supplementation usually combined for optimal results.

Supplementation is programmed to return the animal population to normal copper levels in a controlled situation with repeated monitoring of tissues to prevent toxicity. Supplementation may be delivered by increasing copper in feeds, free-choice minerals, metering devices for water, injections and selective fertilization.

Conclusion:

Farms and ranches with copper deficiency as a primary problem experience poor production and uncontrolled disease syndromes in spite of good management and herd health programs. A sound mineral program can be very rewarding.

It must include laboratory analysis of animal tissues and complete mineral analysis of all material ingested to create an adequate supplementation program. It is necessary to monitor animals clinically and with laboratory analysis at frequent intervals until stability is achieved.

References

1. Cunha, TJ. In: Salt and Trace Minerals for Livestock, Poultry and Other Animals, p 19.
2. Gay, C. Washington State University, Pullman, Washington; (personal communication, 1984).
3. Spiva, CP. Agronomist, A&L Agricultural Labs., Modesto, California; (personal communication).
4. Smart, ME. Can. Bet. J. Dec. 1981, 22, 372.
5. Underwood, EJ. In: Trace Elements in Human and Animal Nutrition, 4th ed.; p 71
6. Anderson, AW. Hoard's Dairyman, Sept. 10, 1985, 992.
7. Sanders, DE. Modern Veterinary Practice, Aug. 1983, p 613-616.
8. Schalm, OW; et al. Bovine Mastitis, p 78
9. Ward, GM. Journal of Animal Science 1978, 46, No.4 1080.
10. Suttle, NF; Angus, KW. J. Comp. Path. 1976, 86, 595.
11. Mills, CF; et al. Brit. J. Nutr. 1976, 35, 309.
12. Pool, DBR. Irish Veterinary Journal 1982, 36, 169.
13. Johnson, JL; et al. Vet. Med. Oct. 1985, 88.
14. Miller, WJ. Dairy Cattle Feeding and Nutrition, 1979, 142.
15. Underwood, EJ. In: The Mineral Nutrition of Livestock, 2nd ed.; p 98.
16. Thomas, JW; Moss, S. J. Dairy Sci. 1951, 34, 929.
17. Ourebo, RG. Private Practitioner, Arlington, Minnesota (personal communication, 1985).
18. Boyne, R. Res. Vet. Sci. 1978, 24, 134.
19. Maas, J. In: Selenium Responsive Diseases in Food Animals, 1985; p 20.
20. Jenkinson, SG; Lawrence, RA; et al. J.Nutr. 1982, 112, 197.
21. Leach, RM. Nutrition Institute on Minerals, 1983, p 3
22. Ensminger and Olentine, Feeds and Nutrition, 1st ed.; p 713.

DISCUSSION

HARRIS: What you found in cattle we have found in thoroughbred horses. Apparently there is a growth plate lesion that affects the fetlock joint. There is total rupture of the articular surface cartilage and its quite traumatic in these animals. They are very lame. This is found in animals that generally are pastured in a Zu rich area, down stream from a smelter, or it occurs spontaneously. It's been suggested that because the animal grows so very rapidly in the appendages that nutrition just can't keep up with it.

ANDERSON: Yes. I have investigated race horses and have not been able to detect any deficiencies.

SMITH: Did you find this condition to be due to low dietary Cu or is it due to a diet that is high in Mo?

ANDERSON: We have both. Most forages run between 8 and 10 ppm Cu. Mo content will run as high as 19.6 ppm. A good ratio is 5 to 1, so we had both circumstances. We didn't find any sulfur. In reflection, we had high Fe levels, levels of 400 to as high as 1240 ppm in some silage. I feel that was involved too.

SMITH: Pica, the eating of abnormal things like manure and earth, has been reported for humans who are Zn deficient. Is this recognized in animals or is this something new? Is this an effort to get Cu?

ANDERSON: I didn't want to infer that pica was a symptom of Cu deficiency, but my personal opinion is that they were looking for Cu. They knew they were deficient in something and they were trying to find it.

SMITH: Those bone abnormalities were dramatic and this is an area that is important for humans. It's been suggested that Cu deficiency might be a model of osteoporosis and that we should study Cu status in osteoporotic humans.

SORENSON: I wonder whether that calf from the cow that was a big milk producer may have died as a result of its consumption of a whole lot more Mo. A calf being nursed by such a cow might need more Cu because of the greater ingestion of Mo.

COUSINS: It seems that some of these problems you've been describing are very similar to problems I had in Scotland with thiomolybdate related changes in Cu status.

OBSERVATIONS ON BIOLOGICAL EFFECTS OF LOW DIETARY COPPER IN NEBRASKA BEEF CATTLE

J. L. Johnson, N. R. Schneider*, A. R. Doster*, M. P. Carlson*, H. W. Leipold**

Veterinary Science Laboratory, University of Nebraska, North Platte, NE; *Veterinary Diagnostic Center, University of Nebraska, Lincoln, NE; **Department of Pathology, Kansas State University, Manhattan, KS.

Nutritional disorders involving the trace elements may arise as either a deficiency or toxicity of particular elements, but more often as deficiencies or toxicities conditioned by the extent to which other organic or inorganic nutrients are present in the diet. Copper is an essential trace element whose presence in biological systems was first reported over 150 years ago,[1] and is utilized in activity of over a dozen enzyme systems. Low serum copper concentrations have been related to disturbances in iron absorption, failures in erythropoiesis, mental deterioration, and death.[2]

Copper is necessary for utilization of iron in hemoglobin synthesis and essential for synthesis of myelin that surrounds peripheral nerves. Copper-containing enzymes in conversion of tyrosine to melanin are the probable cause of failure of normal hair pigmentation when insufficient quantities of copper are fed. Loss of crimp and tensile strength in wool also occurs because of failure to maintain normal disulfide groups that provide cross linkages within the wool fiber. Copper also is involved in bone formation, reproduction, elastin formation and in the prevention of degeneration of bovine myocardium.[3,4,5,6]

There is little placental transfer of copper in the bovine; therefore, the calf is dependent on copper in colostrum. Concentrations of copper in milk drop significantly 48 hours after calving. Copper deficiency in several species has produced anatomical, chemical, and physiological abnormalities similar to those found in ischemic heart disease. A mild microcytic normochromic anemia has been reported by Canadian workers.[7]

Chronic stress or infections can lead to an increase of plasma copper. Plasma or serum concentrations alone are not totally reliable in diagnosis of copper deficiency. Serum and liver copper analyses should be done concomitantly when possible.

A reciprocal antagonism exists between molybdenum and copper. This is important in veterinary medicine because, of farm animals, cattle are the most sensitive to molybdenum followed by sheep, horses, and pigs.[6] A higher rate of ceruloplasmin turnover in cattle than sheep may account for the difference in molybdenum tolerance. A wide tolerance range occurs in various ruminants: 6.2 ppm for cattle compared to 1,000 ppm for mule deer. Dietary copper concentrations of 8-10 ppm protect cattle against excess molybdenum concentrations of approximately 5-6 ppm.[8] Molybdenum, in the presence of sulfates, limits copper retention in cattle and sheep; however, neither molybdenum nor sulfates alone affect copper retention.[5] Effects of excess molybdenum are essentially similar to those of copper deficiency.

OBSERVATIONS

Copper deficiency and excess molybdenum cause similar syndromes.[1,5,6] In Nebraska, copper deficiency has been noted in cattle with abomasal ulcers, congenital defects, nitrate toxicity, diarrheas, infertility, and the alert downer cow syndrome. Copper deficiency has also been observed with abortion.

Ulcers

The incidence of abomasal ulcers and abomasal tympany has been on the increase in western Nebraska and Wyoming over the last 10 years.[9,10] Correlation between infectious agents such as Clostridium perfringens and BVD in association with lower serum and hepatic copper concentrations exist.[9,10,11] The exact role that copper deficiency plays in abomasal ulcers is speculative. Possible pathogenesis includes: as copper-deficient calves are generally anemic, they may be immunologically inferior as copper supplementation or vaccination for the disease agent found will clinically reduce disease morbidity and mortality.[1,10,12,13,14]

Congenital Defects

Various types of albinism have been reported in many breeds of cattle throughout the world. In 1982 a new coat and eye color defect (partial albinism, heterochromia irides) in purebred Black Angus cattle occurred. This condition was reported in several Angus herds in the western states. Field investigations and breeding trials were in conjunction with the Department of Pathology, Kansas State University.

Heterochromia (irides) in Angus calves of both sexes is identified by hair, skin, coat color, and iris color characteristics. Hair of the entire body surface is brown (chocolate) instead of typical black for Angus cattle. The muzzle, hooves, and scrotum in males also are brown. The skin surface is brownish to grey. This is particularly obvious at the glabrous skin, such as around eyelids, ear openings, muzzle, anal and reproductive openings. The most distinguishing factor involves iris color. Angus cattle usually have a dark black iris. Heterochromia irides calves, however, have a light, usually two-colored iris. This gives a double-ringed appearance to the iris when viewed closely: an outer, faintly brown ring and an inner, light blue ring circling the pupil. The pupils always appear constricted in daylight. From a distance, eyes appear white. The fundi appear albinotic.

Melanin, the brown pigment of hair, skin and eyes, is a polymeric material which arises from the oxidation of tyrosine. The enzyme involved in the conversion of tyrosine to melanin is tyrosinase (o-diphenol; O_2 oxidoreductase), a copper containing protein, which catalyzes the conversion of tyrosine to dihydroxyphenylalanine and subsequently to dihydroxyphenylalanine quinone (DOPA reaction). Breeding trials suggest that heterochromia irides is likely transmitted as a simple autosomal recessive.[15]

Congenital angular limb deformity is a periodic, generally isolated observation. Viral isolation of heparinized or EDTA blood has been negative. Copper concentrations in serum have been low, occasionally complicated by low serum selenium.

Neonatal ataxia is a nervous disorder of lambs and calves characterized by inability to suckle. Tissue sections from affected calves reveal amyelination of the brain stem due to lack of cytochrome oxidase, a copper-containing oxidative enzyme necessary for <u>in utero</u> myelin formation.[12]

Nitrates

High incidence of nitrates in beef cattle of west-central Nebraska may be explained partially by the higher molybdenum to copper concentrations in native hays. Hay analyses from several areas revealed molybdenum to copper ratios of 1.5:1 to 11:1. No samples revealed normal 6:1 copper to molybdenum ratios.[16]

Several metalloflavo-protein enzymes are involved in the reduction of nitrate to ammonia in the rumen. These include molybdenum, copper, iron, magnesium, and manganese.[17] Nitrate reductase is a molybdoflavo-protein enzyme. High dietary molybdenum results in lower concentrations of ruminal nitrate, higher nitrite and a higher ruminal ammonia content. The most optimal rumen pH for nitrate reduction is 6.5, while nitrite and hydroxylamine reductions optimize at pH 5.6 to 7.[18]

Subclinical nitrate abortion may relate to complex interactions between maternal-fetal circulation, trace mineral and enzyme interactions controlled by quantity and quality of feedstuffs, absorption time, rumen pH, release of fodder nitrate into rumen liquid and conversion of rumen nitrate to nitrite.[16,19]

Excessive dietary nitrate concentrations may affect copper utilization in cattle, thereby causing copper liver storage aberration. Increased copper liver storage was reported in Canada[7] and liver analysis of nitrate positive stillborn calves at North Platte had a similar correlation. In another copper study in bovine abortion conducted in Nebraska, liver concentrations were lower than expected.[20] This storage aberration could have an effect on maternal-fetal oxidation. The nitrate biological half-life ($t_{1/2}$) for the fetus is exceedingly long (>24 hours) and may explain why the perinate bovine appears more sensitive to the adverse effects of nitrite/nitrate exposure.[19]

Diarrheas

Gastrointestinal irritation is a common entity in trace mineral toxicosis. Diarrhea has been reported with excesses of a number of metallic elements. The most common trace mineral deficiencies that may be associated with diarrhea are copper[14] and selenium.[7,13,21] Interactions are generally involved in calf diarrheas in west-central Nebraska, and these usually include copper, molybdenum, sulfates, iron, selenium, and zinc.

Elemental concentrations of 41 livers from calves with diarrhea, representing 24 different ranches were examined. Animal age varied from 6 to 60 days. Mean concentrations for all elements fell within expected ranges. Thirty-seven individual variations occurred in the 41 livers. Excess storage of elements was recorded for aluminum (1), arsenic (1), copper (3), iron (8), selenium (2), and zinc (3). Deficiencies of copper (1), iron (4), sodium (7), selenium (2), and molybdenum (5) were found. Numerous low and high expected results occurred; these concentrations could cause animal health problems if interaction with other imbalanced trace minerals occurred.

Liver concentrations indicated above normal iron content as the most common trace mineral finding in calves with diarrhea. High iron content (8) was observed on seven different ranches and ranged from 314 to 1540 ppm. Forage analyses generally indicated high molybdenum content in most feedstuffs. Ideal copper/molybdenum ratios should be 6:1 with an adequate ratio 3:1. Ratios of feedstuffs from these Nebraska ranches ranged from 1.5:1 to <1:1. Blood sera from cattle on these ranches routinely indicated marginal to low serum copper concentrations.

No single element appears consistent for neonatal calf diarrhea, but the variety of fluctuations indicate that trace elements play a role in many enteric syndromes. Functions of trace elements are interrelated, and the copper/molybdenum/sulfur interaction has the greatest frequency in the Sandhills of west-central Nebraska. Iron, selenium, and zinc are also important.[1] Diarrhea associated with copper deficiency has been also related to depletion of cytochrome oxidase in the small intestine, resulting in partial villus atrophy.[7]

Downer Cows

An increased incidence of alert downer cows was observed by Nebraska veterinarians during the spring of 1984. Animals affected were predominantly preparturient cows in the third trimester of gestation, often within 2 to 4 weeks of calving. Affected cows were found in sternal recumbency with heads erect, but were weak and reluctant or unable to rise voluntarily. If forced or able to rise, these cows exhibited posterior ataxia; attempts to walk or run often terminated in sudden collapse and death. The animals did not readily respond to intravenous electrolyte therapy. Specific post-mortem findings were minimal or absent, but cows from some herds had evidence of serous atrophy of perirenal fat and little, if any, other body fat remaining. Abortions or stillbirths were reported in some herds.

While cold stress and inadequate nutrition may have contributed to this syndrome, primary analytical findings in cases submitted to the Veterinary Diagnostic Center related to copper deficiency. Concomitant selenium deficiency, hypocalcemia and hypomagnesemia were often present. In some cases, excessive copper or selenium concentrations were found in either liver or kidney. Often there was a deficiency in the other tissues examined. Both maternal and fetal specimens frequently exhibited these findings. Excessive nitrate concentrations were found in maternal and fetal ocular fluid, serum and plasma. The available data suggest that the alert downer cow syndrome most likely is a nutritional/metabolic disorder related to copper availability, absorption, utilization or metabolism; and may be induced by excessive nitrate exposure complicated by stress and inadequate dietary energy.[19]

Infertility and Abortions

Copper deficiency can also cause anestrus and lowered conception in cattle. Fetal resorption and abortions occur in other species.[14]

Herd infertility usually followed a generalized pattern. Fifteen to sixty percent of the animals were reported open at pregnancy examination in the fall. The greatest percent of open animals were first-calf heifers with large frames (exotic and/or crossbred). The previous year these animals were fed only native hay which was low in protein. Supplementation with cake, alfalfa hay and/or mineral was infrequent. Copper and selenium concentrations in serum from these animals were generally lower than expected. Serum profiles (SMAC) recorded variation in the calcium/phosphorus ratio. Total protein, urea, and glucose were low; iron, LDH, SGOT, and SGPT were generally elevated. Overall assessment is indicative of the sparing effect protein has on trace minerals, and native sandhills pasture may not have adequate nutritional levels to support the heavier, higher producing cattle without supplementation. Trace mineral deficiencies were associated with some stillborn calf losses. Evaluation of trace element aberrations and the potential role in maternal-fetal metabolic disorders is continuing.

Anemia

Anemia is one of the most common syndromes associated with copper deficiency. Copper deficiency causes impairment of iron transfer to plasma from the hepatic reticuloendothelial system. Release of iron to plasma depends upon the oxidative enzyme, ceruloplasmin. Copper is the rate-limiting element in production of ceruloplasmin, concentrations of which are directly related to the copper status of the animal. Ceruloplasmin greatly accelerates oxidation of ferrous iron and is the rate-limiting step in the formation of ferric iron in transferrin.[14]

SUMMARY

Copper deficiency exists in Nebraska beef cattle and is manifested by various disease entities and syndromes. Its effects are controlled by management practices, with the third trimester of pregnancy being the most critical period for the brood cow. Interactions with other compounds and elements have not permitted complete characterization of copper metabolism and utilization in cattle. Part of the herd health prenatal program should include monitoring for trace elements. No one analysis is a reliable indicator of the animal's overall copper status. Serum copper determination, multi-elemental liver analyses and complete evaluation of feedstuffs should be conducted. These analyses may indicate possible trace mineral interrelationships that could cause low or deficient copper availability in cattle.

LITERATURE CITED

1. Johnson, JL; Schneider, NR; Anderson, GA; Carlson, MP. Proc. 18th annual conference A.A.B.P., November 19-22. 1985, p 142-146.

2. Hochstein, P; Sreekumar, K; Forman, SJ. Ann. New York Acad. Sci. 1980, 355, 240-247.

3. Ammerman, CB; Goodrich, RD. J.A.S. 1983, 57, 519-533, suppl 2.

4. Baxter, JT. In: Current Veterinary Therapy Food Animal Practice; W. B. Saunders Co.; Philadelphia, London, Toronto, 1981; p 320-330.

5. Hays, VW; Swenson, MJ. In: Minerals Part IV. Physiology of Domestic Animals, Duke's, 9th Ed; Comstock Publishing Associates, Cornell University Press; Ithaca and London, 1977; p 395-410.

6. Underwood, EJ. In: Trace Elements in Human and Animal Nutrition, 4th Ed.; Academic Press; New York, 1977.

7. Smart, ME; Gudmundson, J; Christensen, DA. Can. Vet. J., December. 1981, 22, 372-376.

8. Mineral Tolerance of Domestic Animals: National Academy of Sciences. National Academy Press; Washington DC, 1980.

9. Johnson, JL; Hudson, DB; Bohlender, RE. Proc. 24th annual meeting A.A.V.L.D., October. 1981, p 203-210.

10. Lilley, CW; Hamar, DW; Gerlach, M; Johnson, JL. Vet. Med., October. 1985, p 85-88.

11. Johnson, JL; Lilley, CW; Hamar, DW; Britton, R; Clemens, ET. Proc. Third International Symposium World Association of Veterinary Laboratory Diagnosticians, Ames, Iowa, June 13-15. 1983, p 485-491.

12. Sanders, DE. J.A.V.M.A. 1980, 176, 728-730.

13. Sanders, DE. Mod. Vet. Pract., February. 1984, p 136-138.

14. Sanders, DE; Sanders, JA. Mod. Vet. Pract., August. 1983, p 613-616.

15. Strasia, CA; Johnson, JL; Cole, D; Leipold, HW. Bovine Practitioner. 1983, 18, 147-149.

16. Johnson, JL; Schneider, NR; Kelling, CL; Doster, AR. Proc. 26th annual meeting A.A.V.L.D., October. 1983, p 167-180

17. Nicholls, TJ. Aust. Vet. J. 1980, 56, 95-96.

18. Tillman, AD; Sheriha, GM; Sirny, RJ. J. Anim. Sci. 1965, 84, 1140-1146.

19. Schneider, NR; Johnson, JL. Proc. Nebraska Vet. Med. Assoc., January. 1986, p 75-87.

20. Doster, AR; Schneider, NR. Personal communication.

21. Pottenger, RC. Proc. North Central A.A.V.L.D., June 13. 1984.

(Editor's note: Due to insufficient time there was no discussion of this paper.)

CUPROUS SEBACATE TREATMENT OF MENKES' DISEASE

Darryl M. Williams, MD, John R. Clement, MD,
F. Scott Kennedy, PhD, and Harold Chen, MD
Louisiana State University Medical Center in Shreveport
P. O. Box 33932, Shreveport, Louisiana 71130

In 1962, Menkes *et al.* first described an inherited disorder that was characterized by severe neurologic, integumentary, skeletal, and developmental abnormalities and that was also associated with early death in male infants.[1] In 1972, Danks and his group demonstrated that this disorder was associated with abnormal copper metabolism.[2] Copper absorption was impaired in affected infants, with resultant hypocupremia and hypoceruloplasminemia. Since then, a number of attempts have been made to correct the abnormalities of Menkes' disease with copper administration.[3-11] In general, these attempts have been successful in raising levels of serum copper and ceruloplasmin to normal levels, but convincing changes in the clinical course of treated patients has not been demonstrated. (Table 1)

Danks and his co-workers have studied in detail the copper metabolism of the mottled mutant mouse, a homologous animal model for the disease. As part of these studies, Mann *et al.* demonstrated that treatment with cuprous salts in the presence of sebacic acid resulted in nearly normal growth rates and survival of these animals.[12]

We report preliminary results of similar treatment of a patient with Menkes' disease with copper sebacate.

Table 1
Trials of Copper Therapy in Menkes' Disease

Year	First Author	Preparation	Result
1973	Bucknall[3]	$CuSO_4$, PO Cu acetate, IV Ceruloplasmin, IV	no clinical effect
1975	Grover[4]	$CuSO_4$, IV Cu acetate, IV	"some achievement"
1975	Wehinger[5]	$CuSO_4$, IV	no clinical effect
1975	Garnica[6]	Cu acetate, IV	no clinical effect
1975	Lott[7]	Cu histidinate, PO	increased serum Cu
1977	Williams[8]	Cu histidinate, Sub Q	no clinical effect
1981	Procopis[9]	Cu-EDTA, IM	mild disease/unknown
1983	Sass Kortsak[10]	Cu histidinate	"encouraging progress"
1984	Garnica[11]	$CuSO_4$, IV Plasma, IV	no clinical effect

PATIENT AND METHODS

The patient was the second child of a 24-year-old woman. Two maternal uncles and two male cousins had previously died with documented Menkes' disease. At birth, the patient's serum copper was 30 μg/dl, and serum ceruoplasmin was 8 mg/dl. At one month of age, the patient's values had decreased to 8 μg/dl and 3 mg/dl respectively. At two months of age the patient was referred to the Louisiana State University Hospital. He was noted to be alert, slightly hypotonic, but with good cry and grasp. He was unable to hold his head erect but had good tone and strength in all extremities. Seborrhea was observed, and the hair was noted to be fine and brittle with characteristic microscopic findings. At that time serum copper was 15 μg/dl and ceruloplasmin was 3.8 mg/dl.

At 59 days of age, the patient was begun on copper treatment consisting of cuprous sebacate prepared by the method of Mann *et al*.[12]

Blood for analysis was obtained in acid-washed, heparinized Natelson tubes. Ceruloplasmin was measured as *p* -phenylenediamine oxidase activity, and plasma copper was measured by atomic absorption spectrometry.[8] The patient was tested at intervals for behavioral and motor development using the Denver Development Screening Test.[13]

RESULTS

Copper treatment was begun with 25 µg doses of copper. As tolerance was demonstrated, dosage was increased, and plasma copper and ceruloplasmin levels promptly rose. The most notable side effects were local reactions, including redness, induration, sterile abscesses, and localized lymphadenopathy. In three intervals it was necessary to interrupt treatment because of sterile abscesses and fever. Plasma copper and ceruloplasmin fell quickly following cessation of treatment. (Figure 1) After four months of age, the patient's weight and developmental success stabilized, with little progress after the ninth month. (Figure 2)

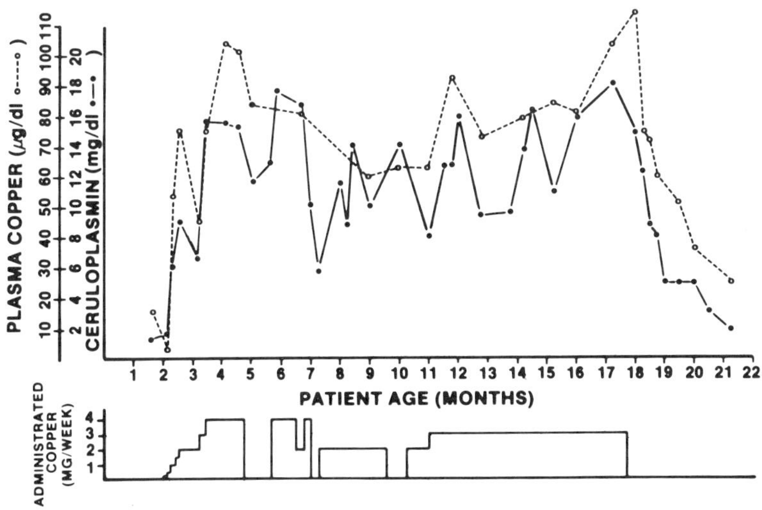

Fig. 1 Responses of plasma ceruloplasmin (.——.) and copper (o---o) levels in a patient with Menkes' disease treated with copper sebacate. Treatment was begun at two months of age with injections of 250 µg Cu. Dosage was increased to the range of 2.0 - 4.0 mg Cu given intramuscularly on a weekly schedule. Treatment was interrupted on three occasions as shown. Both ceruloplasmin and copper levels fell sharply after cessation of treatment.

One year after cessation of therapy, the patient was able to assume the sitting position unassisted. He had developed a vocabulary of intelligible sounds and had become partially toilet trained. Throughout this time the patient remained free of seizures.

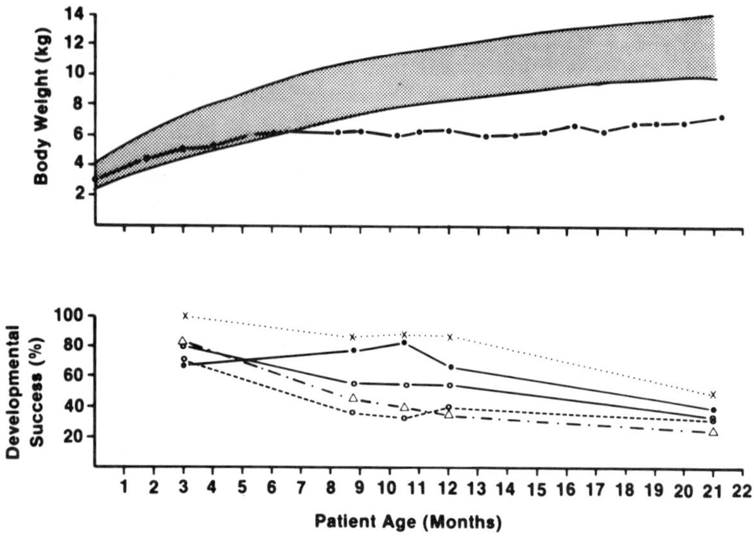

Fig. 2. Changes in body weight and development during treatment with copper sebacate. Panel A. Patient weight (.——.) is compared with growth curves of normal male infants (hatched area includes 5-95 percentile)[13]. Patient weight is within the normal range at birth, but falls below the normal range after 6 months. Panel B. Patient success with the Denver Developmental Screening Test is shown as a percentage of the tests appropriate for the age of testing or younger. Areas of testing included personal and social skills (.——.), fine motor skills (o---o), language (x...x), gross motor (Δ-.-.Δ) and a composite (o——o).

DISCUSSION

Copper treatment of Menkes' disease has characteristically resulted in restoration of normal serum copper and ceruloplasmin levels. At the same time, such treatment has failed to result in meaningful clinical improvement. An early report of one patient described encouraging results, but no further reports have been forthcoming.[4] A second patient treated by Sass Kortsak has been reported to show some improvement.[10] A third patient with a mild form of the disease was treated by the Australian group for an extended length of time, with some suggestion of improvement.[9] Otherwise, treatment has resulted in failure.

The reasons for these failures are not clear. It is possible that the doses of copper are insufficient to meet the tissue needs of the affected individual. No clinical studies are available to address this issue. It is also likely that the time of beginning copper treatment is critical. This appeared to be the case in the animal studies of Mann et al, in which delay of a few days in the institution of copper therapy resulted in failure.[12] It is possible that the form of delivered copper is also crucial. This was also demonstrated in the animal studies of Mann et al. In those studies, only reduced cuprous salts administered with a vehicle such as sebacate resulted in correction of developmental and growth abnormalities. In addition, relatively massive doses of copper were administered in a short interval of time. Such a treatment approach has never been duplicated in human subjects. In large part this is because of the potential toxicity of such large doses of copper.

In the present study, there is little doubt that the patient has shown slow but steady development of motor skills. These skills have been documented by video recordings. Obviously, what is unclear in reporting on a single patient is whether the patient would have achieved such skills with or without the copper. It is unlikely that the patient has a milder form of Menkes' disease, as described by Procopis, since other members of the family have had well documented severe disease resulting in death by the age of two. Thus, it is reasonable to believe that the patient may have had some clinical improvement as a result of the treatment.

More importantly, however, is the demonstration that the patient tolerated cuprous sebacate treatment with no obvious complications except for local sterile abscesses, localized lymphadenopathy, and fever, which resolved with cessation of treatment. Doses of copper that were administered early in the patient's course were relatively small and probably insufficient to meet the replacement needs during the critical developmental period. Subsequently, however, the patient tolerated much larger doses of copper without any additional side effects. For that reason, it appears justified to evaluate the use of higher doses of cuprous sebacate as early as possible in the management of affected individuals.

REFERENCES

1. Menkes, J, Alter, M, Steigleder, GK, Weakley, DR; Sung, JH. Pediatrics 1962, 29, 764.

2. Danks, DM, Campbell, PE, Stevens, BJ, Mayne, V; Cartwright, E. Pediatrics 1972, 50, 188.

3. Bucknall, WE, Haslam, RHA; Holtzman, NA. Pediatrics 1973, 52, 653.

4. Grover, WD; Scrutton, MC. J. Pediatr. 1975, 86, 216.

5. Wehinger, H, Witt, I, Lösel, I, Denz-Seibert, G; Sanders, G. Lancet 1975, 1, 1143.

6. Garnica, AD; Fletcher, SR. Lancet 1975, 2, 659.

7. Lott, IT, DiPaulo, R, Schwartz, D, Janowska, S; Kanfer, JN. N. Engl. J. Med. 1975, 292, 197.

8. Williams, DM, Atkin, CL, Frens, DB; Bray, PF. Pediatr. Res. 1977, 11, 823.

9. Procopis, P, Camakaris, J; Danks, DM. J. Pediatr. 1981, 292, 197.

10. Danks, DM. In: Metabolic Bases of Inherited Disease, 5th ed.; McGraw Hill, New York, 1983; p. 1251.

11. Garnica, AD. Europ. J. Pediatr. 1984, 142, 98.

12. Mann, JR, Camakaris, J, Danks, DM; Walliczek, EG. Biochem. J. 1979, 180, 605.

13. Frankenburg, WK; Dodds, JB. J. Pediatr. 1967, 71, 181.

DISCUSSION

<u>MELETHIL</u>: Have you tried treating with any other organometallics such as Cu(II)(salicylate)$_2$ or anything like that?

<u>WILLIAMS</u>: Yes, a number of years ago we tried Cu(II)(histidine)$_2$.

<u>MELETHIL</u>: You said you might increase the dose in the future and try something more lipid soluble. Is there any suggestion in the literature that part of the problem might be transport of Cu in Menkes' Disease?

<u>WILLIAMS</u>: Yes, there is a lot of thought about that. Ceruloplasmin appears to be normal with regard to protein and amount of Cu. Whether that's caught up in the abnormality is not clear but it's not very likely. For a time, people were enthusiastic about the possibility that metallothioneine was abnormal but now that is in question and the exact nature of abnormal Cu metabolism in these patients is not clear.

<u>BARNEA</u>: I have the impression that it is possible to prevent some of the brain abnormalities in the mouse model if you administer Cu during the first week of life. I'm wondering if it is possible to treat those kids at a much earlier stage, even in utero?

<u>WILLIAMS</u>: The conventional wisdom is that there is a window, although to my knowledge nobody has clearly demonstrated a window experimentally except in Danks' studies that you mentioned and others I briefly summarized in which Cu seems to be most effective in treating mice during a very short time after birth. The conventional explanation is that the blood brain barrier is impervious to Cu after a particular period of time and if it is possible to get Cu across this barrier before it becomes impervious then it may be possible to get Cu levels up high enough to be effective. Again, there are very few data. I know of only a few studies done by Horn and her group comparing fetal brain Cu, which is high, while others have evaluated Menkes' babies after some period of life, where most of the time people find low Cu levels, so that's taken as some evidence that it's a blood brain barrier problem.

As far as treating humans, yes I think the earlier you can get them the better.

BARNEA: Why are those mice given only a single administration within those few days? The first week of life is critical for brain development for many functions and alot of substances administered during this first week of life permanently determine the development of adult brain function. I was surprised to find that they gave only 1 or 2 administrations of Cu. Would this be assumed to have a permanent effect? Wouldn't there be some turnover and loss with time? How much Cu is in the brain?

WILLIAMS: Well, as I said, they gave an enormous amount of Cu.

KEEN: Brain Cu only goes up from about 1.5 to around 2.5 ppm between day 10 and 21 in the postnatal mouse. That's a very small amount. Since most myelination occurs during that time period you really can get away with 1 or 2 doses having a profound effect. Henkin reported 5 or 6 years ago that he had successfully treated a child for several years with Cu-nitrilotriacetate. Do you know what's happened with that child?

WILLIAMS: No I don't. There really are very few follow-up reports on the outcome of these patients.

KEEN: Nina Horn showed quite nicely that you can culture amniotic cells and pick up the Menkes' trait. Since you have the familial history of your patient, did you consider doing amniocentesis in this case or in a future pregnancy and then supplementation at birth? You potentially have a month or so to make your diagnosis.

WILLIAMS: We only became aware of this patient at about the time of the second ceruloplasmin determination. Since we are aware of this family, and there are several women in the family, we will be following their pregnancies.

SMITH: Was your preparation given IV or IM?

WILLIAMS: It was given IM.

SMITH: Why did you want a cuprous preparation?

WILLIAMS: Danks' group tried both Cu(I) and Cu(II) preparations and they suggested that Cu(I) was more effective than Cu(II).

SMITH: Did they give it IM too.

WILLIAMS: I think it was IM.

SMITH: Could you or someone tell me what form Cu(II) sulfate would have in Total Parenteral Nutrition Solutions? We thought the dogma was that it had to be in the Cu(II) state to be absorbed. Here's a case in the literature that suggests that Cu(I) might be more effective.

WILLIAMS: I wish we could say that we knew, in fact, that it was all in the Cu(I) state but that's problematic.

SORENSON: It's great to hear that the notion that these patients should be treated still exists. Other physicians have made the decision to recommend abortion of these children. We will never be able to help these families if all of these children are aborted. So I think your attempt to treat is fanatastic. Regarding the Cu(I) - Cu(II) question, the basis for David Danks' and Jim Camakaris' suggestion that Cu(I) was more effective was never very clear. Their complex or complexes were not well characterized. I'm sure that Cu(I) is made from Cu(II) very fast in vivo, if that is what is needed. Ceruloplasmin becomes blue on standing in air with the oxidation of its Cu(I) to Cu(II). My intuition is that this oxidized form would be reduced if it were again injected into blood.

There is a mixed valence penicillamine complex that has 8 atoms of Cu(I) and 6 atoms of Cu(II). A preparation of it would be easily prepared and it would provide Cu in both oxidation states. Prof. Scheinberg has gotten FDA approval for use of this mixed valence Cu-penicillamine complex to treat rheumatoid arthritis. He tells me that results are good with regard to its use in treating rheumatoid arthritis. If that's the case, then that might be a complex that could be used to provide Cu for any reason. It is water soluble and it could be given by IM or IV injection.

In the film you showed, it looked like this child had had some brain development in that he had progressed beyond stages of development usually seen for Menkes' patients, which are suggested to not have any mental or emotional development at all. This child spoke, he had coordinated movement, and he had hair. Most of these children loose all of their hair and they have no mental development at all. This was a great first effort so you are to be commended.

Biological Aspects of Copper and Iron in Inflammation

Role of Neutrophils in Inflammatory Tissue Injury

Edwin L. Thomas

St. Jude Children's Research Hospital

Memphis, Tennessee 38101

Neutrophilic leukocytes account for about half the number of circulating white blood cells. Their primary role is to protect against infection by killing invading microorganisms. However, the same neutrophil activities important in host-defense can damage other blood cells and the tissues.

Following their release from the bone marrow into the blood stream, neutrophils do not undergo further division, differentiation, or nucleic acid and protein synthesis. They circulate for only a short time, with an average half-life of 7 hours, before being removed from the circulation in the spleen. Neutrophils in blood are quiescent, but can be stimulated to undergo _migration_, _adhesion_, _phagocytosis_, _secretion_ and _oxygen (O_2) metabolism_. These activities are a graded series of responses to increasing concentrations of stimulating agents. Other leukocytes including eosinophils, monocytes, and macrophages express these activities. However, neutrophils are the first of these inflammatory cells to arrive at a site of infection or tissue injury.

In response to chemotactic signals, neutrophils migrate out of the blood stream through the blood vessel walls and into the tissues. Accompanying this chemotactic response is an increase in adhesiveness of the neutrophil surface. Making and breaking adhesive contacts with cell and tissue surfaces provides the traction that enables neutrophils to migrate toward the site where chemotactic signals are generated. In the tissues, neutrophils encounter microorganisms, viruses, and other particles that have been coated with specific antibodies, complement components, and other "opsonizing" plasma components. These surface-bound substances mark the particles as appropriate targets for attack. The opsonized particle binds to the neutrophil through specific receptors on the neutrophil surface, triggering the process of phagocytosis. The neutrophil membrane invaginates and pinches off to form an intracellular compartment, the phagocytic vesicle. The neutrophil cytoplasm is filled with secretory granules, which accounts for the classification of neutrophils together with eosinophils as "granulocytes". The granules consist of tightly packed soluble proteins enclosed by membranes and are a specialized form of the lysosomes present in all nucleated mammalian cells. The cytoplasmic granules merge with the phagocytic vesicle through the process of membrane fusion and release the granule contents into this compartment, forming the phagolysosome. Killing and digestion of microorganisms takes place within phagolysosomes.

Secretion of granule components into phagolysosomes is accompanied by a sudden dramatic increase in O_2 uptake, the "oxidative burst". A latent NADPH-oxidase is assembled and activated in the phagolysosome membrane. This enzyme spans the membrane and is oriented to accept electrons from NADPH in the cytosol, transfer electrons one at a time to O_2, and release the superoxide free radical (O_2^-) into the phagolysosome. Production of O_2^- is the first in a series of reactions that produce toxic O_2 metabolites which contribute to intracellular antimicrobial activity.

Neutrophils contain at least three distinct types of secretory granules, along with a prominent lobed nucleus which accounts for the name polymorphonuclear (PMN) cells. At this time, only the proteolytic enzyme gelatinase has been identified within tertiary granules. However, the membranes surrounding these granules may contain some or all of the flavoprotein, cytochrome, and other proteins that make up the NADPH oxidase.

Secondary granules contain lysozyme and other hydrolytic enzymes, particularly glycosidases which hydrolyze carbohydrate polymers. These granules also contain the protein lactoferrin, which has homology with serum transferrin. Lactoferrin binds iron, copper, and perhaps other metal ions. In acute inflammatory states, lactoferrin-bound metal ions may contribute significantly to the metal content of plasma and inflamed tissues. Because free metal ions can be toxic to cells, lactoferrin may protect neutrophils and the tissues by binding metals released during killing and digestion of microorganisms. Lactoferrin may also have an indirect antimicrobial effect by depriving microorganisms of metals needed for growth. It has been proposed that iron-loaded lactoferrin acts as a catalyst for production of toxic O_2 metabolites. However, iron at the two specific binding sites in lactoferrin appears to be bound in a way that largely deactivates iron as a redox catalyst.

Primary granules contain myeloperoxidase (MPO), lysozyme, and cationic proteins, all of which are involved in antimicrobial activity, and a number of proteolytic enzymes including elastase. Additional lysosomal components not clearly localized to a particular class of granules include the proteolytic enzyme collagenase, phospholipase enzymes, and a family of antibiotic-like peptides called "defensins".

In response to high concentrations of soluble stimulating agents or opsonized particles, cytoplasmic granules fuse with the neutrophil cell membrane, resulting in secretion of granule proteins into the extracellular medium. Similarly, the oxidase becomes distributed and activated over the entire cell surface and releases O_2^- into the medium. It is this release of lysosomal enzymes and toxic O_2 metabolites that contributes to inflammatory tissue injury.

Neutrophils amplify the series of events which may be described as an inflammatory cascade (Fig. 1). Minor alterations in tissue cell functions perhaps related to the influx of calcium convert a xanthine dehydrogenase enzyme to an oxidase that produces small amounts of O_2^- and hydrogen peroxide (H_2O_2). These O_2 metabolites may be released from cells and react with plasma components converting them into chemoattractants. Cell lysis and fragmentation of tissue components releases substances that activate the complement system and result in formation of potent leukocyte chemoattractants. Damage to tissues also exposes collagen, which activates platelets and results in clot formation. Activated platelets release prostaglandins and other arachidonic acid metabolites that may attract neutrophils and also act as vasodilators. Another potent vasodilator is histamine, which is secreted by a special class of tissue cells, the mast cells. Vasodilators stimulate the endothelial cells that make up the walls of blood vessels to undergo a change in shape that opens spaces between the cells and facilitates entry of neutrophils and plasma components into tissues.

Neutrophil influx is followed by stimulation to secrete lysosomal enzymes and produce O_2 metabolites. In addition to damaging tissues, these activities liberate more chemoattractants and vasodilators, resulting in an intensified influx of phagocytic leukocytes. Over a period of seconds to hours, neutrophils predominate at sites of injury and infection. Over a period of days or longer, eosinophils and macrophages become the predominant inflammatory cells.

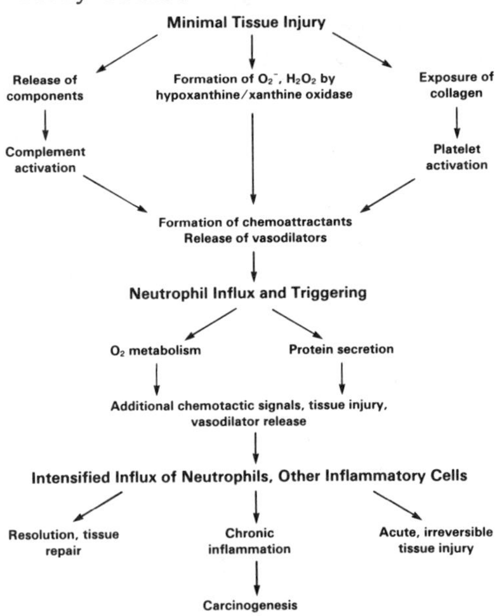

Figure 1. Inflammatory Cascade

The inflammatory cascade has a number of possible outcomes. It is important to remember that inflammation is a normal, healthy response to injury and infection. Phagocytic leukocytes clear the site of microorganisms, damaged tissue components, and leukocyte components, setting the stage for repair and a return to normal function. Inflammation tends to be self-limiting rather than leading to a greater and greater influx of cells and release of inflammatory mediators. Little is known about the signals that control, moderate, and end the inflammatory response.

Chronic inflammation develops when chemotactic signals persist and can't be eliminated by phagocytic leukocytes and tissue repair processes. A slow, progressive loss of tissue function may result. Of particular interest to our laboratory is the relation between inflammation and carcinogenesis. Certain types of tumors tend to arise at sites of chronic inflammation suggesting that attack by leukocytes on normal tissues can cause or promote tumorigenesis.

Another possible but rare outcome is acute, irreversible injury. This type of injury is most often associated with failure of the circulatory system, such as in heart attack, stroke, and other conditions such as ischemic bowel disease. In experimental systems, tissue injury is greatly decreased by blocking neutrophil influx. Another form of acute injury is the result of aggregation. Neutrophils adhere and form multicellular aggregates in the blood stream in response to bacterial toxins and other stimulating agents that enter the circulation. Neutrophil aggregates are deposited in small blood vessels, particularly in the lung, and can block the microcirculation with a sudden disastrous loss of respiratory function. It is tempting to speculate that acute tissue injury results from a failure to control the inflammatory cascade, and that an intact circulatory system including plasma components, blood cells, and endothelial cells of the blood vessels is required for control.

Looking more closely at O_2-dependent mechanisms of toxicity (Fig. 2), O_2^- spontaneously undergoes a dismutation reaction that yields H_2O_2. The rate of this reaction is increased by superoxide dismutase (SOD) enzymes.

O_2^- is short-lived and not highly reactive with biological materials, so that O_2^- is not highly toxic. H_2O_2 is a stabilized molecule that also reacts sluggishly with biological materials. Moreover, all tissues contain catalase and glutathione peroxidase enzymes that protect cells by rapidly eliminating H_2O_2. However, O_2^- and H_2O_2 give rise to toxic agents in the presence of the appropriate catalysts, which are redox-active metal ions such as iron, copper, and other transition metals. For example, O_2^- reacts with certain chelated forms of ferric ion, $[Fe^{3+}]$, to yield a reactive species which may be the perferryl ion, $[FeO_2]^{2+}$. This species acts as an initiating agent for "auto-oxidation" reactions in which electrons are removed from reduced substances and transferred to O_2. Such reactions result in oxidation of readily oxidized substances such as ascorbate, sulfhydryl compounds, and unsaturated moieties of membrane lipids. $[FeO_2]^{2+}$ can also rearrange to yield O_2 and a chelated ferrous ion $[Fe^{2+}]$. The reaction of H_2O_2 with $[Fe^{2+}]$ yields poorly defined iron-oxygen complexes. Such complexes may initiate auto-oxidation reactions and hydroxylate biological materials, or they may release the hydroxyl radical (HO), which causes similar forms of

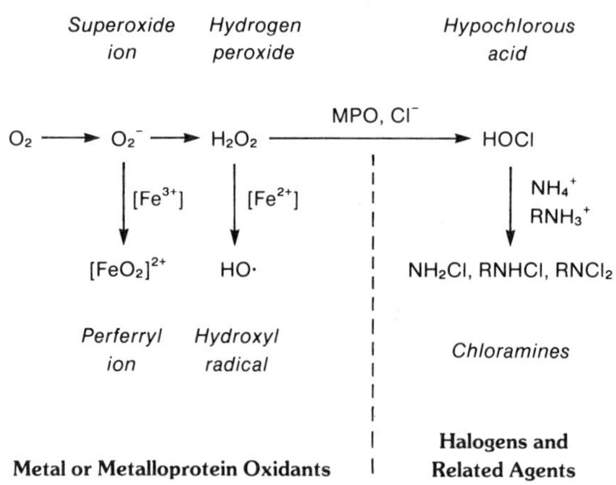

Figure 2. Toxic products of O_2 metabolism

chemical modification. Because HO reacts rapidly and indiscriminately with chemical bonds of all kinds, it is likely to produce damage at or very near the site where it is formed. Though most investigators have focused on the role of iron, copper can also participate in O_2-dependent toxicity. Such reactions may account for the toxicity associated with copper-storage defects.

The physiologically significant metal catalysts of inflammatory tissue injury are unknown. Metalloproteins including transferrin in plasma, hemoglobin and myoglobin released from damaged cells, and lactoferrin and myeloperoxidase from neutrophils have all been implicated. Metal ions and chelates such as iron-heme complexes may be released at sites of inflammation as a result of destruction of metalloproteins by oxidizing agents and hydrolytic enzymes. For example, the iron-hemoprotein myeloperoxidase makes up about 5% of the dry-weight of neutrophils, and a significant fraction of the enzyme is inactivated when neutrophils are stimulated. Therefore, myeloperoxidase may release iron or iron complexes that could catalyze toxic reactions if the iron is not immediately bound and deactivated by lactoferrin. The increase in plasma levels of metal-binding proteins during inflammation may represent an attempt to bind, deactivate, and remove metal ions from the inflammatory site.

The catalytic activity of metal ions in vivo depends on the forms in which they are chelated. For example, iron in catalase or the iron-SOD enzymes of bacteria catalyze reactions that prevent O_2-dependent toxicity, iron in lactoferrin may be inert, and iron in the active form of myeloperoxidase catalyzes a reaction that amplifies the toxicity of H_2O_2. The activity of copper in ceruloplasmin and other complexes in plasma is unclear. Ceruloplasmin has been reported to have SOD activity or to trap free radicals. Copper in this protein does not appear inert as indicated by the oxidase activity of the isolated protein, but isolation may alter its properties.

Metal-catalyzed toxicity of O_2^- and H_2O_2 is not unique to neutrophils. Eosinophils are the most active cells in terms of their capacity to produce O_2^-. In fact, all mammalian cells produce trace amounts of O_2^-

and H_2O_2, and metal-catalyzed toxicity probably contributes to loss of cell functions during aging and the damaging effects of ionizing radiation.

The reaction catalyzed by myeloperoxidase is unique to neutrophils and another class of white blood cells, the monocytes. Myeloperoxidase catalyzes the oxidation of chloride ion (Cl^-) by H_2O_2 to yield hypochlorous acid (HOCl). The toxic properties of HOCl are familiar to everyone; HOCl is widely used as a bleaching and disinfecting agent. However, it's unlikely that HOCl acts directly as a toxic agent _in vivo_. HOCl reacts rapidly with ammonia (NH_4^+) and primary amines (RNH_3^+) to yield chloramines. The total concentration of reactive nitrogen compounds _in vivo_ is very high, so that HOCl has little opportunity to react directly with neutrophils or target cells. Instead, toxicity of the myeloperoxidase/H_2O_2/Cl^- system is mediated by chloramines.

HOCl reacts with NH_4^+ to yield monochloramine (NH_2Cl) and with primary amines to yield mono- and dichloramines (RNHCl and $RNCl_2$). These agents are related to the halogen Cl_2 (chlorine) and are powerful oxidizing and chlorinating agents. They oxidize cysteine, methionine, and tryptophan residues of proteins, chlorinate tyrosine residues, and react with many other readily oxidized substances.

CHEMICAL NAME	STRUCTURE	OPTIMUM CONC. (μM)	COLONIES/PLATE MINUS CONTROL	RATIO (COLONIES/μM)
HISTAMINE-DICHLORAMINE	$C_3N_2H_3(CH_2)_2NCl_2$	20	240 ± 40	12
HISTAMINE-MONOCHLORAMINE	$C_3N_2H_3(CH_2)_2NHCl$	100	100 ± 30	1
MONOCHLORAMINE	NH_2Cl	40	30 ± 20	1
TAURINE-DICHLORAMINE	$^-SO_3(CH_2)_2NCl_2$	50	20 ± 10	0.4
TAURINE-MONOCHLORAMINE	$^-SO_3(CH_2)_2NHCl$	1200	30 ± 20	0.03

Table 1. Relative mutagenic activity

To investigate the possible role of chloramines in inflammatory carcinogenesis, we measured mutagenic activity (Table I). Because most carcinogens have been found to be mutagens, the first event in carcinogenesis is thought to be chemical modification of DNA in normal cells, resulting in expression of an altered regulatory gene-product. Mutagenicity was evaluated by the method of Bruce Ames in which special strains of bacteria are exposed to the potential mutagen, and mutagenic activity is evaluated from the increase in the number of "revertant" colonies. Mutagenesis of the S. typhimurium TA100 strain increased with chloramine concentration up to an optimum concentration. At higher concentrations, killing of the bacteria interfered with the assay. Histamine-chloramines were among the most potent mutagens. Because histamine is released in high concentrations at sites of inflammation, histamine is likely to be chlorinated by neutrophils. The mutagenic activity of histamine chloramines may contribute to the association of inflammation and carcinogenesis.

Turning to the O_2-independent mechanisms of toxicity, it's clear that proteases, glycosidases, and phospholipases can dissolve the components that hold tissues together and probably also modify and damage cell membranes. There's also an extensive interaction of O_2-dependent and O_2-independent mechanisms. For example, the activity of secreted proteases is regulated by plasma proteins that block the active site of proteases. The α_1-protease inhibitor contains an essential methionine residue that is attacked by neutrophil oxidizing agents, rendering the protein unable to block protease activity. On the other hand, collagenase is secreted in a latent form that is activated by agents that attack sulfhydryl groups. In addition, oxidation of tissue components such as hyaluronic acid results in denaturation which increases their susceptibility to hydrolytic attack. Modification of cell membranes by hydrolytic enzymes probably increases their permeability to oxidizing agents, and _vice versa_.

SOD, catalase, and other agents that block the effects of O_2 metabolites may act at many different steps in the inflammatory cascade. Therefore, the ability of such agents to block tissue injury does not necessarily indicate that the reaction of O_2^- or H_2O_2 with tissue components is responsible for toxicity. For example, O_2^- reacts with metal ions to produce agents that initiate the oxidation of ascorbate and other reducing agents by O_2. These auto-oxidation reactions remove protective reducing agents, permitting other agents such as chloramines to inactivate protease inhibitors. This example shows that protection by SOD may be very indirect; SOD can prevent damage by mechanisms in which O_2 metabolites are not involved.

There are many possible sites for pharmacologic intervention to prevent tissue injury. Among the newer approaches under investigation are: (1) preventing formation of chemotactic signals by blocking the hypoxanthine/xanthine oxidase system or by introducing SOD, catalase, or synthetic substances with similar catalytic activities, (2) preventing neutrophil migration with antibodies directed against adhesive glycoproteins on the neutrophil surface, and (3) blocking neutrophil protease activity with genetically or chemically modified forms of the α-1 protease inhibitor that are not inactivated by oxidizing agents. Another possible approach would be to block collagenase activity. Neutrophils probably must express collagenase and other hydrolytic activities in order to migrate through the rather impermeable subendothelial layers of the blood vessel walls. Collagenase inhibitors or agents that prevent oxidative activation of collagenase might block neutrophil influx.

As an emergency treatment to prevent tissue injury immediately after stroke or heart attack, almost any intervention might be considered including use of agents that would completely block neutrophil function throughout the body. Chronic inflammation requires a more conservative and localized approach, so as not to create an immune deficiency with unacceptable risk of infection. To prevent inflammatory carcinogenesis, in which disease may arise long after the inflammatory episode, it will be necessary to know which chemical agents are involved and to very selectively block their formation or action. The best approach to preventing tissue injury may develop from a better understanding of the natural mechanisms for controlling and ending inflammation.

DISCUSSION

KEEN: I was a bit surprised to see that lactoferrin had so much Cu associated with it. Circulating lactoferrin, as I understand it, rarely binds Cu. It binds Fe avidly and Mn but I understand claims for it binding Zn and Cu are somewhat questionable. Do you think this is something unusual about the neutrophil lactoferrin, that it binds Cu?

THOMAS: No. Purified lactoferrin certainly binds Cu. It's easy enough to load it with 2 moles of Cu per mole. One of those being more tightly bound than the other. No one is sure whether it actually binds Cu under in vivo conditions. It does bind Fe more strongly.

SORENSON: A number of investigators study systems composed of "ionic" Cu and Fe along with hydrogen peroxide to suggest the generation of hydroxyl radical in Fenton type chemistry. The conclusion is offered that these reactions represent threats to cells and cellular components. Have you looked at that literature?

THOMAS: Oh sure.

SORENSON: Would you comment with regard to concentrations of those reaction components as to their significance concerning the etiology of some lesion leading to disease.

THOMAS: It's always difficult to extrapolate from model systems to what goes on in vivo. In fact, practically everything we know about functions of neutrophils is based on one or another in vitro system or model system. There's good evidence that the type of chemistry you're discussing, Fenton chemistry, is involved in processes such as normal cell aging, damaging effects of ionizing radiation, etc. It's not so clear that it's important in neutrophil toxicity and in fact there is some very good reasons for thinking that the hydroxyl radical could not be a very effective antimicrobial agent. Hydroxyl radical produced by one cell is not likely to react with another cell. If it has any toxicity it would have to be right at the site where it's produced because it is so reactive. In interpreting that literature one should think of hydroxyl radical as being a name rather than a chemical substance. It describes a particular type of chemistry which is no doubt important in toxicity even if it doesn't involve literally the production of hydroxyl radical.

EGNER: We study the other end of the spectrum with regard to oxygen radical production in terms of the Ames Test and in carcinogenesis. I'm intrigued by your chloramine approach because chloramines are mutagenic in the Ames Test. Does that mean that by implication they become more carginogenic? Do they actually react with cellular DNA or do they cause some change in carcinogen activation?

THOMAS: Chloramines have the potential to react directly with DNA and those that are effective get into cells rapidly. Histamine dichloramine does get into target cells very quickly and probably chlorinates DNA although we haven't demonstrated that directly. That would be considered a primary mutagen and would not necessarily be a carcinogen unless cells are unable to repair the damage or unless a second promoting event takes place.

SORENSON: It appears to me that those who work in this area appear to be unaware of the fact that Cu(I) and Cu(II) as well as Fe(II) and Fe(III) reduce hydroxyl radical at diffusion control rates. When Cu(II) is oxidized to Cu(III) or Fe(III) is oxidized to Fe(IV) the fate of ligands associated with these metalloelements, which have to make up those electron losses, should be studied to account for anything that might happen. It also seems to me that if amounts of hydrogen peroxide used in those in vitro systems were in cells then that would also be devastating to the cell. Is that reasonable?

THOMAS: Hydrogen peroxide added to intact cells is in general not very toxic, depending on the cell, the cell density, and all of that. Cells can frequently tolerate millimolar concentrations of hydrogen peroxide, three orders of magnitude higher than some chloramines, for example. The key is that there has to be a catalyst for the production of toxicity by hydrogen peroxide. In model systems Fe is added up to levels of 0.1 millimolar in order to catalyze those reactions which is obviously not a physiologic condition either. Questions you're raising about Fenton chemistry and hydroxyl radical formation are very good questions and questions that people working in that area have really not answered. If I had another hour

I would present my own view on all of that. Maybe I'll come back in 5 years and do that.

HERNANDEZ: Doug Fox and Neil Granger have shown that the glucose/glucose oxidase system produces large amounts of hydrogen peroxide in our intestinal cat preparation without producing damage. The interaction between O_2^- and hydrogen peroxide is needed in order to produce damage. If we don't have free Fe in the system, how do we get the Fenton reaction to produce hydroxyl radical? Intestinal enterocytes are very rich in Fe deposits so we have enough Fe for this reaction to occur.

THOMAS: One thing to remember about the intestine is that it's the home of the eosinophil. Those tissues are just full of eosinophils which I didn't describe as similar in any respect to neutrophils but with special functions and a separate distinct peroxidase enzyme, the co-factor of which is still questionable. My prejudice would be that the major catalyst for any peroxide-dependent toxicity in the intestine would be the eosinophil peroxidase rather than free Fe or Fe deposits.

SORENSON: Quite a few years ago it was routine to treat the anemia of arthritis with Fe compounds. The resulting Fe storage disease, hemosiderosis, caused dysfunction of synovial membranes to an extent that they utimately required surgical removal. Perhaps this extensive deposition of ferritin represents a situation where there is Fe-mediated degeneration of a particular tissue.

EFFECTS OF CU-DIPS ON ISCHEMIA-REPERFUSION INJURY

Lucrecia A. Hernandez, Matthew B. Grisham, D. Neil Granger
University of South Alabama, Departments of Physiology and Biochemistry, MSB 3024, Mobile, Alabama 36688

A number of mechanisms have been invoked to explain ischemia-induced tissue injury. One of these mechanisms, the production of oxygen-derived free radicals, has received particular attention in recent years. Oxygen-derived free radicals appear to play a major role in the pathogenesis of ischemia-reperfusion injury to digestive organs, such as the intestine,[1,2,3,4] stomach,[5] liver[6] and pancreas.[7] This assertion is largely based on observations that free radical scavengers significantly attenuate the microvascular and parenchymal damage associated with reperfusion of ischemic tissues. Also, when oxygen radicals are generated experimentally, the tissue damage produced mimics that observed after ischemia-reperfusion.[8] Evidence to support the involvement of oxygen-derived free radicals in ischemic tissue injury also comes from studies of the coronary,[9,10,11] renal[12,13] and cutaneous circulations.[14]

The proposed mechanism for oxygen radical formation after ischemia is as follows. During the ischemic period, ATP is catabolized to hypoxanthine[15,16] and the NAD^+-reducing enzyme xanthine dehydrogenase is converted to the oxygen-reducing xanthine oxidase.[17,18] At the time of reperfusion, molecular oxygen is reintroduced into the tissues where it reacts with hypoxanthine and xanthine oxidase to produce the superoxide anion and hydrogen peroxide.[19] These two products then react via the iron-catalyzed Haber-Weiss reaction to produce the hydroxyl radical.[20,21] The highly reactive and cytotoxic hydroxyl radical then causes tissue injury through lipid peroxidation of cell membrane components[19] and by disruption of the interstitial matrix by degradation of hyaluronic acid and collagen.[22,23]

Granger et al[1]. have shown that oxygen radicals are responsible for the increased vascular permeability induced by 1 hour of intestinal ischemia and reperfusion. This conclusion was based on estimates of capillary permeability, where reperfusion after 1 hour of ischemia increased capillary permeability from a control value of 0.08 to 0.41. The increased vascular permeability produced by ischemia was largely prevented (0.14) by intravenous administration of superoxide dismutase (SOD), a specific scavenger of superoxide radicals.

Another compound that is known to have SOD-like activity is Cu(II)-3,5-diisiopropylsalicylic acid (Cu-DIPS), developed by Dr. J.R.J. Sorenson. This copper complex catalyzes the dismutation of superoxide in a manner similar to Cu-Zn SOD. The purpose of the present study was to determine whether Cu-DIPS would provide protection against the ischemia-reperfusion induced increase in intestinal microvascular permeability.

METHODS

The experimental preparation used in this study is similar, except for minor modifications, to that described in detail in previous studies.[1,8] Briefly, cats previously fasted for 18-24 h were initially anesthetized with 50 mg/kg of ketamine-HCl. The right femoral artery and vein were cannulated and anesthesia was maintained by administration of sodium pentobarbital into the femoral vein. Systemic arterial pressure was measured with a Statham P23A transducer (Statham, Oxnard, CA) that was connected to a carotid artery cannula. A tracheotomy was performed to facilitate breathing and as a means of artificial ventilation if the cats failed to breathe spontaneously during the experiment.

A midline abdominal incision was made and a 15-20 cm segment of ileum was isolated; blood and lymph vessels were maintained intact. The remainder of the small and large intestine was extirpated. Immediately after isolation of the ileal segment and before cannulation of the lymphatic and blood vessels, heparin (1000 U/kg) was administered intravenously. Body temperature was maintained at 37°C with a thermistor-controlled infrared lamp. To minimize evaporation and tissue dehydration, all exposed tissues were moistened with saline-soaked gauze and placed in a plastic bag.

A large lymphatic vessel emerging from the mesenteric pedicle was cannulated, and lymph flow was determined by observing the movement of lymph in a calibrated pipette (50-200 µl, full scale). Lymph (C_L) and plasma (C_P) total protein concentrations were measured with a refractometer (American Optical Corp., Buffalo, NY). A large cannula was inserted into the superior mesenteric vein, and venous outflow drained into a reservoir mounted on a vertically positioned pulley system. Blood from the reservoir was returned to the animal via the cannula inserted into the femoral vein. Venous outflow pressure of the intestinal segment was set by adjusting the height of the reservoir and was monitored from a T connector in the venous circuit.

An arterial circuit was established between the superior mesenteric and femoral arteries. Superior mesenteric arterial pressure was measured via a T tube interposed within the arterial circuit using a Statham P23A transducer. All pressure cannulas and associated transducers were positioned at heart level. Heparinized blood from a donor animal was used to prime all the extracorporeal blood circuits. Systemic and superior mesenteric arterial pressures and venous pressure were continually recorded with a Grass physiologic recorder (Grass Instrument Co., Quincy, MA). At the end of each experiment the intestinal segment was weighed and lymph flow was normalized to milliliters per minute per 100 g tissue.

In all experiments, control values for intestinal lymph flow, and lymph and plasma protein concentrations were obtained at a normal local arterial pressure and at a venous pressure of 0 mmHg. Local arterial pressure was reduced to decrease blood flow to 15-20% of control, using an adjustable clamp interposed within the local arterial circuit. Forty-five minutes after reducing arterial pressure, additional measurements of lymph flow and of lymph and plasma protein concentrations were made. After 60 min of local arterial hypotension, the arterial clamp was released and all measured parameters were allowed to reach a new steady state. Venous pressure was then elevated in 10 mmHg increments up to 40 mmHg. Intestinal venous pressure was maintained constant at each pressure level until all parameters were in a steady state, at which time lymph and plasma samples were acquired for protein concentration determinations. This procedure was used in all experiments except the control experiments. The control experiments involved the same procedure as described above, with the exception that the intestine was not subjected to 1 h of ischemia. Cu-DIPS (25 mg/kg) was dissolved in cat plasma and administered intravenously 15 minutes prior to induction of ischemia in 5 cats.

The osmotic reflection coefficient (σ) was estimated using the steady-state relationship between the lymph-to-plasma protein concentration ratio (L/P) and lymph flow. As lymph flow is increased, L/P rapidly decreases (filtration rate-dependent) and then becomes relatively constant at a minimal value (filtration rate-independent) when lymph flow is high. At a normal portal pressure, the exchange of macromolecules across the intestinal capillary wall occurs by both diffusion and convection. Elevation of venous pressure increases the convective movement of macromolecules across the capillary wall while at the same time the diffusive contribution to total exchange is reduced to a neglegible level. Theoretical and experimental evidence suggests that $\sigma = 1-L/P$ when L/P is filtration rate-independent, i.e., when diffusive exchange is neglegible.[24]

Conventional statistical methods (mean ± SEM, unpaired t-test) were used to analyze the data. Statistical comparisons of vascular permeability for the various experiments involved an unpaired t-test between filtration rate-independent values of lymph-to-plasma concentration ratio, i.e., $1-\sigma$.

RESULTS

Figure 1 summarizes the vascular permeability data obtained in the control, ischemia-reperfusion and Cu-DIPS treated ischemia-reperfusion groups. In control intestinal preparations, vascular permeability was 0.08 ± 0.005. Vascular permeability increased significantly to 0.40 ± 0.03 ($p < 0.05$) in preparations subjected to 1 hour of ischemia and reperfusion. The control and ischemia-reperfusion values were not significantly different from those published in previous reports.[1,8] Vascular permeability measured in the ischemia-reperfusion group treated with Cu-DIPS (0.19 ± 0.0033) was significantly lower ($p < 0.05$) than the values obtained in the untreated ischemia-reperfusion experiments, but were significantly higher ($p < 0.05$) than the control value. These results indicate that the increased intestinal vascular permeability induced by 1 hour of ischemia (and reperfusion) is significantly attenuated by pretreatment with Cu-DIPS.

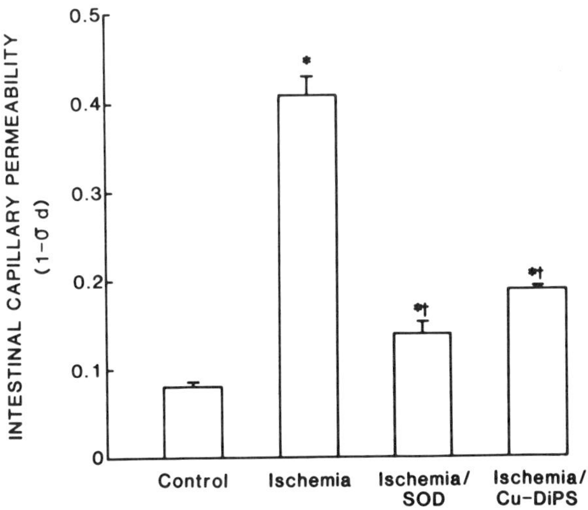

* $p < 0.05$ to control
† $p < 0.05$ to ischemia

DISCUSSION

In recent years there has been growing interest concerning the role of reactive metabolites of oxygen in the tissue injury associated with inflammation and ischemia. Most mammalian cells possess the ability to produce very reactive oxygen metabolites such as the superoxide anions (O_2^-), hydrogen peroxide (H_2O_2) and the hydroxyl radical ($OH^\cdot$), by the sequential univalent reduction of molecular oxygen. A variety of intracellular defense mechanism, namely antioxidant enzymes and scavengers, detoxify the reactive oxygen metabolites produced under basal conditions and protect the tissue against oxidant-induced injury. However, if the rate of production of oxygen free radicals exceeds the capacity of the endogenous antioxidant mechanisms, then serious cellular injury may ensue.

The gastrointestinal tract is particularly well endowed with the biochemical machinery necessary for the production of potent cytotoxic oxidants. Several lines of evidence suggest that the enzyme xanthine oxidase is a major source of oxidants during reperfusion of the ischemic bowel: 1) the intestinal mucosa is a very rich source of the enzyme;[25] 2) NAD^+-reducing xanthine dehydrogenase (D) is rapidly converted to the O_2^- and H_2O_2-producing oxidase (O) form during intestinal ischemia (a 6-8 fold increase in oxidase activity occurs during one hour of ischemia);[26] 3) hypoxanthine, the major substrate for xanthine oxidase, increases 10-20 fold during intestinal ischemia;[27] 4) xanthine oxidase inhibitors (allopurinol, pterin aldehyde) afford significant protection against the ischemia-reperfusion induced increase in microvascular permeability[2,28] and mucosal lesion formation;[2] 5) trypsin inhibitors, which prevent xanthine dehydrogenase-to-oxidase conversion,[26] also protect against ischemia-reperfusion injury;[29] and 6) local intra-arterial infusion of hypoxanthine and xanthine oxidase (at rates which produced blood levels comparable to those measured in the ischemic bowel) increases vascular permeability in the normal (non-ischemic) bowel to an extent comparable to that observed in the postischemic intestine. The hypoxanthine-xanthine oxidase infusion-induced increase in vascular permeability is largely prevented with superoxide dismutase or dimethylsulfoxide.[30] The results described above support the hypothesis that the tissue hypoxia which occurs during intestinal ischemia leads to intracellular accumulation of hypoxanthine and proteolytic conversion xanthine dehydrogenase to xanthine oxidase. When the intestine is reperfused (reoxygenated), molecular oxygen reacts with hypoxanthine and xanthine oxidase to form a burst of superoxide and hydrogen peroxide. The O_2^- and H_2O_2 then react, via the Haber-Weiss reaction, to form the hydroxyl radical, an extremely powerful oxidant. The highly reactive hydroxyl radical mediates the ischemia-reperfusion induced increase in microvascular permeability and hermorrhagic mucosal lesions through lipid peroxidation of endothelial and epithelial cell membrane components and degradation of collagen and hyaluronic acid, components of vascular and epithelial basement membranes.

The increased intestinal vascular permeability produced by ischemia-reperfusion is largely prevented by the administration of compounds which either scavenge oxy-radicals (SOD, catalase, dimethylsulfoxide), inhibit xanthine oxidase (allopurinol, pterin aldehyde, folic acid), prevent xanthine dehydrogenase-to-oxidase conversion (trypsin inhibitor), or chelate iron (deferoxamine, apotransferrin) (see Table 1). In the present study we found that Cu(II)-DIPS also provides significant protection (0.19 ± 0.033) against the increased vascular permeability produced by ischemia-reperfusion (Fig. 1). These results are not entirely surprising inasmuch as Cu-DIPS, a small lipophilic complex, possesses SOD-like activity.[31] The dismutation of superoxide by Cu-DIPS is similar to that observed with SOD, however, Cu-DIPS is not as efficient as SOD in catalyzing this reaction. Cu-DIPS is known to provide protection in different models of inflammation,[32] promote wound healing and to act as a radioprotectant.[33] The results from this study suggest that copper combination complexes (like Cu-DIPS) may prove to be useful in the treatment of ischemic disorders and in preservation of organs for transplantation.

Table 1: Effects of ischemia-reperfusion on intestinal vascular permeability.[25,34]

Experimental Group	Intestinal vascular permeability
Controls	0.08 ± 0.005
Ischemia-reperfusion (I-R)	0.41 ± 0.02
I-R + superoxide dismutase	0.14 ± 0.013
I-R + allopurinol	0.18 ± 0.013
I-R + folic acid	0.16 ± 0.04
I-R + pterin aldehyde	0.15 ± 0.02
I-R + catalase	0.19 ± 0.02
I-R + dimethylsulfoxide	0.19 ± 0.023
I-R + trypsin inhibitor	0.16 ± 0.012
I-R + deferoxamine	0.15 ± 0.009
I-R + apotransferrin	0.17 ± 0.002

REFERENCES

1. Granger, DN; Rutili, G; McCord, JM. Gastroenterology 1981, 81, 22-29.

2. Parks, DA; Bulkley, GB; Granger, DN; Hamilton, SR; McCord, JM. Gastroenterology 1982, 82, 9-15.

3. Schoenberg, MH; Muhl, E; Sellin, D; Younes, M; Schildberg, FW; Haglund, U. Acta Chir. Scand 1984, 150, 301-309.

4. Younes, M; Schoenberg, MH; Jung, H; Fredholm, BB; Haglund, U; Schildberg, FW. Res. Exp. Med. 1984, 184, 259-264.

5. Itoh, M; Guth, PH. Gastroenterology 1985, 88, 1162-1168.

6. Adkison, D; Hollwarth, ME; Benoit, JN; Parks, DA; McCord, JM; Granger, DN. Acta Physiol. Scand. 1986, 126 (Suppl. 548), 101-108.

7. Sanfey, H; Bulkley, GB; Cameron, JL. Ann. Surg. 1984, 200, 405-413.

8. Parks, DA; Shah, AK; Granger, DN. Am. J. Physiol. 1984, 247, G167-G170.

9. Chambers, DE; Parks, DA; Patterson, G; et al. Fed. Proc. 1983, 42, 1093.

10. Hess, ML; Okabe, E; Kontos, H. J. Mol. Cell. Cardiol. 1981, 13, 767-772.

11. Lefer, AM; Araki, H; Okamatsu, S. Circ. Shock 1981, 8, 273-282.

12. Chatterjee, SN; Berne, TV. Am. J. Surg. 1976, 131, 658-660.

13. Hansson, R; Gustafsson, B; Jonsson, O; et al. Trans. Proc. 1982, 14, 51-58.

14. Del Maestro, RF; Bjork, J; Arfors, KE. Microvasc. Res. 1981, 22, 255-270.

15. Cunningham, SK; Keaveny, TV. Ir. J. Med. Sci. 1977, 146, 136-143.

16. Jones, CE; Crowell, JW; Smith, EE. Am. J. Physiol. 1968, 214, 1374-1377.

17. McCord, JM; Roy, RS. Can. J. Physiol. Pharmacol. 1982, 60, 1346-1352.

18. Roy, R; McCord, JM. Fed. Proc. 1982, 41, 767.

19. Fridovich, J. Science 1978, 201, 875-880.

20. McCord, JM; Day, ED. FEBS Lett. 1978, 86, 139-142.

21. Parks, DA; Granger, DN. Am. J. Physiol. 1983, 245, G285-G289.

22. Brown, K; Fridovich, J. Acta Physiol. Scand. 1980, 492, 9-18.

23. McCord, JM. Science 1974, 185, 529-531.

24. Granger, DN; Taylor, AE. Am. J. Physiol. 1980, 238, H457-H464.

25. Granger, DN; Hollwarth, ME; Parks, DA. Acta Physiol. Scand. 1986, 126 (Suppl. 548), 47-65.

26. Roy, RS; McCord, JM. In: Proc. Conf. Superoxide and Superoxide Dismutase; Elsevier/North Holland Biomedical Press; New York, 1983; pp 145-153.

27. Schoenberg, MH; Fredholm, BB; Haglund, U; Jung, H; Sellin, D; Younes, M; Schildberg, FW. Acta Physiol. Scand. 1985, 124, 581-589.

28. Granger, DN; Parks, DA; Hollwarth, ME; McCord, JM. Gastroenterology 1986, 90, 80-84.

29. Parks, DA; Granger, DN. Gastroenterology 1985, 89, 6-12.

30. Parks, DA; Granger, DN. Am. J. Physiol. 1983, 247, G131-G138.

31. de Alvare, LR; Goda, K; Kimura, T. Biochem. Biophys. Res. Commun. 1976, 69, 687-694.

32. Sorenson, JRJ. J. Med. Chem. 1976, 19, 135-148.

33. Sorenson, JRJ. J. Med. Chem. 1984, 27, 1747
Marcel Dekker, Inc.; New York and Basel, 1982.

34. Hernandez, LA; Grisham, MB; Granger, DN. The Physiologist (in press) 1986.

ACKNOWLEDGEMENTS

Supported by a grant from the National Institutes of Health (AM26441).

DISCUSSION

LOVEN: Were your drugs given before or after ischemia?

HERNANDEZ: We gave Cu(II)(3,5-dips)$_2$ before ischemia. We gave Cu-Zn SOD 15 minutes before reperfusion. Deferoxamine and transferrin were given 5 minutes into ischemia and 10 minutes into reperfusion. Allopurinol was given for 2 days prior to inducing ischemia.

LOVEN: I wanted to be sure you were producing ischemia and not anoxia. Jones suggests that enzymes like xanthine oxidase may still be active at relatively low oxygen concentrations. You could be generating some O_2^- even during the ischemic stage and have a larger flux on reoxygenation.

HERNANDEZ: Experiments using Cu-Zn SOD throughout the ischemia period and into reperfusion have not shown any more protection than found with treatment during reperfusion only. Maybe O_2^- is being produced throughout the period of ischemia but its production is not as significant as after reperfusion.

CROOK: Can you explain how you made your antineutrophil serum?

HERNANDEZ: We have an immunologist working with us. White blood cells are isolated from cats, purified, and suspended in Freund's adjuvant prior to injection into rabbits. Two weeks later we reboost these rabbits with the same adjuvant suspension of neutrophils and then serum is withdrawn from these rabbits. We had 4 rabbits. All of these rabbits were wonderful producers. We obtained alot of very potent antineutrophil serum from them. In my cat studies I use one ml of this antineutrophil serum and in 1 hour all of these cells are absent from intestinal tissues.

CROOK: I ask that because I'm always concerned about immune defenses when the vascular supply is interrupted and reestablished. I'm also curious as to when circulation is reestablished whether or not other substances are released into the tissue that may provide protection in the proposed scheme?

HERNANDEZ: What other substances are you referring to that might be released from the tissue?

CROOK: I am uncertain as to which specific substances might be coincidentally associated with the administration of the drug but they may produce an effect that is being attributed to the drug. Maybe another way to do this kind of experiment is to induce hemorrhagic shock, which follows bleeding as a result of decreased vascular prefusion due to decreased blood volume, and then determine if allopurinol, for example, indeed has any protective effect.

HERNANDEZ: We and others have shown that allopurinol and SOD protects against rat gastric lesions induced by shocking animals with a lowering of blood pressure to around 20 mm of mercury for 30 minutes. Shocked rats given allopurinol or SOD were protected. Protection by allopurinol and SOD support a free radical mechanism for tissue damage.

BHATHENA: What proportion of your white cell infiltrate are neutrophils and how is this related to neutrophil enzyme activity?

HERNANDEZ: I really don't know. Matt Grisham has been trying to separate myeloperoxidase (MPO) derived from eosinophils and he actually can tell that it's the MPO derived from neutrophils that increases. In the original report the number of white blood cells increased based upon count. This was then compared to the MPO activity. This is the evidence we have that we are actually observing an influx of neutrophils in our ischemia model.

BHATHENA: In other words, there is a quantity of enzyme going in but not the activity per se?

THOMAS: I can offer comment on that. To measure the MPO activity of neutrophils total activation requires the addition of detergent because the enzyme is essentially in a latent or cryptic form within granules. The way that this group is measuring activity is by adding detergent and looking at the total expression of activity. It really is a measure of total MPO activity which is a measure of the number of neutrophils.

I'd also like to commend this group for the work they are doing. Some of their approaches that are being used such as antineutrophil serum don't have a lot of hope for direct therapeutic application but it's clear that diseases that are of greatest concern can quite possibly be treated, based upon the kind of research strategies this group has developed.

FIELDS: Intravenous administration of fructose to experimental animals and humans increases uric acid production in the liver. The increase in uric acid is associated with decreased ATP, increased IMP, increased xanthine, and increased hypoxanthine. A lot of oxygen radicals and H_2O_2 are produced along with this uric acid in the liver and the liver is loaded with Fe. Would you expect the same type of damage in the liver when fructose is fed?.

HERNANDEZ: It is really difficult to compare what might happen in a normal tissue and a tissue that becomes ischemic. The normal cell maintains a balance between its antioxidants and its oxidants. We measure normal SOD, catalase, GSH, MPO, and xanthine oxidase activity in normal tissue. I suspect that if we produce ischemia in the liver and reperfuse it, we will probably be able to use the Fe in the same way that it's been used in the intestine. I think we need the presence of ischemia for xanthine dehydogenase to be converted to the oxidase while in normal circumstances it would not occur.

FIELDS: Couldn't it occur with Cu deficiency without ischemia because in Cu deficiency SOD is reduced?

HERNANDEZ: In Cu deficiency the availability of Cu for SOD synthesis is decreased. On the other hand, Cu-Zn SOD is needed everyday by all cells, there is always that 5% of the O_2^- that is being produced in normal cells that escapes from mitochondria. This is one rationale as to why all cells have Cu-Zn SOD, to get rid of this free O_2^-. The symptomalogy of Cu deficiency as evidenced in animals and humans is consistent with an inability to resynthesize Cu-Zn SOD. Such cells are then at a disadvantage with regard to their ability to dismutate O_2^- that is normally being produced in the cell.

WILLIAMS: Were those animals that were partially protected with deferoxamine or apotransferrin infusions perfused with whole blood? Did you ever measure serum transferrin Fe and did you ablate protection by infusing animals with Fe loaded transferrin?

HERNANDEZ: These experiments were intended to measure normal Fe content in the cell before and after ischemia, during reperfusion, to study cellular Fe metabolism. We wanted to measure free Fe but these levels are to small to measure. This is something we will eventually have to do if we want to study the role of Fe.

SORENSON: Measurable Fe contents represent some sort of complexed form of Fe. The free ionic form is extremely low and not measurable with existing equipment.

HERNANDEZ: Can you measure all those different forms of Fe? Can the percent of Fe in each complex be measured? Can you measure how much Fe is being transported by transferrin, in all the different stages that Fe goes through from the time it's absorbed to the time it's utilized?

SORENSON: Yes, I think so. One can measure Fe-dependent enzyme activity, transferrin Fe, apotransferrin and fillerable Fe complexes. Changes in total Fe content in your system would be a good initial investigation.

MODULATION OF LYMPHOCYTE FUNCTION BY COPPER AND THIOLS

Peter E. Lipsky
Harold C. Simmons Arthritis Research Center
and the Department of Internal Medicine
UTHSCD Southwestern Medical School
Dallas, Texas 75235

INTRODUCTION

D-Penicillamine is effective in the treatment of rheumatoid arthritis. Moreover, D-penicillamine can modify the disease process as evidenced by the capacity to decrease acute phase reactants[1]. Despite this well-documented evidence of clinical efficacy, the mechanism by which D-penicillamine achieves this therapeutic result in rheumatoid arthritis remains unknown.

We have previously presented evidence from in vitro studies suggesting that D-penicillamine may have immunosuppressive capabilities[2-6]. Thus, D-penicillamine was shown to inhibit in vitro T lymphocyte proliferation and the activity of helper T cells in supporting the generation of antibody-forming cells in vitro. Specificity for this inhibitory effect was shown by the finding that D-penicillamine had no apparent effect on the function of monocytes or B lymphocytes. The inhibitory effect of D-penicillamine was markedly accentuated by the presence of copper ions. A variety of other divalent cations were ineffective at increasing the inhibitory action of D-penicillamine on T lymphocyte function. Finally, ceruloplasmin also augmented the inhibitory effect of D-penicillamine on T cell function in a manner similar to that exerted by $CuSO_4$[5].

Recent clinical studies have indicated that a variety of other drugs may suppress rheumatoid inflammation in a manner that is analogous to the effect of D-penicillamine. These drugs, which include 5-thiopyridoxine[7], 5-mercaptopropionylglycine[8,9], and captopril[10] are all sulfhydryl containing compounds. Moreover, these compounds appear to have in vitro effects that are similar to those caused by D-penicillamine, suppressing T cell function in a copper dependent manner[4,6].

The data therefore suggest that D-penicillamine and related drugs may be active in rheumatoid arthritis by virtue of their capacity to inhibit T lymphocyte function specifically. Additional experiments demonstrated that the mechanism by which the mixture of D-penicillamine and copper or ceruloplasmin altered T cell function involved the generation of hydrogen peroxide[5]. Cupric ions had the capacity to oxidize the free thiol of D-penicillamine leading to the rapid production of H_2O_2, which was markedly inhibitory of T cell function[5]. These results suggested that the therapeutic action of D-penicillamine in rheumatoid arthritis may result from the production of H_2O_2 locally in the synovial tissue, which suppresses the T cell mediated immunoreactivity propagating the chronic inflammation.

IMMUNOSUPPRESSION BY CAPTOPRIL

Captopril is a sulfhydryl containing angiotensin converting enzyme inhibitor that has been shown to be effective in treating rheumatoid arthritis[10]. It appears that its therapeutic effect in rheumatoid arthritis is related to its free thiol group and not its capacity to inhibit angiotensin converting enzyme activity, as enalapril, another angiotensin converting enzyme inhibitor does not appear to have therapeutic efficacy in rheumatoid arthritis[11]. It was therefore of interest to examine the spectrum of activity of these two angiotensin converting enzyme inhibitors on in vitro assays of human lymphocyte function.

For these studies, the mitogenic lectin phytohemagglutinin (PHA) was used to stimulate human T lymphocyte

DNA synthesis that was assayed as ^{3}H-thymidine incorporation. It has previously been found that copper ions or ceruloplasmin markedly enhance the inhibitory effect of D-penicillamine on T lymphocyte DNA synthesis[5]. Similar results were found in these studies (Table I). Thus, D-penicillamine itself had only a modest inhibitory effect on PHA-induced ^{3}H-thymidine incorporation. Similarly, $CuSO_4$ had no significant inhibitory effect. However, when $CuSO_4$ was added to cultures containing D-penicillamine, a markedly enhanced inhibitory effect was noted as has been previously described[2 6]. Ceruloplasmin could substitute for $CuSO_4$ in these experiments[5], but other divalent cations did not increase the inhibitory effect of D-penicillamine[2]. The importance of the reduced thiol in mediating inhibition was indicated by the finding that oxidized D-penicillamine disulfide did not cause inhibition of lymphocyte responsiveness even in the presence of $CuSO_4$[2]. Finally, as previously reported[5], inhibition resulted from the generation of hydrogen peroxide and was completely prevented by catalase (Table I).

TABLE I
Effect of D-Penicillamine and $CuSO_4$
on Mitogen-Induced Human T Cell DNA Synthesis

D-penicillamine (μg/ml)	Addition	T cell ^{3}H-Thymidine Incorporation No $CuSO_4$	 $CuSO_4$-2μg/ml
		(Δ cpm x 10^{-3})	
0	0	51.6 ± 0.5	68.2 ± 1.2
10	0	45.2 ± 1.4	64.8 ± 1.6
50	0	49.6 ± 3.0	9.7 ± 1.0
100	0	44.6 ± 2.4	0.1 ± 0.0
100	catalase	30.9 ± 0.2	51.0 ± 0.6

Human peripheral blood mononuclear cells were cultured in microtiter wells (1 x 10^{5}/well) and stimulated with phytohemagglutinin. After a 3 day incubation, ^{3}H-thymidine incorporation was determined.

The effect of captopril was comparable to that of D-penicillamine (Table II). Thus, captopril did not significantly inhibit mitogen-induced T cell proliferation alone but was markedly inhibitory in the presence of $CuSO_4$. All the inhibitory effects of captopril and $CuSO_4$ were completely prevented by catalase (data not shown). It should be noted that although captopril was qualitatively similar to D-penicillamine, it was somewhat less active, requiring approximately twice the concentration to achieve the same degree of inhibition (data not shown).

TABLE II
Effect of Captopril and $CuSO_4$ on
Mitogen-induced Lymphocyte DNA Synthesis

Compound	PHA-Induced Lymphocyte DNA Synthesis No $CuSO_4$	Plus $CuSO_4$
	^{3}H-Thymidine Incorporation, Δ cpm x 10^{-3}	
Nil	43.3	43.1
Captopril	35.1	2.3
S-methyl captopril	52.7	48.2
α-Methylbutyryl-L proline	47.0	51.3
Enalaprilic acid	43.7	44.9

Human peripheral blood mononuclear cells were cultured in microtiter wells (1 x 10^5/well) and stimulated with phytohemagglutinin. After a 3 day incubation, ^{3}H-thymidine incorporation was determined. The various compounds were added at a final concentration of 100ug/ml. $CuSO_4$ was added where indicated at a final concentration of 2ug/ml (8uM Cu^{++}).

Two captopril analogues, S-methyl captopril and α-methylbutyryl-L proline, which lack free sulfhydryl groups, did not inhibit T lymphocyte proliferation and did not become suppressive in the presence of $CuSO_4$ (Table II). Moreover enalaprilic acid, another angiotensin converting enzyme inhibitor that lacks a free sulfhydryl, was also unable to inhibit T lymphocyte proliferation, even when $CuSO_4$ was present. These results indicate that suppression of T lymphocyte proliferation requires the presence of a free sulfhydryl group and cannot be explained by inhibition of angiotensin converting enzyme.

INHIBITION OF NATURAL KILLER CELL FUNCTION BY THIOLS AND $CuSO_4$

Natural killer cell function has been reported to be suppressed by the action of H_2O_2[12,13]. It was therefore of interest to examine whether thiols and $CuSO_4$ generated adequate H_2O_2 to inhibit NK function. For these experiments a previously described[3] preincubation protocol was employed in which lymphocytes were exposed to the putative inhibitors for a brief period of time, washed and then tested for residual function. As can be seen in Table III, a brief exposure to the mixture of D-penicillamine and $CuSO_4$ inhibited subsequent NK function and mitogen-induced T cell proliferation. Inhibition was completely prevented when catalase was included in the incubation mixture. Similar results were observed when cells were exposed to the mixture of captopril and $CuSO_4$, whereas enalaprilic acid and $CuSO_4$ caused no inhibition of either NK or T cell function. These findings indicate that hydrogen peroxide generated from the mixture of $CuSO_4$ and a thiol containing compound such as D-penicillamine or captopril can inhibit both NK and T cell function.

DISCUSSION AND CONCLUSIONS

It has become clear from this and other studies[2-6] that thiol containing compounds such as D-penicillamine and captopril in the presence of $CuSO_4$ or ceruloplasmin inhibit T cell and NK function by generating H_2O_2. It apprears that H_2O_2 generated in this system is directly inhibitory of T cell function, as scavengers of superoxide or hydroxyl radicals do not protect T cells[5]. Of interest, the inhibitory effect of H_2O_2 can be further increased by myeloperoxidase[13]. The exact mechanism whereby T cells and NK cells are damaged has not been delineated, but may involve the formation of toxic lipid peroxides[14].

TABLE III
Effect of D-Penicillamine, Captopil and Enalaprilic Acid on Human Natural Killer Cell Function and Mitogen Induced T Cell Proliferation

Preincubation			Natural Killer	PHA-Induced T Cell
Compound	$CuSO_4$	Catalase	Cell Function	DNA Synthesis
			(% specific lysis)	(Δ cpm x 10^{-3})
0	0	0	29.9	31.6
	+	0	34.0	25.7
	0	+	34.4	26.6
D-peni-	0	0	30.2	30.5
cillamine	+	0	1.0	0
	+	+	28.9	31.1
Captopril	0	0	29.0	30.1
	+	0	0.8	0
	+	+	32.5	29.2
Enalaprilic	0	0	28.4	29.6
Acid	+	0	32.3	28.2
	+	+	29.3	25.3

Non-adherent cells were prepared by incubating peripheral blood mononuclear cells on a glass petri dish. The non-adherent cells were then passed over a nylon wool column. The resulting population contains about 3% monocytes and is enriched in T cells and NK cells. These cells were preincubated for 1 hour at 37°C with the indicated compound at a final concentration of 100 µg/ml in the presence or absence of $CuSO_4$ (2 µg/ml) and/or catalase. After the preincubation, the cells were washed extensively, suspended in fresh medium and assayed for NK activity (the ability to cause ^{51}Cr release from K-562 cells at an E:T ratio of 25:1) and PHA-induced ^{3}H-thymidine incorporation.

The chemical reactions leading to the generation of H_2O_2 in this system have not been completely delineated but the superoxide anion appears to be an important intermediary[5,15,16]. As indicated below, there are at least two mechanisms whereby the superoxide intermediate can be generated after the oxidation of the sulhydryl anion by cupric ions[5,15,16,17]. Either cuprous ions or the thiyl radical generated from this reaction may reduce molecular oxygen to superoxide. Subsequent dismutation catalyzed by penicillamine-copper complexes[18] will lead to the rapid formation of H_2O_2.

Generation of Hydrogen Peroxide from Copper Mediated Oxidation of Thiols

$$RS^- + Cu^{++} \longrightarrow RS^{\cdot} + Cu^+$$

$$1. \quad Cu^+ + O_2 \longrightarrow Cu^{++} + O_2^{\cdot -}$$

$$2. \quad RS^{\cdot} + O_2 \longrightarrow RS^+ + O_2^{\cdot -}$$

$$2O_2^{\cdot -} + 2H^+ \xrightarrow[Cp]{Pen\text{-}Cu} H_2O_2 + O_2$$

The results suggest that the therapeutic effect of D-penicillamine, captopril and other thiols in rheumatoid arthritis may relate to the capacity of these compounds to generate H_2O_2 locally in the synovial tissue and thereby suppress the activity of T cells and NK cells that propagate the immunologically-mediated inflammation. A number of features of the rheumatoid synovium may modulate this effect. Thus, for example, the rheumatoid synovium is characterized by diminished perfusion and intense inflammation with resultant acidosis[19]. Since the maintenance of D-penicillamine and related thiol compounds in the reduced form is favored at acid pH, the local environment of the rheumatoid synovium may facilitate the immunosuppressive capacity of these agents. Similarly, rheumatoid synovial fluid contains markedly increased concentrations of ceruloplasmin[20], that may also enhance the capacity of the reduced thiols to generate H_2O_2. An additional feature of the synovium can alter the immunosuppressive capacity of the H_2O_2 produced. Thus, mononuclear phagocytes are able to protect T cells from the injurious effects of H_2O_2 by virtue of their abundant content of catalase[5]. These results suggest that the functional competence of mononuclear phagocytes within the inflamed synovium may also play a critical role in determining the therapeutic effectiveness of D-penicillamine and other thiol-containing disease-modifying drugs.

Thiol-containing compounds cause considerable toxicity in treated patients[21]. Among the toxic manifestations are a variety of autoimmune side effects. It is possible that some of these are caused by the inappropriate generation of H_2O_2 at various tissue sites with subsequent immunization to peroxide denatured self-antigens. This would imply that the therapeutic and toxic effects of D-penicillamine and other thiol-containing compounds may result from the capacity of these agents to generate H_2O_2 after oxidation by copper ions. Both possibilities require experimental verification.

Acknowledgments

The thiol compounds and analogues were graciously provided by Merck Sharp and Dohme Research Laboratories. The author is grateful to Dr. Ron Estabrook, Dept. of Biochemistry, UTHSCD for his many insightful suggestions. This work was supported by USPHS program project grant AM-09989.

References

1. Multicentre Trial Group. Lancet. 1973, 1, 275.
2. Lipsky, P.E; Ziff, M. J Immunol. 1978, 120, 1006.
3. Lipsky P.E; Ziff, M. J Clin Invest. 1980, 65, 1069.
4. Lipsky P.E. Agents Action. 1981, Suppl. 8, 95.
5. Lipsky P.E. J. Clin. Invest. 1984, 73, 56.
6. Lipsky, P.E. In: Inflammatory Diseases and Copper; Humana Clifton, NJ. 1982; p. 581.
7. Huskisson E.C; Jaffe I.A; Scott, J; Dieppe P.A. Arthritis Rheum. 1980, 23, 106.
8. Amor, B; Mery, C; Degery, A. Arthritis Rheum. 1982, 25, 698.
9. Pasero, G; Pellegrini, P; Ambanelli, U; Ciompi, M.L; Colamussi, V; Ferraccioli, G; Barbieri, P; Mazzoni,

M.R; Menegale, G; Trippi, D. Arthritis Rheum. 1982, 25, 923.
10. Martin, M.R.F; McKenna, F; Bird, H.A; Surrall, K.E; Dixon, J.S; Wright, V. Lancet. 1984, 1, 1325.
11. Jaffe, I. Arthritis Rheum. 1984, 27, 840.
12. Seaman, W.E; Gindhart, T.D; Blackman, M.A; Dalal, B; Talal, N; Werb, Z. J. Clin. Invest. 1982, 69, 876.
13. El-Hag, A; Lipsky, P.E; Bennett, M; Clark, R.A. J. Immunol. 1986, 136, 3420.
14. Tien, M; Bucher, J.R; Aust, S.D. Biochem. Biophys. Res. Comm. 1982, 107, 279.
15. Misra, H.P. J. Biol. Chem. 1974, 249; 2151.
16. Starkebaum, G; Root, R.K. J. Immunol. 1985, 134, 3371.
17. Feldman, S.L; Hunter, J.S.V; Zgirski, A; Chidambaram, M.V; Frieden, E. J. Inorg. Biochem. 1982, 17, 51.
18. Younes, M; Weser, U. Biochem Biophys Res Comm. 1977, 78, 1247.
19. Wallis, W.J; Simkin, P.A; Nelp, W.B. Arthritis Rheum. 1985, 28, 1096.
20. Scudder, P.R; McMurray, W; White, A.G; Dormandy, T.L. Ann Rheum Dis. 1978, 37, 71.
21. Jaffe, I.A. Amer. J. Med. 1986, 80, 471.

DISCUSSION

BHATHENA: Is cysteine also effective?

LIPSKY: Cysteine itself has not been tested in rheumatoid arthritis. There is some conflicting evidence in the rat which is hard to interpret. Cysteine is interesting in that cysteine levels in plasma of patients with rheumatoid arthritis are very much depressed as are all sulfhydryl containing substances. It's one hallmark of the disease which returns to normal when they begin to go into remission.

THOMAS: I would like to offer a comment as to why peroxide might be especially toxic to T-cells. Tom Oni and Carl Pierce have shown that suppressor T-cells release a soluble factor which they call SIRS which is activated by peroxide, it's not a catalyst for peroxide toxicity. SIRS requires one mole of peroxide for activation and then it's a very effective suppressor substance which they have tested on a number of types of cells.

D-penicillamine would not be expected to be taken up into cells. What's known about it's ability to actually get into synovial fluid?

LIPSKY: Levels of penicillamine have only been measured in blood, not in synovial fluid. Presumably it gets in but we really don't know. I think the situation with regard to the SIRS work is of interest but probably not comparable to what we are measuring. We have no production of SIRS in this system. The SIRS itself once activated by peroxide really surpresses T-cells and B-cells and a number of other cell types and lacks the specificity that we find. This is really a unique toxic effect on T-cells. Why this effect should be manifest we really don't know. You mentioned earlier that all cells had catalase. I'm not sure that's actually been measured in resting T-lymphocytes. I can't find that data for resting T-lymphocytes. They may well have it but my guess is that if they do it would be remarkably less than the amount, for example, in a blood monocyte, and perhaps, a B-lymphocyte.

THOMAS: Is it clear that your target cells are undergoing some sort of specific inhibition or are they simply being killed? Are they losing all their functions?

LIPSKY: One can use lower concentrations and lose certain functions, one at a time, with great specificity. As to whether the biochemical activity of peroxide on T-cells and B-cells is the same and one just lacks the capacity to repair that damage, we don't know at this time.

THOMAS: But, it's possible that the target for toxicity might be right on the cell membrane so it's not necessary for peroxide to get in where the catalase would be.

LIPSKY: Right.

BARNEA: I noticed that the molar ratio of penicillamine to Cu was about 1000 to 1. Could you share with us how this whole scheme involving peroxide and hydoxyl radical formation occurs, especially with regard to ionization of the sulfhydryl group. The chemical state you presented was based almost on one-to-one ratio of thiol to Cu.

LIPSKY: Obviously these are not the only reactions going on in these tissues and there's a lot of oxidation-reduction not involving Cu. Perhaps John (Sorenson) can help us with regard to why Cu is acting in this sort of catalytic way.

SORENSON: You're suggesting that some Cu or Cu-penicillamine complex disproportionates O_2^- to give hydrogen peroxide. Hydrogen peroxide plus O_2^- or Cu(I) were suggested to produce hydroxyl radical which might account for some chemical damage that occurs in your system. That's reasonable Haber-Weiss and Fenton chemistry but I am not convinced that occurs in vitro or in vivo. The stoichiometry has not as yet been worked out. The question I have is: Are these cells dying?

LIPSKY: If you use a high enough concentration of peroxide they will die. If you use a lower concentration of peroxide they are functionally inactive and they don't die at least for the first few days of culture.

SORENSON: Copper may have a role in modulating the cell growth cycle to stop lymphocytes in the S phase. If that's what's happening, then mitogens would appear to be ineffective.

LIPSKY: I'm not aware of that data. We have looked at that quite carefully and have not found anything related to that.

SORENSON: Have you looked for slowing or stopping of the cell growth cycle?

LIPSKY: Following cells with cell cycle analysis using acridine orange we don't see any growth retardation that is Cu specific at the G_1/S interphase. These cells, after they have been affected with peroxide or thiol and Cu, sit in G_0 and don't get out of G_0. In addition, using normal cells, we are not able to see a role for Cu in any part of the cell cycle.

SORENSON: In the penicillamine-Cu system you see stoppage of cells in G_0?

LIPSKY: Ninety percent of resting human lymphocytes come out of the blood in G_0. When one stimulates them with a mitogen they leave G_0 enter G_1 and then progress through the cell cycle and cycle many times. If you treat them with penicillamine and Cu, you're treating them in G_0. Now when you stimulate them with a mitogen, they stay in G_0. You don't get any entry, that we can detect, of these cells into the cell cycle. That's one observation. The second observation is that if you take normal cells and stimulate them with a mitogen, they'll enter G_1. If you do this experiment in the presence of say 50 uM EDTA, which is not enough to chelate all of the Ca or Mg in the medium, and then you do that with 50 uM EDTA plus Zn because they need Zn, then the cells will progress through the cell cycle normally although presumably all Cu will be chelated.

SORENSON: Yes, but by the fact that the Cu is chelated doesn't mean that it's not useful and it may in fact be more useful.

LIPSKY: It may be more useful. It doesn't help these cells. We can find no evidence of a role for ceruloplasmin or Cu. Now it's possible that a trace of Cu is necessary but we certainly don't need ceruloplasmin. We don't think we need Cu in order to get these normal untreated cells to cycle.

SORENSON: But, they may not cycle very long.

LIPSKY: If we treat these cells with penicillamine plus Cu or peroxide and wash them and then stimulate them with mitogen, as far as we can see these cells just don't move, they stay in G_0.

SORENSON: Does Cu-Zn SOD do that?

LIPSKY: No. Why? Would you expect it to?

SORENSON: It would provide hydrogen peroxide.

LIPSKY: How much, to these particular cultures?

SORENSON: Superoxide dismutase is as effective as a hydrogen peroxide generator as any other complex that's ever been studied.

LIPSKY: As long as you have O_2^-.

SORENSON: Yes.

LIPSKY: So we would need O_2^- there.

SORENSON: Maybe I missed where the hydrogen peroxide was coming from in the penicillamine-Cu mixture.

LIPSKY: What you are asking is if we mix penicillamine and Cu plus O_2^- do we get the same degree of inhibition.

SORENSON: Isn't that what is being proposed for the system?

LIPSKY: Right. Our problem was if we use suboptimal amounts of penicillamine and Cu and add in O_2^-, we thought we got enhanced inhibition. We've never been able to convince ourselves as to whether that was because we had enhanced generation of peroxide or that some of the SOD itself was denatured and was providing more Cu.

SORENSON: Is your penicillamine-Cu solution colored purple or blue?

LIPSKY: It's light bluish.

SORENSON: That's the color of the mixed valence complex and you could be forming that complex in your mixture.

LIPSKY: Data with the mixed valence complex are the following. The complex itself does not inhibit at all at any concentration that we can get into solution.

SORENSON: None of those complexes I sent you caused any inhibition?

LIPSKY: None of yours.

SORENSON: Good.

LIPSKY: Interestingly, the presence of penicillamine or another thiol will inhibit just as though you were adding $Cu(II)SO_4$ rather than mixed valence complex, so it looks as though the mixed valence complex can act as a very effective donator of Cu for this response but it, in and of itself, is not active. The pale yellow Cu(I) complex behaved the same way. The same thing is true for the cysteine complex and for one other that I can't recall.

SORENSON: Okay. Then in answer to Prof. Barnea's question, there's likely to be more than one process going on to account for the overall effects. One is Cu dependent and another is, perhaps, Cu-penicillamine dependent.

LIPSKY: And many that are probably irrelevant to this process. Alot of oxidation-reduction, especially oxidation of penicillamine to form the mixed disulfide, is going on which rapidly takes it out of the picture.

BARNEA: You said the pH was lower?

LIPSKY: Right, in vivo. We didn't do careful concentration curves with altered pH in the in vivo situation.

BARNEA: But the pK_a for cysteine or glutathione would be around 6.0 or so. It might be protonated and you might not get a complex.

LIPSKY: I don't think that would make a difference. The complex is not critical here.

COPPER REGULATION OF MACROPHAGE EICOSANOID AND CYCLIC-AMP SYNTHESIS

G.R. Elliott, M.J.P. Adolfs, M. van Batenburg and I.L. Bonta
Department of Pharmacology, Erasmus University Rotterdam, P.O.Box 1738, 3000 DR Rotterdam, The Netherlands

INTRODUCTION.

Copper has been shown to be anti-inflammatory in a number of acute and chronic models of inflammation, as assessed by oedematous changes (1). However, its effect on other parameters associated with chronic reactions, e.g. tissue damage and growth, are equivocal (2,3). Copper can effect a number of reactions known to occur at the inflammatory site. For example it can inhibit chondrocyte growth and mitogen induced lymphocyte proliferation and it is necessary for lysyl oxidase activity and hence collagen synthesis (4-6). Copper could therefore, stimulate tissue repair and limit immune activation, and the associated stimulation of other cell types, at the inflammatory site. The macrophage plays an important role in regulating chronic inflammations releasing not only lysosomal enzymes and active oxygen species which could damage tissue but also secreting mediators such as interleukin-1, which activates the immune system (7,8). The macrophage is an important source of, and target for, metabolites of arachidonic acid (AA) such as the anti-inflammatory prostaglandin (PG) E_2 and the pro-inflammatory leukotrienes (LT) (9). For example LTC_4 stimulates the secretion of lysosomal enzymes from macrophages whilst PGE_2 blocks their release (10). In fact PGE_2 has been shown to be anti-inflammatory in a number of models of inflammation e.g. the carrageenan sponge granuloma model (11). PGE_2 is thought to work by stimulating intracellular cyclic-AMP (cAMP) formation and stable cAMP compounds and agents elevating intracellular cAMP levels, for example phosphodiesterase inhibitors, mimick many of the effects of PGE_2 (7). Interestingly copper has been reported to inhibit both PGE_2 synthesis by ram seminal vesical homogenates and cAMP production by rat cerebellar homogenates, (12-14). If copper similarly effected macrophage eicosanoid and cAMP synthesis then it could profoundly modify the course of an inflammation, the net effect depending on the relative importance of the various reactions occuring at the time. As part of our studies into the role of copper in chronic inflammatory conditions, and in order to be able to assess its potential for modifying such reactions, we have investigated the effect of copper on macrophage cAMP formation and eicosanoid and lysosomal enzyme release in vitro.

MATERIALS AND METHODS.

Male Wistar rats (175-200g) were purchased from TNO, Zeist. Indomethacin was obtained from the Pharmacy Dept., Dijkzigt Hospital, Rotterdam and isobutylmethylxanthine (IBMX) from Jansen Pharmaceuticals. Radiolabelled standards were ordered from New England Nuclear and standard eicosanoids, chloroquine and cAMP from Sigma Ltd. Antisera against PGE_2 and TxB_2 were bought from Institute Pasteur, Paris and antisera against 6ketoPGF_1 from Seragen Inc. Boston, USA. Other chemicals were of standard analytical grade.

Macrophages were isolated from Geys Balanced Salt Solution (GBSS) washes of rat peritonea by centrifugation on a Ficoll gradient, 24h after injecting 5ml of a 1% starch suspension, and suspended at 3×10^6 cells /ml GBSS. Viability, assessed by trypan blue exclusion, was greater than 95% and the macrophage preparation was greater than 80% pure. Cells (1ml alliquot) were incubated for 1h at 37^oC with, or without, copper (10^{-4}), chloroquine, indomethacin, IBMX or PGE_2. The cells were centrifuged and the supernatants stored at -70^oC for later analysis of eicosanoids (by radio immunoassay), lactate dehydrogenase (LDH) and beta-glucuronidase (GUR). LDH was assessed by the method of Wroblewsky and La Due (15) and GUR activity was quantified using the conversion of phenolphthalein glucuronide to phenolphthalein, which was then measured spectrophotometrically at 555nm. Enzyme activity is given as a percent of that released from 3×10^6 macrophages incubated with 0.1% triton X100 in GBSS for 20min at 37°C. The cell pellet was heated at 95°C for 5min in 150 μl Tris/EDTA buffer and the medium cAMP content assayed using the protein binding method of Gilman (16). The release of radiolabel from macrophages prelabelled with ^{14}C AA was performed as previously described (17).

RESULTS

The compounds used had no effect on the assay systems. Values given represent the mean ± S.D. of triplicate incubations. Each experiment is representative of at least two others. Copper stimulated the synthesis of PGE_2, TxB_2 and 6keto$PGF_{1\alpha}$ but induced a decrease in the intracellular cAMP content of rat, starch elicited, peritoneal macrophages (table 1).

Table 1. The effect of copper on macrophage eicosanoid and cAMP synthesis.

treatment	cAMP(pmol)	PGE_2(ng/ml)	TxB_2(mg/ml)	6keto$PGF_{1\alpha}$ (ng/ml)
control	1.9 ± 0.57	0.21 ± 0.03	0.28 ± 0.01	0.10 ± 0.01
10^{-4}M copper.	< 0.30	6.00 ± 0.25	5.20 ± 0.30	3.00 ± 0.50

The stimulation of eicosanoid synthesis was associated with an increase in GUR, but not LDH, secretion, (table 2).

Table 2. The effect of copper on macrophage lysosomal enzyme release.

treatment	enzyme activity as a % of that released using triton X100. LDH	GUR
control	7.0	2.5
10^{-4}M copper	10.5	15.5

Similar effects on eicosanoid and cAMP synthesis were observed when cupric nitrate was used, data not shown. The increase in eicosanoid release was almost completely inhibited by 30 μM indomethacin (table 3), but only partly reduced by the phospholipase inhibitor, chloroquine (100 μM) (table 4).

Table 3. The effect of indomethacin on copper induced stimulation of macrophage eicosanoid synthesis

treatment	PGE_2(ng/ml)	TxB_2(ng/ml)	6keto$PGF_{1\alpha}$(ng/ml)
control	0.13 ± 0.01	0.25 ± 0.02	0.11 ± 0.05
30 μM indo.	0.00	0.00	0.00
10^{-4}M copper	1.50 ± 0.03	1.70 ± 0.03	1.00 ± 0.04
indo.+ copper	0.00	0.00	0.00

Table 4. The effect of chloroquine on copper induced stimulation of macrophage eicosanoid release.

treatment	PGE_2(ng/ml)	TxB_2(ng/ml)	6keto$PGF_{1\alpha}$(ng/ml)
control	0.30 ± 0.02	0.47 ± 0.03	0.70 ± 0.05
100 μM chloroquine	0.11 ± 0.06	0.18 ± 0.05	0.26 ± 0.04
10^{-4} M copper	2.70 ± 0.07	3.00 ± 0.14	3.50 ± 0.60
chloroquine + copper	1.50 ± 0.35	3.10 ± 0.14	2.80 ± 0.60

Copper also stimulated the release of radioactivity from macrophages prelabelled with ^{14}C AA (control release 13% of maximal, 10^{-4}M copper 19% of maximal, $p < 0.05$ experimental vs control). Neither stimulating the macrophage adenylate cyclase with PGE_2 nor inhibiting phosphodiesterase activity with IBMX protected the cells against the depletion of cAMP induced by copper (table 5).

Table 5. The effect of PGE_2 and IBMX on copper induced changes in macrophage cAMP levels.

	control	PGE_2(100 ng)	control	IBMX(100 μM)
control	3.2 ± 0.5	5.3 ± 1.2	4.7 ± 0.5	8.7 ± 1.1
10^{-4}M copper	0.3	0.3	0.7 ± 0.4	0.6 ± 0.03

cAMP values are expressed as pmol/$3x10^6$cells.

Discussion.

Copper (10^{-4}M) stimulated the release of eicosanoids and GUR from rat, starch elicited, peritoneal macrophages but inhibited cAMP formation. It appears that copper activates phospholipase (PL) activity, possibly PLA_2, as shown by the increased release of radiolabel from prelabelled cells. The fact that there was no interaction between indomethacin and copper, i.e. indomethacin completely inhibited the copper induced stimulation of eicosanoid release, indicates that copper has no effect on the cyclooxygenase. It is possible that copper had an effect on individual enzymes further along the cyclooxygenase path, e.g. PGE_2 isomerase, and that we did not detected this. Copper, however, partly reversed the inhibition due to chloroquine indicating that this compound and copper competed for the same enzyme, i.e. a PL, and supporting the view expressed above. These results are interesting for several reasons. First, it appears that ionic copper could influence macrophage functions, at least in vitro. Second we have shown that copper, as reported for other systems, can induce a decrease in macrophage cAMP synthesis by apparently directly inhibiting the adenylate cyclase. Third, we have demonstrated that copper, at a concentration which reportedly inhibited PGE_2 synthesis by seminal vesical preparations (12), stimulated PGE_2, TxB_2 and 6keto$PGF_{1\alpha}$ synthesis. These points, and the possible mechanism(s) by which copper could stimulate macrophage eicosanoid and enzyme release, will be discussed in turn.

It is unlikely that free copper is found in vivo. 90% has been reported to be bound to ceruloplasmin and the rest to copper dependent enzymes and amino acids (1). Where ionic copper has been reported to be active there was always the possibility that it had formed a complex with some molecule, either in vitro or in vivo, e.g. a serum component or some molecule in a tissue homogenate (18,1). In fact it has been suggested that copper has to be complexed before it can exert an effect (19). We used a basic salt solution (GBSS) in our experiments so that copper would have had to have formed a complex at the cell surface or, if taken up, within the cell. It has been shown that Mycoplasma accumulates copper better if it is added as a phenanthroline complex rather than as copper sulphate. Copper sulphate inhibited enzyme activities better than the complex however, when cell homogenates were used indicating that copper was the active component and that the complexing agent aided its uptake (19). We do not know at this moment if copper had to enter the macrophage to exert its effect or if it stimulated membrane changes which then triggered the chain of events. It is also possible that copper complexes could stimulate eicosanoid release at much lower concentrations. The inhibition of cAMP synthesis observed in the presence of copper was not reversed by inhibiting phosphodiesterase activity (using IBMX) or by stimulating cAMP synthesis with PGE_2 indicating that copper had a direct inhibitory effect on the adenylate cyclase. This suggestion is supported by the fact that we could not detect an increase in medium cAMP levels when macrophages were incubated with IBMX and copper, (data not shown). It appears that copper has the ability to inhibit the adenylate cyclase of a number of different systems. Several authors have reported that copper inhibits PGE_2 synthesis from exogenous AA (12,13). We have clearly shown that it stimulated PGE_2 release from endogenous AA. Whether this discrepancy is due to differences in the tissue/cells used, to the fact that we used intact cells and other authors tissue homogenates, or that they used exogenous AA, is not clear and further experiments are being carried out to explore these possibilities. We have shown earlier, however, that at least one compound has apparently differing effects on AA metabolism depending on the source of the substrate, i.e. exogenous or endogenous (17). Although the data is insufficient to allow us to state with any certainty the mechanism involved in copper activation of macrophages there is sufficient data to suggest two possible pathways which could result in an increase in eicosanoid and enzyme release. The first pathway involves the decrease in cAMP levels. cAMP is known to regulate phospholipase activity, possibly via lipocortin phosphorylation (20). A decrease in cAMP levels could deregulate PLA_2 so indirectly stimulating AA release from phospholipid and hence PG formation. The second pathway could be via an effect on macrophage metabolism in general. Copper has been shown to inhibit oxygen consumption, as assayed by an accumulation of pyruvate, in Mycoplasma (21). It is possible that such an effect in macrophages could render the cell glycolytic resulting in a stimulation of mitochondrial activity (21) and hence fat metabolism. If AA turnover was also affected then there could be an increase in PG release. When discussing possible mechanisms by which copper could exert an effect it is essential to bear in mind that it (or its complexes) has super oxide dismutase activity (18). Although it is unlikely that oxygen radical induced cell damage, and hence PL activation, played a role in the changes observed (we found no increase in LDH release or trypan blue uptake) it is possible that by scavenging active oxygen radicals copper could have stimulated eicosanoid formation. This is because, at least in cell free systems, the cyclooxygenase/peroxidase enzyme complex is self destructive. The destructive agent is probably an active oxygen species generated during the conversion of PGG_2 to PGH_2 (22). If copper acted as a scavenger intracellularly then it could have protected the enzyme complex and indirectly stimulated eicosanoid formation. In summary, we have shown that copper can affect macrophage cAMP formation and eicosanoid and lysosomal enzyme release. It is doubtful if such changes could be considered to be antiinflammatory but could help explain some of the effects observed in conditions such as Wilson's disease, where metallic copper has been shown to accumulate and cause toxic changes (23), and the alterations in uterine function observed in women using copper containing intrauterine devices (24).

(Editor's note: Metallic copper is not the form of copper that accumulates in Wilson's disease and copper containing intrauterine devices have been shown to be antiinflammatory, N.E.J.M. 1985, 312, 937, 941, and 984.)

References

1) Lewis, AJ: Smith, WE: Brown, DH. In: Inflammatory diseases and copper; Humana Press, 1982; p303.

2) Lewis, EJ: West, GB. In: Agents and Actions Suppliments vol 14; Birkhauser, 1981; p339.

3) Rainsford, KD: In: Ref.1); p375.

4) Kirkpatrick, CJ: Mohr, W: Haferkamp, O. Expl.Cell.Biol. 1982, 50, 108.

5) Lipsky, PE. In: Ref.1); p581.

6) Sorenson, JRJ. In: Ref.2); p305.

7) Bonta, IL: Parnham, MJ. Int.J. Immunopharmac. 1982, 4, 103.

8) Oppenheim, JJ: Gery, I. Immunology Today. 1982, 3, 113.

9) Bonney, RJ: Davies, P. In: Contemporary Topics in Immunobiology. 1983; p199.

10) Schenkelaars, EJ: Bonta, IL. Int. J. Immunopharmac. 1986, 8, 305.

11) Bonta, IL: Adolfs, MJP: Parnham, MJ. Prostaglandins. 1981, 22, 95.

12) Lee, RE: Lands, WEM. Biochem. Biophys. Acta. 1972, 260, 203.

13) Maddox, IS. Biochem. Biophys. Acta. 1973, 306, 74.

14) Nathanson, JA: Bloom, FE. Mol.Pharmac. 1976, 12, 390.

15) Wroblewsky, F: LaDue, JS. Proc. Soc. Exptl.Biol.Med. 1955, 20, 210.

16) Gilman, AG. Proc.Nat.Acad.Sci. (USA). 1970, 67, 305.

17) Elliott, GR: Braquet, P: van Batenberg, M: Bonta, IL. Biochim. Biophys. Acta. 1985, 837, 213.

18) Smit, H: van der Groot, H: Nauta, WT: Timmerman, H: de Bolster, MWG: Stouthamer, AH: Vis, RD. Antimicrob. Agents and Chemother. 1982, 21, 881.

19) Sorenson, JRJ. J.Med.Chem. 1976, 19, 135.

20) Hirata, F: Matsuda, K: Notsu, Y: Hattori, T: Del Carmine, R. Proc.Nat.Acad.Sci. (USA) 1984, 81, 4717.

21) van der Goot, H: Pijper, PJ: Smit, H. In: Ref 1), p409.

22) Keuhl, FA: Egan, RW: Humes, JL. In: Progress in Lipid Research, Pergamon Press, 1981; p97.

23) Ludwig, JC: Chvapil, M. In: Ref 2), p65.

24) Mercado, E: Aznar, R: Gallegos, AJ: Dominguez, R: Rosado, A Contraception. 1977, 16, 229.

DISCUSSION

LOVEN: Do you think that Cu is acting on those free sulfhydryl groups of adenylate cyclase or by some different mechanism?

ELLIOTT (Graham): At the moment I've got no idea as to how it's acting. It's possible that if we use low concentrations of ATP we would also see an effect of Cu. I don't know. All these concentrations we have been using are saturating.

SORENSON: How do these results correlate with results in bovine homogenates where additions of Cu(II)SO_4 or chloride decrease E_2 synthesis and increase F_2 alpha synthesis?

ELLIOTT (Graham): They don't correlate at all. This could be because of the tissue difference or it could be the fact that the original workers were using radioactive arachidonic acid. Now we've found that if a tissue is incubated with radiolabeled arachidonic acid, one doesn't always get the same effect found with cold arachidonic acid and radioimmuno assay. One possible explanation is that stimulation of very active cyclooxygenase metabolizes the endogenously released substrate in preference to exogenous substrate. It has been reported that added radiolabeled arachidonic acid causes an apparent decrease in the release of PGE_2 although radioimmuno assay measurement shows an increase. I don't really trust studies using radiolabeled substrates.

SORENSON: So these cells were just exposed to these experimental conditions and then ...

ELLIOTT (Graham): These cells were incubated in a balanced salt medium with $Cu(II)SO_4$ for 1 hour, then spun down, and cyclic AMP and release of eicosanoids were measured.

SORENSON: I have suggested that if one puts $Cu(II)SO_4$ into any biological system and that causes some effect then that effect is mediated by some complex. This suggestion is open to challenge.

ELLIOTT (Graham): I'm not quite sure what complex the Cu formed in this system, at least in the medium, because there is nothing there.

SORENSON: You have no amino acids?

ELLIOTT (Graham): No, no serum, nothing at all. Just a salt solution with Ca and Mg etc.

SORENSON: Then perhaps it's a cellular mediated event, and that's why you found an adenylate cyclase affect.

ELLIOTT (Graham): Yes. It could also happen if Cu forms a complex inside the cell.

SORENSON: It may not be able to pass through any bilipid cellular membrane which requires lipophilic character for passive transport.

ELLIOTT (Graham): I don't know. Prof. Barnea, did Cu get in by itself in your system?

BARNEA: Essentially your results are exactly opposite ours. First of all we don't get anything with $Cu(II)Cl_2$ which is essentially the same as $Cu(II)SO_4$. I was surprised to see that you observed an effect with ionic Cu.

I have a comment about adenylate cyclase. I presume you used membrane preparations for your measurement of the fluoride effect on GTP synthesis since this cannot be done with intact cells. However, membrane preparations lose their adenylate cyclase activity very rapidly. It has been reported that membranes prepared from brain tissue, for example, rapidly lose about 70% of their adenylate cyclase activity within a few minutes. This loss actually can be reversed by GTP and by fluoride. Essentially one can protect against that loss and reverse inactivation of enzyme activity if you don't wait too long. If membrane preparations are held for some time, some of the activity is lost. Effects seen with cells may not be the result of a direct effect on adenylate cyclase but an alteration of the conformation of other constituents around adenylate cyclase.

ELLIOTT (Graham): Our macrophages can be kept for 2 or 3 days as a pellet at -70°C without loss of activity.

BARNEA: Yes, but these are whole cells?

ELLIOTT (Graham): Yes, we store these cells and then we use them to make membrane preparations on the day we plan to use them.

BARNEA: I've just suggested that you may not be able to reproduce reported effects of Cu with your membrane preparations because they have lost a lot of their catalytic activity.

ELLIOTT (Graham): We have activity without GTP and it's enhanced with GTP.

SORENSON: The bonding between Cu and the cell membrane is complexation, coordinate covalent bonding, so that would be the complexation that then may be involved in triggering events that you've been able to measure.

Antiinflammatory Activities of Copper Complexes

ANTIINFLAMMATORY AND ANTIULCER ACTIVITIES OF NON-STEROIDAL ANTIINFLAMMATORY AGENT COPPER COMPLEXES.

John R.J. Sorenson

Department of Biopharmaceutical Sciences,

College of Pharmacy, University of Arkansas for

Medical Sciences, 4301 W. Markham, Little Rock,

Arkansas 72205 U.S.A.

To date, over 70 copper complexes have been studied as antiinflammatory agents. Results of these studies have been reviewed recently[1] and confirm as well as extend original observations that copper complexes of inactive ligands and active antiinflammatory drugs are more active than the parent ligand or inorganic copper. The followng is a brief presentation of some data from the original report suggesting that copper complexes are the active metabolites of antiarthritic drugs[2].

As shown in Table I, $Cu(II)_2(acetate)_4(H_2O)_2$ was found to be active in the initial test (carrageenan paw edema) for antiinflammatory activity but inactive in the two follow-up antiinflammatory screens (cotton wad granuloma and adjuvant arthritis). Copper chloride dihydrate had no activity in any of these models of inflammation using large initial screening doses of 1.2, 0.6, or 0.2 mmol/kg, respectively. Ligands such as anthranilic acid and 3,5-diisopropylsalicylic acid(3,5-dips) which were anticipated to be inactive were found to be inactive. However, their copper complexes were found to be potent antiinflammatory agents in all three models of inflammation. These observations and results obtained with other complexes prepared with inactive ligands supported the notion that complexed copper was a more active antiinflammatory form of copper and led to the suggestion that copper complexes of active antiinflammatory agents might be more active than the parent antiinflammatory drugs. Representative data taken from the original publication[2] for aspirin, D-penicillamine, and their copper complexes are also compared in Table I. These data show that these complexes are much more effective than their parent drugs and support the hypothesis that active metabolites of antiarthritic drugs are their copper complexes. Since amounts of copper in these complexes do not appear to correlate with activity of these complexes, it is suggested that pharmacologic activity is due to the physicochemical properties of these copper complexes.

Two general chemical mechanisms are immediately apparent possibilities. Small molecular weight copper complexes may serve as transport forms of copper that allow activation of copper-dependent enzymes or they may have chemical reactivities that facilitate correction of the chemical problem that led to the disease state.

Table I. ANTIINFLAMMATORY ACTIVITIES OF SOME COPPER COMPLEXES

Compound	Carrageenan Paw Edema	Cotton Wad Granuloma	Adjuvant Arthritis	Copper (%)
$Cu(II)_2(acetate)_4(H_2O)_2$	A at 0.02	I at 0.30	I at 0.089	31.8
anthranilic acid	I at 1.46	NT	I at 0.219	
3,5-dips acid	I at 0.90	NT	I at 0.135	
$Cu(II)(anthranilate)_2$	A at 0.02	A at 0.07	A at 0.003	18.9
$Cu(II)(3,5,dips)_2$	A at 0.02	A at 0.01	A at 0.002	12.5
Aspirin	A at 0.36	A at 1.11(i.g.)	A at 0.033	
$Cu(II)_2(aspirinate)_4$	A at 0.01	A at 0.01	A at 0.001	15.0
D-Penicillamine	I at 1.34	I at 0.67	I at 0.201	
$Cu(I)(D\text{-}pen)(H_2O)1.5$	A at 0.03	A at 0.04	NT	26.7
$Cu(II)_2(D\text{-}pen\ disulfide)_2(H_2O)_2$	A at 0.01	A at 0.03	A at 0.040	15.4

A = lowest active dose tested; I = inactive; NT - not tested.

All compounds were given by subcutaneous injection unless indicated as intragastric (i.g.) and expressed as millimole per kilogram of body weight (m mol/kg).

Activitation of lysyl oxidase, which is ultimately required to catalyze cross-linking of connective tissue components in the tissue repair phase of antiinflammatory action, is an important mechanistic consideration in accounting for antiinflammatory activity of copper complexes[3]. Activation of the copper-dependent superoxide dismutases (Cu-ZnSOD), which are required to prevent accumulation of superoxide in the cellular fluid matrices is another important mechanistic consideration since accumulation of superoxide as a result of reduced Cu-ZnSOD activity has been suggested as the cause of arthritic and other inflammatory diseases[4-6]. Small molecular weight copper complexes have been shown to disproportionate superoxide[7-8] and this chemical reactivity has attracted a great deal of attention in accounting for antiinflammatory activity of copper complexes[9]. Other less well understood possible mechanisms involve modulation of inflammation mediators such as histamine, lysosomal membrane stabilization, and prostaglandin syntheses[1].

Acute toxicity studies were done early in the course of this work[2,10]. These studies demonstrated that antiinflammatory copper complexes were less toxic than inorganic forms of copper as well as their parent antiarthritic drugs. The oral LD_{50} values for copper aspirinate were found to be 1.06 $\pm$ 0.26 and 1.16 $\pm$ 0.33 mmol/kg, respectively, for male and female Sprague-Dawley rats. Chronic toxicity studies using 0.12 mmol/kg given orally to male Sprague-Dawley rats 5 days per week for 3 months did not affect growth, survival, or copper and zinc concentrations in 15 tissues including skin and fur[11]. Histopathologic examination of all tissues, except skin and fur, at the light microscopic level revealed no evidence of pathologic changes except for liver. An increase in Kupffer cell number was noted for those animals that were sacrificed at the end of the 3 month treatment period. Kupffer cell number progressively decreased in livers of animals sacrificed during the subsequent 2 months of this 5 month study. The projected human dose is 0.003 to 0.006 mmol/kg daily with a decrease to some maintenance dose if remission results. This regimen seems to be safe enough to correct what may be, in part, a copper deficiency or inadequacy disease.

Antiinflammatory activities of copper complexes explain earlier observations by Fenz, Forestier et al., Kuzell et al., and Hangarter et al. that copper complexes are effective in treating arthritic and other human degenerative diseases. Their clinical results have been reviewed in detail elsewhere[12].

Antiulcer Activities of Copper Complexes

Pharmacologic studies also revealed that copper complexes are unique antiinflammatory agents when it was demonstrated that they were potent antiulcer agents[2]. To date, antiulcer data for over 60 copper compounds have been reported and these data confirm as well as extend this remarkable observation[1]. Representative data from the original report[2] are presented in Table II.

Table II. ORAL ANTIULCER ACTIVITY OF SOME NON-STEROIDAL ANTIINFLAMMATORY AGENT COPPER COMPLEXES

Compound	Antiulcer Activity (ED_{50}) (μmol/kg)	% Copper
Cu(II)$_2$(acetate)$_4$(H_2O)$_2$	564	32
Cu(II)(anthranilate)$_2$	13	19
Cu(II)(3,5-dips)$_2$	5	13
Cu(II)$_2$(aspirinate)$_4$	13	15
Cu(II)$_2$(niflumate)$_4$	4	10
Cu(II)$_2$(D-pen disulfide)$_2$·3H_2O	6	16
Cu(II)$_2$(fenamole)$_2$(acetate)$_4$	7	19
Cu(II)(fenamole)$_2$(HCl)$_2$	10	14
Cu(II)(salicylate)$_2$·4H_2O	11	15
Cu(II)$_2$[1-(p-chlorobenzoyl)-5-methoxy-2-methyl-3-indolyl-acetate]$_4$(CH_3COCH_3)$_2$	3	8
Cu(II)(4-n-butyl-1,2-diphenyl-3,5-pyrazolidinedione)$_2$	7	9

Copper(II)$_2$(acetate)$_4$ had only very weak antiulcer activity while copper complexes of the non-steroidal antiinflammatory agents were found to be potent antiulcer agents. There are no more potent antiulcer agents in this model of ulcer. In addition, some of these complexes were also shown to be effective in preventing corticoid-induced ulcers[2].

Comparison of the relative antiulcer activities of three different penicillamine complexes in the Shay ulcer model revealed that the water soluable mixed valence complex, $Na_5Cu(I)_8Cu(II)_6$ (Penicillamine)$_{12}$ Cl, was the most effective[13], as shown in Table III,

Many copper complexes of amino acids have also been found to be effective in preventing and reducing the severity of existing Shay ulcers[14,15]. The copper complex of glycine was the most effective of all amino acid complexes studied, its ED_{50} was 23 μmol/kg for reduction of number of ulcers and severity of existing ulcers[14].

TABLE III. ORAL ANTIULCER ACTIVITIES OF PENICILLAMINE COPPER COMPLEXES, PENICILLAMINE, COPPER CHLORIDE, AND COPPER ACETATE, IN THE SHAY RAT

Compound	ED_{50} (μmol/kg)		
	Antiulcer	Antiseverity	%Cu
$Na_5Cu(II)_6Cu(I)_8(Pen)_{12}Cl$	10	4	32
[Cu(II)Penicillamine Disulfide]$_2$	>56	20	18
Cu(I)(Pen)	156	57	30
D-penicillamine	ULCEROGENIC		
$Cu(II)Cl_2(H_2O)_2$	18	35	37
$Cu(II)_2(Acetate)_4(H_2O)_2$	53	55	32

In our original report of antiulcer activity for copper complexes we also reported that these copper complexes reduced gastric acid secretion in the Shay rat[2]. The influence of copper complexes on histamine activity was examined by studying the activity of the copper complex of histamine. Cu(II)(histamine)$(Cl)_2$ was found to be a potent antiulcer and antisecretory agent, its ED_{50} for anti Shay ulcer activity is 37 µmol/kg and for reduction in severity of existing ulcers the ED_{50} is 17µmol/kg. This observation was confirmed and extended by Marletta et al.[16] and Alberghina et al.[17] with the reports that the mixed copper complex of phenylalanine and tryptophan inhibited histamine-, pentagastrin-, and bethanacol-induced gastric acid secretion (Table IV).

Antiulcer activity for copper complexes has been confirmed and extended further by West and his colleagues with reports showing that $Cu(II)_2$$(aspirinate)_4$ and Cu(II)$(salicylate)_2$ were more effective in the aspirin-exacerbated Shay ulcer model than either the H_1(Mepyramine) or H_2(Metiamide) histamine blockers as shown in Table V[18]. It was also shown that Cu(II)$(salicylate)_2$ was more effective than these histamine blockers in the Cold-Stress ulcer model (Table VI)[18].

TABLE IV. ANTISECRETORY ACTIVITY OF Cu(II)(TRYPTOPHAN)-(PHENYLALAMINE) COMPLEX IN THE PENTAGASTRIN, HISTAMINE, OR BETHANECHOL STIMULATED SHAY RAT

Compounds	ΔµEqH+/hr*	inhibition
Pentagastrin	23.68±11.23	87%
Pentagastrin + Cu-Tryp-Phe	2.98± 1.30**	
Histamine · 2HCl	16.54± 1.91	82.5%
Histamine · 2HCl + Cu-Tyrp-Phe	2.90± 0.82**	
Bethanechol Chloride	28.78± 5.14	82%
Bethanechol Chloride + Cu-Tyrp-Phe	5.30± 1.65**	

*µEqH+(M±SD) secreted above basal value, **$p<0.01$

TABLE V. ORAL ANTIULCER ACTIVITY OF $Cu(II)_2(ASPIRINATE)_4$, $Cu(II)(SALICYLATE)_2$, AND HISTAMINE BLOCKERS IN THE ASPIRIN PRETREATED SHAY RAT

Drug	Dose (µmol/kg)	% inhibition
Aspirin	278	+ 23.0±8.0 (ulcerogenesis)
	555	+ 49.7±8.7* (ulcerogenesis)
$Cu(II)_2(aspirinate)_4$	6	41.7±5.9*
	12	69.5±9.0*
$Cu(II)(salicylate)_2$	29	25.8±8.3
	59	52.4±8.1*
Mepyramine(Pyrilamine)	35	10.7±5.3
	88	18.1±5.1
	350	21.6±7.8
Metiamide	18	8.3±1.7
	45	12.8±2.0
	111	9.2±2.5
	446	19.9±5.9

*$p<0.05$

TABLE VI. ORAL ANTIULCER ACTIVITY OF $Cu(II)(SALICYLATE)_2$ AND HISTAMINE BLOCKERS IN THE COLD-WATER STRESSED RAT

Drug	Dose (μmol/kg)	% inhibition
$Cu(II)(salicylate)_2$	29	71.1± 6.7*
	59	84.6± 5.9*
Mepyramine(Pyrilamine)	35	28.0± 6.7
	88	77.0± 5.0*
	350	79.8±10.2*
Metiamide	18	44.6± 2.9*
	45	84.6± 5.2*
	111	87.2± 8.8*
	446	98.4± 7.7*

*$p<0.05$

$Cu(II)(salicylate)_2$ was also shown to be more effective than cimetidine in preventing ulcers induced by the antiinflammatory drugs aspirin, and indomethacin, as well as Cold-Stress (Table VII)[19].

Finally, Townsend has demonstrated that $Cu(II)$-$(tryptophan)_2$ and $Cu(II)_2(aspirinate)_4$ increased the rate of healing of surgically placed glandular gastric ulcers using histochemical methods[20]. These complexes prevented wound regression and increased the rate of reepithelization. The appearance of replaced connective tissue components could not be distinguished from tissue of normal non-operated rats.

Potent antiulcer activity without ulcerogenic activity[21] coupled with potent antiinflammatory activity has distinguished copper complexes as a unique class of antiarthritic drugs since all antiarthritic drugs do not have antiulcer activity and they are ulcerogenic. Known copper-dependent biochemical processes can be cited as possibly accounting for these antiinflammatory and antiulcer activities, and the pharmacologic use of copper complexes to treat ulcers offers a physiologic approach to treatment.

TABLE VII. COMPARISON OF $Cu(II)(SALICYLATE)_2$ AND CIMETIDINE, GIVEN ORALLY, IN PROTECTING AGAINST GASTRIC LESIONS PRODUCED BY ORAL ULCEROGENIC DOSES OF ASPIRIN (2.22 mmol/kg), INDOMETHACIN (56 μmol/kg), AND COLD-STRESS IN RATS

Ulcerogen	Antiulcer Agent	Dose (μmol/kg)	% inhibition (M±SE)
Aspirin	Cimetidine	148	19 ± 5
		297	16 ± 4
	$Cu(II)(Salicylate)_2$	22	21 ± 6
		44	61 ± 10*
Indomethacin	Cimetidine	148	10 ± 3
		297	24 ± 6
	$Cu(II)(Salicylate)_2$	22	37 ± 7*
		44	50 ± 5*
Cold-Stress	Cimetidine	148	38 ± 8*
		297	61 ± 9*
	$Cu(II)(Salicylate)_2$	22	58 ± 8*
		44	79 ± 7*

*$p<0.05$

References

1. Sorenson, JRJ. In: Metal Ions in Biological Systems-14 (ed. H. Sigel), Marcel Dekker, New York, 1976, p 77.
2. Sorenson JRJ. J. Med. Chem. 1976, 19, 135.
3. Harris ED; Di Silvestro RA; Balthrop JE. In: Inflammatory Diseases and Copper (ed. J.R.J. Sorenson) Humana Press, Clifton, New Jersey, 1982, p 183.
4. McCord JM. Science, 1974, 185, 529.
5. Rister M. In: Trace Elements in the Pathogenesis and Treatment of Inflammation (eds. K.D. Rainsford, K. Brune and M.W. Whitehouse) Birkhauser Verlag, Basel, 1980, p 137.
6. Rister M; Bauermeister K; Gravert U; Gladtke E. Eur. J. Pediatr., 1979, 130, 1217.
7. DeAlvare LR; Goda K; Kimura T. Biochem. Biophys. Res. Commun., 1976, 39, 687.

8. Weser U; Richter C; Wendel A; Younes M. Bioinorg. Chem., 1978, 8, 201.
9. Sorenson JRJ; Oberley LW; Crouch RK; Kensler TW. In: Oxygen Radicals in Chemistry and Biology (eds. W. Bors, M. Saran, and D. Tait) Walter de Gruyter and Co., Berlin, 1984, p 821.
10. Sorenson JRJ. Prog. in Med. Chem. 1978, 15, 211.
11. Sorenson JRJ; Rolniak TM; Chang LW. Inorg. Chim. Acta, 1984, 91, L31.
12. Sorenson JRJ; Hangarter W. Inflammation, 1977, 2, 217.
13. Sorenson JRJ; Ramakrishma K; Rolniak TM. Agents and Actions, 1982, 12, 408.
14. Kishore V; Rolniak TM; Ramakrishma K; Sorenson JRJ. In: Inflammatory Diseases and Copper (ed. J.R.J. Sorenson) Humana Press, Clifton, N.J., 1982, p 363.
15. West GB. In: Inflammatory Diseases and Copper (ed. J.R.J. Sorenson) Humana Press, Clifton, N.J., 1982, p 319.
16. Marletta F; Rizzarelli E; Mangiameli A; Alberghina M; Brogna A; Sammartano S; Blasi A. Rendic. Gastroenterol. 1977, 9 (Suppl.1), 35.
17. Alberghina M; Brogna A; Mangiameli M; Marletta F; Rizzarelli E; Sammartano S. Il Farmaco, 1982, 12, 805.
18. Hayden LJ; Thomas C; West GB. J. Pharm. Pharmacol. 1978, 30, 244.
19. West GB. J. Pharmacol. Methods 1982, 8, 33.
20. Townsend SF; Sorenson JRJ. Agents and Actions, Supplement 8, (eds. K.D. Rainsford, K. Brune, and M.W. Whitehouse) Birkhauser Verlag, Basel, 1981, p 389.
21. Williams DA; Walz DT; Foye WO. J. Pharm. Sci., 1976, 65, 126.

DISCUSSION

HARRIS: I was wondering why some of these compounds are effective and others are less effective. It occurred to me that you are attributing a lot of your efficacy to the ablilty of the compound to find its target and modulate events that are occurring right there at that site. My question is, why aren't these other compounds effective? Is it possible that when you're dealing with these less effective compounds you're really dealing with compounds that are taken up by other tissues in a rather non-discriminate way and lost? It would seem that a compound, to be effective, would have to have that particular property of showing a much longer half-life in the system and being essentially ignored or not taken up by the liver and other tissues. Does that seem to be reasonable and when you design these drugs are you looking for something that the liver is going to ignore?

SORENSON: Yes. Delivery without losses is likely to be most important. We pay attention to the physicochemical properties of those complexes that are more effective than others. The Cu amino acid complexes still offer a useful approach to treating or preventing ulcer disease even though they're less effective than some others. The recommended daily intake of Cu is 50 uM/day. Amounts of Cu in these complexes represent much less Cu than this so one could use any of those amino acid complexes to get a part of what is recommended daily intake and I'm sure get beneficial antiinflammatory and antiulcer effects. Copper complexes of drugs probably have physicochemical properties that facilitate target delivery.

There is a big question as to what is the form of these complexes once they're absorbed. We simply show the complex that we administer or at least put into suspension prior to administration. Most or essentially all of that material exists as complexes in the stomach (pH 6.0). If one takes say 10 mg of a drug to get some pharmacological effect, it's likely that only femtomolar quantities ever get to the site of the lesion. There are a lot of lost processes. Some of all drugs in the first pass through the liver are taken up by the liver, metabolized, and eliminated, only a part gets through, that part that gets through goes to all tissues. Fortunately there's an increase in vascular permeability at the site of inflammation or ulcer so there's a lot of delivery to that site and the drug concentration is higher at that site than it would be elsewhere. We would like to study absorption, tissue distribution, and things like that to better understand what are the important physicochemical properties and then synthesize more effective compounds. However, we are straped for money so we just measure pharmacologic effects as a means to justifying the other work and perhaps, in the future, we may be able to study these parameters as well as other mechanistic aspects.

MOSTAFA: Did you study stereochemical aspects of Cu complexes?

SORENSON: Just a little. Not enough to make any claims. I'm sure there are important stereochemical relationships for the ligand that would contribute to the physicochemical properties that influence activity.

MOSTAFA: You mentioned Dr. West. Is Dr. West at ISU?

SORENSON: No, Geoff West is at North East London Polytechnic in London.

CROOK: I would be reluctant to place a whole lot of credence in the activity of a compound in treating acute rheumatic fever. I know it has the magic word rheumatic in there but the etiology and everything else makes it a completely different disease and not at all similar to either rheumatoid arthritis or osteoarthritis. Am I correct in presuming that there were not any placebo-treated groups in those studies?

SORENSON: Yes. When all of these studies were done placebo treatment was unknown.

CROOK: Do you have any feeling for what it truly means to have only 50% or 60% inhibition of ulcer formation? Have you studied chronic or just acute treatment with Cu complexes? Has it been demonstrated that there are 50% less ulcers with treatment? Have all these studies been acute? Has all the animal work been done with just a single dose?

SORENSON: No. Doses used to get those ED_{50} doses ranged from 0 to 40 mg/kg. Activity is measured at each dose and the dose needed to produce a 50% reduction in number of ulcers and severity is calculated. It is then possible to compare the effectiveness of one compound with that of another based upon their ED_{50} values.

CROOK: Was all of the antiinflammatory activity observed following subcutaneous administration?

SORENSON: Early on that's what we did. Yes.

CROOK: Are they just as efficacious orally?

SORENSON: No. No drug that I know of is as effective orally as it is subcutaneously. I didn't have time to present some of our newer data evaluating copper aspirnate given orally. It was reported by Williams, Waltz, and Foye (J. Pharm. Sci. 1976, 65,126) that $Cu(II)_2(aspirnate)_4$ was 1.7 times as effective as aspirin when it was given orally. However, they didn't take into account the fact that the molecular weight of $Cu(II)_2(aspirinate)_4$ is about 5 times greater than aspirin. As it turns out, $Cu(II)_2(aspirnate)_4$ is 8 times as effective as aspirin on molar basis. We repeated that study and we found that to be true both with regard to acute and chronic stages of polyarthritic inflammation following oral administration. We would also like to know whether or not absorption is different for aspirin and $Cu(II)_2(aspirinate)_4$. Absorption would also influence the relative activities of these two compounds.

CHIDAMBARAM: You have an impressive list of Cu compounds there. Which are the more effective complexes, Cu(II) or Cu(I)?

SORENSON: I don't know. We don't have enough anti-inflammatory data for Cu(I) compounds to offer a meaningful comparison. We do have other data that we had gotten with a series of penicillamine complexes where we had Cu(I) and Cu(II) complexes. The most effective penicillamine complex was the mixed valance complex where there's both Cu(I) and Cu(II). My inclination is that if you give Cu(II) and Cu(I) is needed, the conversion from Cu(II) to Cu(I) will be made in vivo.

CHIDAMBARAM: What's the half-life of these compounds?

SORENSON: I don't know.

MELETHIL: Can you give me some references concerning the structure of Cu(II)(salicylate)$_2$ and the number of molecules of water found in solvates?

SORENSON: Yes. Chemical abstracts should be helpful.

ELLIOTT (Gary): When I look at these chelates I have the impression that their formation constants are, perhaps, in the lower twenties?

SORENSON: Less than that, perhaps in the range of 10^{15} to 10^{20}.

ELLIOTT (Gary): I'm wondering if we go to a more stable chelate such as N^1, N^8 bis-2,3-dihydroxybenzoyl spermidine which has a formation constant of 10^{31}, do you anticipate lower biological activity with these more stable Cu chelates? Is part of their activity due to their lability and release of Cu?

SORENSON: It's possible. They may be a better transport form of Cu. I can't rule out the possibility that they have some chemical reactivity that is unique and promotes some useful biochemical process. It's worth studying some very stable complexes to determine whether or not they have any beneficial effects.

RISK: Are you hypothesizing that these Cu complexes are just convenient transport molecules that allow Cu to get into the cell or is the complex acting directly? I personally favor the latter hypothesis. I believe there's a lot of insight one can get with regard to structure-activity-relationships that you've done, in effect, with these different amino acids complexes. I assume that those amino acids differ tremendously in their ability to permeate cells but they do not vary tremendously in their efficacy. There's only a 2 or 3 fold variation. This could suggest something significant about the mechanism of action.

SORENSON: Yes, both transport and direct action are

possible. A better understanding of structure-activity-relationships in accounting for the ultimate physiologic or pharmacologic effects will be helpful in developing more effective complexes in either case.

RISK: Are you willing to speculate on what are the mechanisms of action?

SORENSON: I hold open two general possibilities. That is, that they are just transport forms of Cu or they are able to cross cell membranes and perhaps induce a particular Cu-dependent enzyme such as Cu-Zn SOD or lysyl oxidase. These complexes may also modulate some O_2^--dependent process or some redox process.

RISK: Would you care to speculate as to why Cu(II)(3,5-dips)$_2$ is so effective?

SORENSON: It is a very lipid soluble complex. That physicochemical property is probably important in accounting for its effectiveness. As to why it seems to be the only one effective in many different disease models, it's the one we suggest investigators study initially and many investigators are so pleased with its activity they want to spend all of their time studying it as opposed to studying other complexes that we would like to send them.

COMPARATIVE PHARMACOKINETICS OF SALICYLIC ACID AND COPPER SALICYLATE IN RATS

Bhasker V. Shetty and Srikumaran Melethil[1]

University of Missouri, Schools of Pharmacy and Medicine, Kansas City, MO 64108-2792

ABSTRACT

The objective of this study was to provide a basic understanding of the effect of complexation with copper on the pharmacokinetics of salicylic acid (SA). Male Sprague-Dawley rats (250-300 gms) received an I.V. dose of either SA or copper-salicylate complex, (Cu-SA) (10 mg/kg of SA) in a randomized balanced crossover fashion. Based on free copper ion measurements, the amount of intact complex in the vehicle used to prepare the drug solution for injection ranged between 68.4 and 76.6%. Serial blood samples (100 µl) were drawn for 12 hours post administration. Plasma samples were analyzed for SA by reversed phase HPLC on a C-18 column with U.V. detection at 237 nm. The SA concentration-time profiles were similar up to the first 90 minutes for both treatments. For the remainder of the sampling period, SA concentrations were consistently lower where the animals had received Cu-SA. In rats (n=5) injected with Cu-SA, SA declined monoexponentially with an elimination rate constant (mean ± SD) of 0.0108 ± 0.00146 min^{-1}, a plasma clearance of 2.15 ± 0.621 ml/(min kg) and a Vd of 188.2 ± 38.67 ml/kg. In contrast, rats injected with SA showed

[1] Correspondence Author

a biexponential decline of the drug with a β of 0.00475 ± 0.00144 min^{-1}, plasma clearance of 1.08 ± 0.153 ml/(min·kg) and Vd(β) of 243.4 ± 44.96 ml/kg. Non-compartmental analysis of SA concentration-time data following Cu-SA administration gave Vd(SS) and MRT values of 190.5 ± 31.81 ml/kg and 92.29 ± 9.279 min, respectively. The corresponding values after SA administration were 232.9 ± 40.17 ml/kg and 211.7 ± 73.88 min. Differences in the terminal rate constants, volumes of distribution (Vd(β)), clearances and MRT between the two treatments were significant ($p<0.05$).

INTRODUCTION

Copper complexes of drugs are gaining attention as potential therapeutic agents. An extensive body of literature exists on the pharmacological effects of a wide variety of copper complexes. Some major effects of these complexes include anti-inflammatory (1), anti-cancer (2,3), anti-diabetic (4), anti-ulcer (5,6), and anti-convulsant (7) properties. Most of the publications to date have focused on pharmacological and biochemical aspects of these complexes. An extensive literature search failed to reveal any comparative pharmacokinetic studies of these drugs and their copper complexes. Since copper complexes of drugs in general are more lipophilic in character than the ligand, it can be speculated that the pharmacokinetic properties of a drug and its copper complex may vary. The objective of this study was, therefore, to provide a basic understanding of the effect of complexation with copper on the pharmacokinetics of salicylic acid (SA) in rats.

EXPERIMENTAL

Synthesis

Copper-salicylate complex (Cu-SA) was prepared according to published methods (1) with minor modifications. A 5% sodium salicylate solution prepared in 10% sodium hydroxide was filtered and its pH adjusted to 7.8 with 10% hydrochloric acid. One hundred ml of this solution was then added to 300 ml of a stirred aqueous solution of copper sulfate pentahydrate (0.0336 M). After stirring for 30 min, the solution was refrigerated overnight when pale blue needle-like crystals separated. These crystals were separated, washed with cold water, and recrystallized from warm water (40°C). The precipitate was then dried at 40°C to constant weight.

Composition

Carbon and hydrogen contents of Cu-SA were obtained through a commercial source (Spang Microanalytical Laboratory, Eagle Harbor, MI). Copper content of the complex was determined by flameless atomic absorption spectrophotometry (8). The SA content of Cu-SA was determined by an HPLC method (see "Analytical Methods" Section). Results on the composition of Cu-SA are given in Table 1.

Animal Studies

In preparing a solution of Cu-SA for I.V. administration, it was decided to select a vehicle in which the complex was mostly present in the intact form. The extent of intact copper complex in solution was determined as the difference between the total and free (ionic) copper. Ionic copper was measured by means of an ion specific electrode for copper (Model 901 Ionanalyzer, Orion, Cambridge, MA). Vehicles tested were normal saline, aqueous acetic acid buffer (pH 6 and 8) methanol-water (50:50), 1% Tween 80, aqueous nitric acid (pH 2.0 - 5.0) and boric acid (0.1 M - 0.5 M). The maximum extent of intact Cu-SA was measured in 0.2 M boric acid adjusted to a pH of 5.5 (mean ± SD:73.6 ± 3.17%, n=4). This vehicle was used for preparing SA and Cu-SA solutions for I.V. administration.

Pharmacokinetic studies were done in male, Sprague-Dawley rats (250-300 gms) with indwelling jugular catheters surgically implanted at least 60 hours prior to the study. Each animal received an I.V. dose of either SA or Cu-SA (10 mg/kg of SA) in a randomized crossover fashion. A washout period of two days was allowed between the two treatments. Following administration of the drugs, 100 μl blood samples were drawn via the catheter at various time intervals up to 12 hours. The sampling times were 1, 3, 5, 10, 15, 20, 30, 45, 60, 90, 120, 150, 180, 210, 240, 270, 300, 360, 420, 480, 540, 600, and 720 min. To minimize changes in blood volume or hematocrit, blood sampling was followed by immediate replacement with fresh blood from donor animals. Plasma was separated immediately after collection and frozen until analysis.

Analytical Methods

Salicylic Acid. Plasma samples (50 μl aliquots) were analyzed for SA by a previously reported reverse phase HPLC method (9) with minor modification. Briefly, the procedure involved the addition of 50 μl of 1% nitric acid (to precipitate proteins) and internal standard (m-anisic acid, 500 ng) to the plasma sample followed by extraction with 8 ml of ethyl acetate:butyl chloride (50:50) mixture. Following extraction, the organic layer was dried at 37°C in a vortex evaporator (Haakebuchler Instruments, Inc., Saddlebrook, NJ). The residue was reconstituted in a suitable volume (0.5 - 2 ml) of the mobile phase (3% acetic acid:acetonitrile, 85:15) and chromatographed on a C-18 column (5 μ particles, 15 cm x 4.6 mm, Altex Scientific, Inc., Berkeley, CA) with UV detection at 237 nm (Spectraflow 783, Kratos Analytical, Ramsey, NJ). Peak areas were obtained directly by an integrator (Model C-R3A, Shimadzu Corporation, Kyoto, Japan) from which peak area ratios (SA/internal standard) were computed. These ratios were converted to SA concentrations using linear calibration plots prepared from standard solutions (1-80 μg/ml SA) in rat plasma. Correlation coefficients for the calibration plots over the 3-month period of analysis ranged from 0.9960 to 1.0. The minimum detectable concentration of this method is 1 μg/ml.

Plasma samples obtained from the study could contain SA in both free and complexed (Cu-SA) forms. Separate experiments, where Cu-SA was added to plasma, showed that this method accurately measures SA when present in either of the two forms.

Data Treatment

The exponential equation that best characterizes a given SA concentration-time profile was determined using non-linear regression analysis (10), the F test (11), and Akaike's Information Criteria, AIC (12). The coefficients and exponents of this equation were used to estimate model dependent parameters like volume of the central compartment, Vc; volume of distribution, Vd or Vd(β); and elimination half-life, t 1/2 (13). Model independent parameters, like steady state volume of distribution, Vd(SS); mean residence time, MRT; and clearance, CL; were also calculated by established procedures (13).

RESULTS

The elemental composition of the copper-salicylate complex is shown in Table 1. The formula for the complex is consistent with $(C_6H_4OH\ CO_2)_4 \cdot Cu_2$, reported earlier (14).

Table 1. Composition of Copper-Salicylate

Component	Found (%)	Calculated[a] (%)
C	49.19, 49.24	49.78
H	3.31, 3.43	2.98
Cu	18.7 ± 0.422[b]	18.8
SA	82.4 ± 2.71[c]	81.2

[a] based on formula weight $(C_6H_4OH\ CO_2)_4 \cdot Cu_2$

[b] n=6 [c] n=8

In 4 out of 5 animals injected with Cu-SA, visual examination of the concentration-time profiles showed a clear monoexponential decline of SA. In one animal (rat #6), there was some uncertainty because of a sharp decline in concentration values between the 1 and 3 min samples, after which, the decline was log-linear. Statistical evaluation of non-linear regression data (F test and AIC) indicated a monoexponential decline for all animals except for rat #6. For this animal, a biexponential equation was found to give the best fit. In contrast, when the same group of animals were injected with SA, the concentration-time profiles appeared biphasic in all cases except rat #4. In this animal, the loss of the 1 min sample and scatter of concentration values observed in the first 30 min de-emphasized the biphasic decline. Statistical evaluation supported the visual observation by confirming a biexponential decline for all rats except for rat #4, for which a monoexponential decline was indicated. A comparison of the concentration-time profiles obtained in each rat following the two treatments (SA and Cu-SA) showed that SA concentrations were essentially equal during the first 90 min. After that, SA concentrations declined more rapidly in animals that received Cu-SA, as shown in Fig. 1.

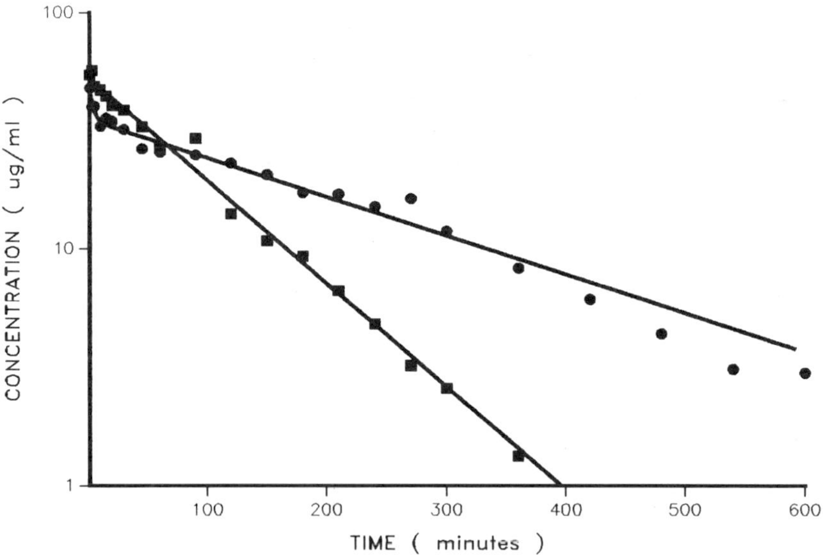

Fig. 1. Plasma Salicylic Acid following Cu-SA (■) and SA (●) in Rat #7.

Details of individual pharmacokinetic values are given in Tables 2 and 3. The terminal rate constant for SA (mean ± SD) was 0.0108 ± 0.00146 min^{-1} following Cu-SA. The corresponding value following SA injection was about 3 times lower (0.00475 ± 0.00144 min^{-1}) and was significantly different ($p < 0.05$). The corresponding model independent parameter, the mean residence time (MRT), also showed a significant difference. Clearance values for SA following Cu-SA administration (2.15 ± 0.621 ml/(min•kg)) were also significantly higher than those obtained after SA administration (1.08 ± 0.153 ml/(min•kg)).

Table 2. Pharmacokinetic Parameters of SA following Injection of Cu-SA

Rat #	ke (min^{-1})	Vd (ml/kg)	Cl (ml/(min•kg))	Vd(SS) (ml/kg)	MRT (min)
1	0.00991	250.0	2.48	216.5	82.65
2	0.00948	150.0	1.42	149.3	103.9
4	0.0118	161.8	1.91	172.6	91.71
6	0.0129[a]	190.6[b]	3.03	227.1	84.03
7	0.0100	188.4	1.89	187.2	99.15
Mean	0.0108	188.2	2.15	190.5	92.29
SD	0.00146	38.67	0.621	31.80	9.279

[a] β value, biexponential decline (see text). [b] Vd(β)

Table 3. Pharmacokinetic Parameters of SA following Injection of SA

Rat #	α (min^{-1})	β (min^{-1})	Vc (ml/kg)	Vd(β) (ml/kg)	Cl (ml/(min•kg))	Vd(SS) (ml/kg)	MRT (min)
1	0.144	0.00295	148.4	289.4	0.852	285.2	330.2
2	0.328	0.00513	122.1	223.0	1.14	216.2	185.5
4	-[a]	0.00668	-	179.7[b]	1.12	189.4	155.6
6	0.364	0.00522	154.1	243.3	1.27	210.5	153.0
7	0.308	0.00377	192.0	282.0	1.06	264.1	234.2
Mean	0.286	0.00475	154.4	243.4	1.08	232.9	211.7
SD	0.0975	0.00144	29.18	44.96	0.153	40.17	73.88

[a] Monoexponential decline (see text). [b] Vd

The volume of distribution for SA following administration of Cu-SA (188.2 ± 38.67 ml/kg, Table 2) was significantly lower than that obtained for the drug following SA (243.4 ± 44.96 ml/kg, Table 3). While the Vd(SS) values showed a similar trend, the differences were not significant.

DISCUSSION

Pharmacokinetics of SA in rats have been reported to be dose dependent (15). Following a 10 mg/Kg I.V. dose of SA, values reported by Bakar and Niazi (16) for the terminal half-life, Vd, and clearance were 142.8 ± 6.6 min, 146.8 ± 5.6 ml/kg, and 0.757 ± 0.04 ml/(min·kg)., respectively. These values are consistent with results obtained in this study (Table 3). In contrast, they have reported a monoexponential decline for SA. This is primarily due to the fact that their first plasma sampling time was 15 min and, consequently, they missed the relatively short distribution phase.

Copper complexes of many drugs have been shown to be pharmacologically more active than their parent drug. For example, copper-aspirinate was found to be more active as an anti-inflammatory agent than aspirin in three different models of inflammation (1). This increase in activity has been attributed to the lipophilic nature of the complex which enables it to penetrate into the tissue much better than the drug. *In vitro* partitioning studies have confirmed that copper-aspirinate complex is 10 times more lipophilic than the parent drug aspirin (17).

The *in vivo* stability of Cu-SA is unknown and extremely difficult to examine. Assuming Cu-SA remains intact, differences in lipid solubilities of SA and Cu-SA may partially explain the differences observed in clearances and plasma concentration-time profiles. The higher lipid solubility of Cu-SA (as compared to SA) which permits more rapid diffusion of the complex from plasma to tissue would result in a higher plasma clearance (lower half-life) for SA following Cu-SA. The lipophilicity of Cu-SA is also likely to make it achieve distribution equilibrium more rapidly than SA, thus minimizing tendencies to exhibit multicompartment behavior. This is consistent with the profiles observed in this study where concentration-time profiles for SA are monophasic following Cu-SA while the corresponding profiles are biphasic after SA. It is also possible that for reasons discussed above the uptake of SA and Cu-SA by red blood cells may be different, thus contributing to the differences observed in plasma clearance and concentration-time profiles. A higher uptake of the more lipid soluble Cu-SA by RBC is consistent with the results obtained in the study. Contrary to expectations based on increased lipophilicity of Cu-SA, the volumes of distribution, Vd(β) and Vd(SS) for SA are lower following administration of the complex than after SA. Since protein binding also influences volumes of distribution, differences in binding may account for the results. Currently, comparative studies are in progress to determine differences in both the lipophilicity and plasma protein binding characteristics of Cu-SA and SA.

It is also possible that Cu-SA may dissociate *in vivo*. In that case, the observed results may also be due to the effect of Cu on SA disposition. Since Cu-SA is 18% copper, a 10 mg/Kg dose of this compound represents a copper dose of 0.54 mg in a 300 gm rat. Assuming normal plasma copper concentration is about 1 mg/L (18) and using a blood volume of 0.021L (7% of body weight (19)), copper concentration can increase 20-30 fold upon injection, particularly if Cu-SA rapidly dissociates into its components, perhaps due to interaction with plasma proteins. Such increases in copper concentrations have been reported earlier (18). Experiments are in progress to understand the effect of copper on SA disposition. The results from experiments currently in progress (lipophilicity, protein binding, and effect of copper on SA disposition) will enable a better understanding of the complicated pharmacokinetics of drug-metal complexes.

ACKNOWLEDGEMENTS

This work was supported in part by funds from the Weldon Spring Endowment Fund, University of Missouri. The authors wish to thank Mrs. Jeanne Caldwell for the excellent assistance provided in the preparation of this manuscript.

REFERENCES

1. Sorenson JRJ. J. Med. Chem., 1976, 19, 137.
2. Lumme P; Elo H. Inorganica Chim. Acta, 1984,92, 241.
3. Leuthauser, SWC; Oberley, LW; Oberley, TD; Sorenson, JRJ; Ramakrishna, K. J. Nat. Cancer Inst., 1981, 66, 1077.
4. Gandy, SE; Buse, MG; Sorenson, JRJ; Crouch, RK. Diabetologia, 1983, 24, 437.
5. Sorenson, JRJ, ed., In: Inflammatory Diseases and Copper, Humana Press, Inc., Clifton, NJ, 1982, pp. 363.
6. Sorenson, JRJ; Ramakrishna, K; Rolnaik, TM. Agents and Actions, 1982, 12, 408.
7. Hemphill, DD, ed., In: Trace Substances in Environmental Health-XIV. University of Missouri, Columbia, MO, 1980.
8. Anon. Analytical Methods for Furnace Atomic Absorption Spectroscopy, Perkin-Elmer, 1981.
9. Mason, WD; Gillilan, R. Anal. Letters, 1983, 16, 903.
10. PCNONLIN by Metzler, CM; Weiner, D. Statistical Consultants, Lexington, KY.
11. Boxenbaum, HG; Riegelman, S; Elashoff, RM. J. Pharmacokin. Biopharm., 1974, 2, 123.
12. Yamaoka, K; Nakagawa, T; Uno, T. J. Pharmacokin. Biopharm., 1978, 6, 165.
13. Gibaldi, M; Perrier, D. In: Pharmacokinetics 2nd Ed., Marcel Dekker, Inc., NY, 1982.
14. Kishita, H; Kubo, M. Acta Crystallographica, 1963, 16, 699.
15. Nelson, E; Hanano, M; Levy, G. J. Pharmacol. Exptl. Therap., 1966, 153, 159.
16. Bakar, SK; Niazi, S. J. Pharm. Sci., 1983, 72, 1030.
17. Williams, DA; Walz, DJ; Foye, WO. J. Pharm. Sci., 1976, 65, 126.
18. Brown, DH; Dunlop, J; Smith, WE; Teape, J. Agents and Actions, 1980, 10, 465.
19. Baker, HJ; Lindsey, JR; Weisbroth, SH, eds., In: The Laboratory Rat - Volume I (Biology and Diseases), Academic Press, New York, NY, 1979, pp. 108.

DISCUSSION

HARRIS: The first slide you showed us was rather shocking. You showed that only 67% of the Cu-salicylate remained intact.

MELETHIL: When I put it in normal saline, I measure 100% as free Cu. When I put it in nitric acid, pH 2 or 3, the complex is destroyed based on ion specific electrode measurements.

HARRIS: Could Prof. Sorenson comment on whether he has conducted similar measurements with his complexes in these vehicles? (Prof. Sorenson had left the lecture hall.)

MELETHIL: I have no idea what happens with other complexes but, as I suspected a long time ago, the pharmacologist or whoever is doing the work is not paying attention to this or its implications. We tried 1% Tween 80. We had a hard time with it. It wouldn't go into solution.

LOVEN: To follow that up, has anyone used a double isotope labeled Cu(II)(salicylate)$_2$ complex?

MELETHIL: We thought about it. I would welcome any opportunity to do that with somebody who can produce the labeled complex.

BHATHENA: I have one difficulty understanding your data. Your data shows that when you give salicylic acid more than 50% of it disappeared in 3 minutes. When you give Cu-salicylate the rate of disappearance was slower. I wonder what this means since it should be removed from blood faster than salicylic acid?

MELETHIL: I don't know how you arrived at that.

BHATHENA: Your data indicated that more than 50% of the free salicylic acid disappeared in 3 minutes.

MELETHIL: The point I was making was that this is disappearance of drug from plasma. It's going into tissues.

BHATHENA: So you think it's better if the drug goes into tissues faster?

METETHIL: In principle yes. Concentrations fell from about 60% or 70% to 30% or 40% with time. In principle the more rapid it gets into tissues the faster the onset. That does not say that salicylic acid is more effective.

FIELDS: If the idea behind all these drugs is to either accelerate the entry of Cu into tissues or to induce some type of Cu-containing enzymes, then did you measure serum Cu before and after the administration of your Cu-salicylate and did you measure the concentration of Cu-salicylate in tissues? The fact that it clears from the plasma doesn't really mean that it has gone into tissue. It could have been cleared through the kidneys and excreted in urine.

MELETHIL: We did not measure Cu concentrations but it is something we are going to do. With regard to your second question, keep in mind that what I am measuring in plasma is salicylic acid. I take plasma and add nitric acid to break up everything. Measuring salicylic acid in all tissues would take a very long time.

FIELDS: Unless you use radioactive materials.

MELETHIL: We could measure radioactivity. I always try to do things with cold materials so I can develop methods suitable for future clinical studies. There's nothing wrong with measuring radioactivity and double labeling. We thought about it but we don't have anybody to help us with it as yet.

CHIDAMBARAM: Is salicylic acid in plasma bound to something or is it free?

MELETHIL: Within the concentration range we used, approximately 80% to 90% is bound to albumin.

CHIDAMBARAM: As Dr. Fields pointed out, I think salicylic acid is a strong Cu chelating agent, like ceruloplasmin, in blood.

MELETHIL: In this first experiment we wanted to see if there were any measurable changes at all.

CHIDAMBARAM: I was not clear as to how you arrived at the

selection of buffer you used to dissolve the complex.

MELETHIL: We tried various ranges of boric buffer from 0 to 0.5 molar. We wanted something that was safe so we tried acetate, nitrate, and borate. We were told to stay away from phosphates because phosphates might bind to and change it. When we did this experiment we found that with 0 to 0.5 borate we got the lowest amount of free ion so that's what we used. The idea was to find a physiologically and pharmacologically non-toxic system in which the complex was most stable, if you will.

SORENSON: I understand Prof. Melethil felt he needed to have a solution. That's a problem. It is not possible to dissolve this complex and still maintain the integrity of the complex. We've always given everything in suspension and we use suspending agents that do not affect complex stability. With this procedure we know we're giving a certain dose of complex. When you give anything intravenously it must be in solution. I can appreciate that. However, it would have been much more meaningful, in the pharmacologic sense as well as in the pharmacokinetics sense, to have given this complex subcutaneously as a suspension and then follow the distribution from the site of injection into plasma and its disappearance from plasma. This will have to be part of the pharmacokinetic description since these complexes are not soluble in common vehicles.

MELETHIL: John, I'm sure you recognize that whether you give it subcutaneously or not, for the drug to be absorbed, it has to first go into solution at the site of administration because it won't go through the membrane unless it is in solution.

SORENSON: Yes, but those solution phenomena must occur at the site of injection with interstitial fluid or the compound must first be suspended in serum or plasma before giving it intravenously.

HARRIS: I will rephrase the question regarding the slide we saw in the first part of your presentation. When this material was dissolved in buffer medium almost 33% of the compound disintegrated and only 67% remained intact. Is it a possibility that these compounds are less stable and do not retain their Cu in the complexed form during the course of administration?

SORENSON: Yes. I think the integrity of the complex was compromised by having put it in borate.

HARRIS: How routinely are these complexes analyzed for their stability prior to their administration?

SORENSON: The suspending agents we use do not compete with the complex ligand for the Cu in the complex. Borate can compete and apparently the original salicylate complex was, in part, destroyed. There is no need to determine stability of the complex in our suspensions.

IAMIN: A HUMAN GROWTH FACTOR WITH MULTIPLE WOUND-HEALING PROPERTIES

Loren Pickart

Bioheal Inc.

15232 S.E. 48th Dr., Bellevue, WA 98006

Wound-healing remains a major medical problem. Yearly in the United States alone there are 29 million in-patient hospital surgical procedures (plus many more handled on an out-patient basis), 2.3 million burns requiring medical attention, approximately 6 million bedsores (decubitus ulcers) at any given time, 1.1 million severe fractures, and 2 million chronic diabetic ulcers (leading to 40,000 leg amputations yearly) [1].

The mechanisms of wound-healing and tissue repair are often inadequate and partial. Healing is impaired after severe burns and frostbite, in bedsores in the elderly, in cancer patients after chemotherapy and radiation treatments (radiation-bed ulcers and osteoradionecrosis), and in persons with diabetes and fistulas. The lesions caused by auto-immune diseases such as Krohn's disease or osteoarthritis also are characterized by deficient healing.

Delayed healing markedly increases patient discomfort and hospitalization costs. Such wounds may persist as a chronic sore that requires continuing medical care to control infection and tissue necrosis. Even when such wounds finally heal, the wound area is often devoid of feeling and sensory perception and filled with scar tissue.

Iamin (from "iama", classical Greek for a healing cure for wounds) or Gly-L-His-L-Lys:

copper(II), a human blood factor, possesses multiple wound-healing properties. We have found iamin to markedly accelerate the healing of superficial wounds in rats, mice, and pigs. The biochemical basis for this action appears to reside in iamin's tissue-protective superoxide dismutase-like activity and its potent and specific in vivo chemoattractant actions on cells critical in healing (macrophages, monocytes, mast cells, capillary endothelial cells). In mice, iamin increases levels of the red blood cell inducing hormone, erythropoietin. In vitro, it induces axonal and dendrite outgrowth from neurons and inhibits platelet aggregation and thromboxane production [2-4].

METHODS OF PROCEDURE AND RESULTS

Preparation of iamin

Optimal healing occurs with a ratio of 0.50 to 0.75 copper atoms per tripeptide molecule and a ratio of 0.50 was routinely used in all experiments described below. Most procedures to crystallize the peptide-copper complex produce a molecule with a slight excess of copper with ratios from 1.03 to 1.10. Copper in molar excess to the peptide is loosely bound and may delay healing since there are reports that free copper salts attract inflammatory cells such as neutrophils. However, the tripeptide without copper has no wound-healing action.

Crystalline Gly-L-His-L-Lys:copper(II) was prepared by published methods then diluted with an equi-molar amount of free tripeptide to obtain a ratio of 0.50 [5]. This mixture is diluted into water containing 0.9% benzoyl alcohol (Abbott) or phosphate buffered saline to obtain the desired concentration, then sterilized by cold filtration.

Healing of superficial wounds

The rate of healing of superficial wounds was accelerated by treatment with iamin. In rats, a circular (2 cm diameter) patch of skin was excised from the back. The outer circumference of the wound was injected with either 5 micrograms of iamin in phosphate buffered saline (controls received saline only) immediately after wounding, a second time at 24 hours, and a third time at 48 hours. The wounds were inspected on days 5, 10, 15, 20 and 25 by debridement of the wound to a sharp wound margin, but with preservation of the proliferating granulation tissue, then photographed. Data was analyzed by measurement of the wound circumference by computerized digitization of the perimeter of the wound in the photograph, then corrected with a scale bar to compensate for variations in magnification among the photographs. Timepoint and group information were coded at the time of digitization to eliminate subjective bias. Histological analysis found the treated wound area to be similar to normally-healing skin. Qualitatively similar results were observed in mice and pigs. Longer injection schedules do not significantly improve healing.

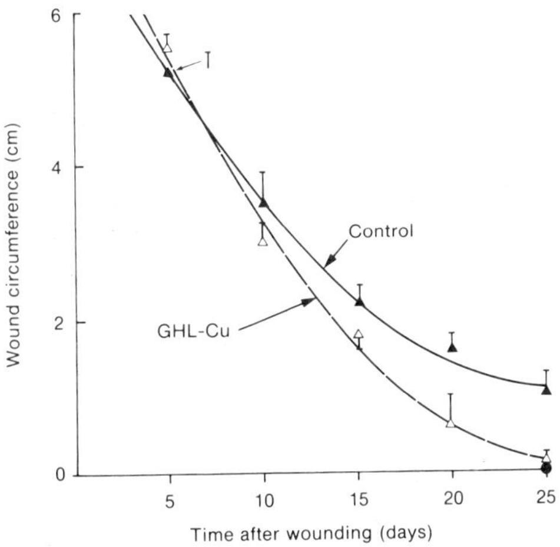

Figure 1. Effect of iamin on healing of 2 cm diameter surgical defects (12 rats each group).

The direct, one-time, application (swabbed on) of iamin to the open wound also produces a healing response but a high concentration (4 mg/ml) must be used. Similar treatment with copper(II) complexes of Gly-L-His-Gly,

Gly-L-His, and L-histidine or with copper sulfate had no significant effect.

Recovery of burned tissue

The optimal amount of iamin for recovery of scald-burned tissue is between 10 and 100 micrograms per square centimeter in combination with 1 to 3% dimethylsulfoxide (DMSO) to facilitate uptake through the skin. Optimal injections are given three times - after burning and 24 and 48 hours later.

Burns were induced in Swiss-Webster mice by applying a 1.5 centimeter diameter brass cylinder at 100° C for 5 seconds to the shaved backs of the anesthetized animals. After the burning, and at 24 and 48 hours later, the burn area was infiltrated with 20 or 200 micrograms of iamin (27 gauge needle, in 0.03 ml volume) in water containing 0.9% benzoyl alcohol as a preservative plus 3% DMSO. Controls received similar volumes of injection water or 3% DMSO.

Initial effects of compounds can be determined by the effect on retarding the post-burn skin hardening - effective compounds keep the burned skin softer and more pliable. Burns treated with iamin show less necrosis (darkening and hardening) at 24 hours post-burn.

In most control animals, the burned skin necroses and hardens, then falls off in 7 to 9 days. This is followed by a healing from the periphery and the wound closes in about 25 days. Infection is not a problem in this procedure.

The effectiveness of iamin is determined by the length of time the burned skin is retained and the percentage of burns in which a major portion of the burned skin recovers and is re-established as healthy skin. Experiments are terminated at 30 days and, at appropriate intervals, skin samples are taken for histology.

By 10 days post-burn, in controls the skin necrosed and in only 1 of 14 animals was there partial recovery of the burned skin. In contrast, in the iamin-treated group the original skin recovered from the burn trauma with normal appearance and hair growth in 13 of

15 animals treated with 200 micrograms of iamin and 9 of 15 treated with 20 micrograms. DMSO alone was of little value. Burned skin retained for 30 days is fully recovered. Lower dosages of iamin (2 to 5 micrograms per square cm) do not stop skin loss but enhance burn-wound closure by about 60%.

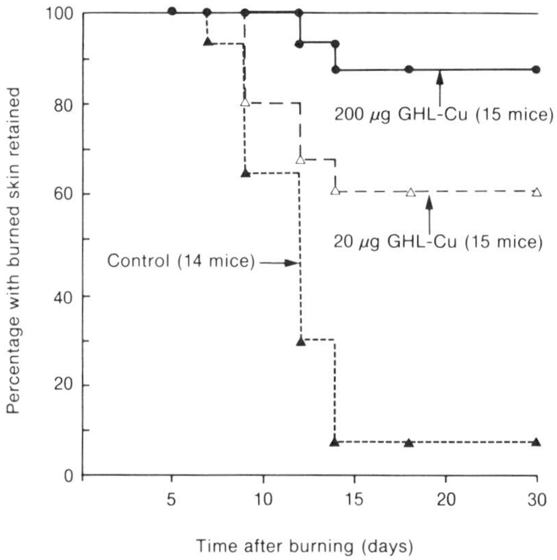

Figure 2. Recovery of burned tissue. Scald-burned skin normally falls off. However, high concentrations of iamin inhibit burn necrosis and often the burned skin fully recovers.

These results indicate that the development of effective therapeutic methods for the use of iamin would be of great value in the treatment of burns. The high percentage of recovery of burned skin was unexpected. Such treatment could minimize the extent of burn trauma in patients.

Potential problems in use of iamin

Treatment of surgical wounds and burns with iamin produces a greater liquid (serum-like) exudate over wounds. This is noticeable within 30 minutes after treatment of a surgical wound. After 1 day, treated wounds have a more moist appearance and the scabby layer that forms over the wound is somewhat thicker. This occurs after localized injection and when high concentrations (4 mg/ml) are topically applied to a fresh surgical defect. This is likely to be due to iamin's mild anti-thrombotic actions. At 1 microgram/ml, the compound inhibits collagen-induced platelet aggregation by about 50% and platelet production of the vasoconstrictive hormone thromboxane by about 80%.

There is disagreement in the literature on wound-healing, and among surgeons I have talked with, as to the effect of the "wetness" of the wound on the rate of healing. More critical may be the type of white cells appearing in the exudate. A high population of macrophages and monocytes might accelerate healing while a high proportion of neutrophils could delay healing. In theory, such a phenomenon could complicate the healing of burn wounds by increasing eschar formation over the wound. Yet, our best results have been observed on the recovery of burned (100 degree C) tissue.

DISCUSSION

Iamin (Gly-L-His-L-Lys:Cu(II)) represents a new approach to a critical unfilled need - that of accelerating healing processes. The molecule has healing-associated properties present in no other known factor.

Iamin was the first discovered representative of a newly emerging family of biological modulators. In addition, bursin (L-Lys-L-His-Gly-amide) has been isolated from the bursa of Fabricius of chickens and induces the differentiation of avian and mammalian B-cells while the tripeptide L-His-Gly-L-Lys which inhibits neuronal firing has been isolated from the cat spinal cord and the blood of coma patients [6,7]. Neither of these second two compounds mimic iamin's wound-healing actions. While these very small peptides can form somewhat similar three-dimensional structures, neither bursin or the coma factor can form the type of tridentate copper binding structure that gives Gly-His-Lys its high copper affinity and superoxide dismutase-like activity.

Iamin acts as a trigger to induce the critical second phase of healing. Proper wound healing and tissue repair requires a sequence of orderly steps. Initially after wounding, the body seals off blood flow into the area and a white cell type, called neutrophils, secrete toxic superoxide anion into the wound to kill bacteria and induce a general inflammation. In the second phase of healing, the production of superoxide anion and the inflammatory response is suppressed, while macrophages invade to clean up tissue debris, secrete angiogenic factors, and open a path for healing-associated cells [8]. Endothelial cells form capillaries while mast cells serve as "helper cells" to amplify capillary growth [9]. Monocytes secrete fibroblast-activating factors [10].

Nerve axons regrow back into the area to restore feeling and motor control. Once the capillary and new blood vessel networks are functioning, the third phase of healing begins and there is a great proliferation of granulation tissue and of the fibroblasts that secrete the structural protein collagen into the wound area. In poorly-healing wounds, the initial inflammatory phase is prolonged while the second phase is usually deficient and blood vessels do not adequately invade the wound.

Iamin acts as (1) an anti-inflammatory agent and (2) as a chemoattractant for cells critical in healing. The tripeptide's copper binding affinity is equivalent to that of the copper binding site on plasma albumin and the addition of free tripeptide at hormonal concentrations to biological systems is likely to result in a significant portion being converted to the copper complex. The complex has high superoxide dismutase activity (constant = 5 x 10 exp 8) [11] - an activity associated with tissue survival and detoxification of oxygen radicals produced in the initial phase of healing and thus reduces the inflammatory response. Analogs of Gly-L-His-L-Lys, such as Gly-L-His and Gly-L-His-L-Gly, form a similar copper(II) complexes [12] that also possess significant superoxide dismutase-like activities and in vivo anti-inflammatory actions [13]. However, these analogs lack iamin's specific chemoattractant effects on white cells [14,15].

The second source of iamin's healing actions resides in its cellular chemoattractant effects. It is essential that adequate numbers of macrophages, monocytes, mast cells, and capillary cells enter the wound area. Iamin is a potent chemoattractant (at hormonal levels) in animals for all four of these cell types and is the only known factor capable of attracting these specific cell types [3,15]. However, it does not attract other non-essential white cells types such as eosinophils, lymphocytes, and neutrophils [15].

Iamin induces new capillary growth in rabbits without causing inflammation. Immediately prior to new capillary ingrowth (76 to 80 hours after application), iamin induces the synthesis of a family of six proteins of molecular weights ranging from 35,000 to 66,000 plus another group of chemoattractant proteins of about 700,000 daltons [16].

In vitro, iamin induces axonal and dendrite outgrowth from cultured chick and rat cerebral, central, and peripheral neurons and may serve to re-establish nerve function in the wounded area [17,18]. When a large amount of tissue is being regenerated, nerves must grow into the wound before the surrounding cells begin to divide and organize to repair the damage. Some have postulated that the nerves may generate electrical fields that promote the sequence of transformations that culminate in tissue replacement.

The anti-thrombotic (inhibition of platelet aggregation and thromboxane production) and erythropoietin stimulating action of the molecule [19] are only seen at pharmacologic dosages and their physiological significance is uncertain. Yet their general effect would be to improve blood flow and tissue oxygenation to the wound area.

The enhancement of the rates of wound-healing observed in immature animals is highly significant since these animals have high endogenous rates of growth and healing. For example, rats and mice grow at about 70-times the human rate and their tissue repair ability is also enhanced. In contrast, tissue repair processes are greatly diminished in adult and elderly humans, patients with diabetes, and after burns, chemotherapy and radiation therapy. Hence, iamin may have a greater potency for healing in humans with these conditions.

Sorenson has emphasized the anti-inflammatory and healing actions of a variety of low molecular weight synthetic copper complexes [20]. Iamin represents the first small biological molecule with similar attributes.

ACKNOWLEDGEMENT

I would like to thank Dr. John Majnarich and the staff of the BioMed Research Laboratories for their help and encouragement in this work. This project was supported in part by United States Public Health Service Grant AG-06270 from the National Institute of Aging.

REFERENCES

1. Sources: National Center for Health Statistics and the Health Care Financing Administration, U.S. Dept. Health and Human Services; the American Cancer Society and the American Diabetes Association.

2. Pickart, L. Lymphokines 1983, 8, 425.

3. Pickart, L.; Downey, D.; Lovejoy, S.; Weinstein, B. In: Superoxide and Superoxide Dismutase in Chemistry, Biology and Medicine; Elsevier, Amsterdam, 1986, p. 555.

4. Pickart, L. In: Peptide Growth Factors, Methods in Enzymology, Academic Press, New York, In press.

5. Perkins, C.M.; Stenkamp, R,; Rose, N.J.; Weinstein, B.; Pickart, L. Inorg. Chim. Acta 1984, 67, 93.

6. Audhya, T.; Kroon, D.; Heavner, G.; Viamontes, G.; Goldstein, G. Science 1986, 231, 997.

7. Lote, C.J.; Gent, J.P.; Wolstencroft, J.H. and Szelke, M. Nature 1976, 264, 188. Also C.J. Lote, personal communication.

8. Hunt, T.K.; Knighton, D.R.; Thakral, K.K.; Goodson, W.H.; Andrews, W.S. Surgery 1984, 96, 48.

9. Folkman, J. Ann. N. Y. Acad. Sci. 1982, 410, 212.

10. Dohlman, J.G.; Payan, D.G.; Goetzl, E.J. Immunology 1984, 52, 577.

11. Beilski, B.H.J.; Rose, N.J.; Pickart,L. Work in progress.

12. Kwa, E.Y.; Lin, B.S.; Rose, N.J.; Weinstein, B,; Pickart, L. Peptides: Structure and Function 1983, 8, 805.

13. Weser, U,; Schubotz, L.M. In: Trace Elements in the Pathogenesis and Treatment of Inflammation, Birkhauser, Basel, 1980, p 103.

14. Poole, T.; Zetter, B.R. Cancer Res. 1983, 43, 5857.

15. Zetter, B.R. Rasmussen, N.; Brown, L. Lab. Invest. 1985, 53, 362.

16. Raju, K,; Alessandri, G.; Gullino, P. Cancer Res. 1984, 44, 1579.

17. Sensenbrenner, M.; Jaros, G.G.; Moonen, G.; Meyer, B.J. Experientia 1980, 36, 660.

18. Lindner, G.; Gross, G.; Halle, W.; Mandel, P. Zeit. Mikro. Anat. Forsch. 1979, 93, 820.

19. Naughton, B.; Naughton, G.; Lui, P. J. Surg. Oncol. 1982, 21, 97.

20. Sorenson, J.R.J. Comprehesive Therapy 1985, 11, 49.

DISCUSSION

BARNEA: Has the structural specificity of GHL been studied? There are other tripeptides like GHL found in the brain that are quite similar in that they have histidine in the second position. There could be some kind of relationship.

PICKART: The specificity is fairly rigid. Cu(II)(glycylhistidylglycine) has SOD-like activity. It binds Cu in the same tridentate fashion as SOD but it doesn't bring in the same white cells. SOD will bring in neutrophils but it won't bring in other cells that are important in healing. If you remove the terminal lysine you totally modify immune functions. One cannot change this peptide much and retain activity.

BARNEA: Is GHL produced in the liver like albumin or is it produced in other tissues?

PICKART: I have no idea where it's produced. The reason I've never gone after that is that those groups that tried to determine the source of Thyrotrophin Releasing Hormone ran into extreme difficulty. At the hormonal level it's really hard to find out where something is made. It may be made like carnosine and anserine. Those compounds are made by a series of enzymes. They are fairly high in plasma. When I went back and read the original papers on that work those authors suggested that an enzymatic system made a series of di- and tri-peptides with glycine, histidine, and lysine so Iamin is probably synthesized by an enzymatic system.

HARRIS: You isolated GHL from Fraction V of the blood. Was it a Cu-GHL complex at that point?

PICKART: Yes. We simply were not aware of it because we were adding Cellex to our solution to get better recovery. We put Cellex in plasma, lowered the pH to 4.5, and boiled the mixture.

HARRIS: It's your understanding now that in blood this compound is a Cu complex of GHL?

PICKART: We isolate it in the highest concentration from Cohn Fraction V. This is an impure albumin. It's not found in crystalline albumin but Cohn Fraction V is where the Cu is being carried on albumin. Maybe it is co-chelating in some manner to the Cu transport site.

HARRIS: What precisely constituted the control in those very dramatic patch studies?

PICKART: Buffer served as control.

HARRIS: Did you evaluate simple Cu salts?

PICKART: Cu salts are not effective. We've also looked at Cu(II)(histidate)$_2$. The critical factor is that these substances do not possess the specific peptide structure needed to interact with the receptor and they will not bring in the correct mix of immune mediating white cells.

HARRIS: What did you mean when you said the Cu(II)-(glycinate)$_2$ didn't work? Does it mean that the patch never adhered?

PICKART: It didn't accelerate wound healing. Some of the Cu that we injected tended to be inflammatory.

JOHNSON: Was your mode of injection intradermal or subcutaneous?

PICKART: Basically, just a little bleb under the skin something like an allergy test.

JOHNSON: Did you find a lot of granulation tissue in your controls compared to none with faster healing?

PICKART: We have had histology done at the University of Washington and those histological reports keep coming back with the comment that the tissue looks very normal. It's simply just accelerated healing. More fat vacuoles seem to develop in the healing, treated, area but we don't know what that means.

JOHNSON: Have you tried a salve or topical application?

PICKART: We are working on that right now. We have had some success after putting it on the wound. We are also trying to put fats into the molecule. We also use it with

DMSO at times. DMSO itself doesn't have much of an effect but it does seem to help general healing, especially with burns.

JOHNSON: We used DMSO for equine wounds quite a bit. We use a lot of Cu(II)SO_4 to remove necrotic flesh and I was wondering whether or not subcutaneous injection might work in equines. There is a big market in equine wound healing.

FIELDS: In Cu deficiency the immune response is depressed. Do you think your drug could help Cu deficient animals?

PICKART: Yes. I've closely followed those experiments. What amazes me is that these animals tend to look very good after they're treated. We don't run into infections. We could be turning on a good deal of the immune system. Of course, they use this with Bursin. The immune functions have not been looked at in depth but it's a hormonally active chemoattractant for these cells so it may be interacting at a very fundamental level.

CHIDAMBARAM: Dog serum albumin doesn't contain histidine at position three. What would be the likely peptide in dogs, in your opinion, that transports Cu?

PICKART: I really don't know and I haven't studied dogs. Dogs handle Cu differently.

REINERS: I was very interested in the neuron studies you cited. Does Iamin actually act as a mytotic agent?

PICKART: There have been a couple different studies. One suggests stimulation of embryonic cerebral neuronal growth. At the same time it seems to suppress glial cell growth. There's been other studies done with different tissues showing dendrite out growth. I don't know if this is really mytosis. We've done some experiments in animals and I really can't state with certainty that there is growth stimulation.

PEROXIDE SCAVENGING BY Cu(II) SULFATE AND Cu(II)(3,5-DIISOPROPYLSALICYLATE)$_2$

Gregory A. Reed and Cherukury Madhu

University of Kansas Medical Center

Kansas City, Kansas 66103

INTRODUCTION

The production of reactive oxygen species (ROS) in biological systems provides key mediators of physiological and pathological processes. Arguably the most important of this class are the partially reduced oxygen species, which include the superoxide anion radical ($O_2^{\bar{\cdot}}$), hydrogen peroxide (H_2O_2), and the hydroxyl radical (•OH). These species have been proposed as mediators of both inflammation[1-3] and in various stages in carcinogenesis.[3-5] One line of evidence supporting a role for ROS in these processes drives from the ability of exogenous scavengers for ROS to inhibit the process.

The exogenous scavengers of ROS which have been employed include both enzymes and small molecular weight compounds. Among the small molecular weight species is the complex Cu(II)(3,5-diisopropylsalicylate)$_2$ [CuDIPS]. This complex, originally prepared and characterized by Sorenson, was found to possess potent anti-inflammatory activity.[6] Later studies demonstrated the ability of CuDIPS to inhibit the initiation[7] and the promotion[8-10] stages of carcinogenesis, and to possess anti-tumor activity as well.[11-12] The closely related complex Cu(II)(salicylate)$_2$ was shown to be an effective scavenger of superoxide[13], and subsequent studies demonstrated the same activity for CuDIPS.[11,14]

Given the roles of ROS as mediators of inflammation and of carcinogenesis, the ability of CuDIPS to catalyze the dismutation of superoxide may explain in part the observed biological activities of the complex. Two considerations, however, raise doubt that this superoxide dismutase (SOD) activity alone explains the biological activities. The first is due to the nature of the SOD reaction:

$$2\ O_2^{\bar{\cdot}} + 2H^+ \longrightarrow O_2 + H_2O_2 \qquad \text{Eq. 1}$$

As this reaction yields hydrogen peroxide as a product, it represents only partial scavenging of ROS. Hydrogen peroxide itself may promote inflammation[1,3] or carcinogenesis[1,3,5]; thus the SOD reaction itself offers incomplete removal of these mediators. A second rationale for suspecting additional activities of CuDIPS with ROS is based on the known reactivities of transition metal ions and their complexes with peroxides. Specifically, some Cu(II) complexes are known to react readily with hydrogen peroxide.[15] We have examined the ability of CuDIPS and $CuSO_4$ to scavenge hydrogen peroxide via either catalatic or peroxidative reaction using standard assay systems for these activities. The results are considered in regard to the noted biological activities of CuDIPS and the role of ROS in these activities.

EXPERIMENTAL

The peroxidase substrate N,N,N',N'-tetramethylphenylenediamine (TMPD) was supplied by Aldrich Chemical Co., Milwaukee, WI. All other substrates, cofactors, and proteins were purchased from Sigma Chemical Co., Saint Louis, MO. CuDIPS was provided by Dr. John Sorenson, Little Rock, AR. All other reagents were from Fisher Scientific.

The activity of bovine erythrocyte SOD and the SOD-mimetic activities of CuDIPS and $CuSO_4$ were measured by their effect on the xanthine-xanthine oxidase-dependent reduction of horse heart cytochrome c. The procedure, based on that of Flohe and Otting[16], employed 45 μM xanthine and 18 μM cytochrome c in 0.1 M potassium phosphate buffer, pH 7.4. Both SOD and $CuSO_4$ were added in buffer, whereas CuDIPS was added as a solution in DMSO. The DMSO content of incubations was kept constant at 1% (v/v), a level which had no effect on the rate of cytochrome c reduction. Reactions were initiated by the addition of xanthine oxidase. Results are expressed as the per cent of the control initial rate of cytochrome c reduction observed in the presence of inhibitors.

Catalase activity was determined by the measurement of oxygen evolution using a Clark-type electrode. Incubations were carried out in a sealed, stirred cell maintained at 37°C. Studies using purified bovine liver catalase established the linearity and the accuracy of this assay using from .01 to .1 units of catalase/ml. All reactions were carried out in 0.1 M potassium phosphate, pH 7.4, and were initiated by the addition of hydrogen peroxide.

The peroxidation of TMPD was monitored by following the absorbance of the TMPD radical cation at 602 nm. The extinction coefficient of this species is 12 mM^{-1} cm^{-1}.[17] Incubations contained 0.1 M potassium phosphate, pH 7.4, and 80 μM TMPD, and were initiated by the addition of hydrogen peroxide.

RESULTS AND DISCUSSION

The goal of this study was to determine the activity of CuDIPS as a scavenger of hydrogen peroxide via both catalatic and peroxidative pathways. Moreover, we wished to compare these activities to the SOD activity of CuDIPS on a quantitative basis in order to consider the relative importance of the various ROS-scavenging activities of the complex. The SOD activity was determined based on the inhibition of the reduction of ferricytochrome c utilizing the xanthine-xanthine oxidase system for generation of superoxide. The amount of xanthine oxidase was adjusted to yield a control rate of reduction of about .02 A.U. min^{-1} in accordance with the standard assay procedure.[16] We determined that the concentration of CuDIPS exerting a 50% reduction in the rate of cytochrome c reduction (IC_{50}) was 23 μM (Table 1). This value is somewhat higher than the 1 to 5 μM values previously reported.[14,19] Previous workers have reported that serum albumin substantially

Table 1. IC_{50} Values for $O_2^{\bar{}}$ dependent Cytochrome c Reduction

Conditions	Bovine SOD	CuDIPS	$CuSO_4$
Control	4.0 U/ml	23 μM	24 μM
BSA, 1 mg/ml	0.9	28	24
BSA, 10 mg/ml	2.9	>> 100[a]	ND[b]

[a] Less than 15% inhibition was observed at 100 μM CuDIPS, the highest concentration used.

[b] ND = not determined.

lowers the SOD activity of CuDIPS.[18,20] We observed no effect of 1 mg/ml bovine serum albumin (BSA), whereas 10 mg/ml BSA virtually abolished the SOD activity. Neither concentration of BSA affected the activity of bovine erythrocyte SOD in this assay. It is somewhat surprising that 1 mg/ml BSA did not effect the activities of CuDIPS and $CuSO_4$. This is equivalent to 15 μM BSA, a concentration comparable to that of the catalysts, and yet no effect was seen.

The dismutation of superoxide yields hydrogen peroxide which is also proinflammatory[1,3] and procarcinogenic.[1,3,5] One means of scavenging hydrogen peroxide is via a catalase reaction:

$$2\ H_2O_2 \longrightarrow O_2 + 2\ H_2O \qquad \text{Eq. 2}$$

Whereas SOD catalyzes a 1-electron dismutation reaction, catalase catalyzes a 2-electron dismutation of hydrogen peroxide. Both CuDIPS and $CuSO_4$ catalyze hydrogen peroxide-dependent oxygen evolution as would be expected for a catalase reaction. The concentrations of catalyst and of hydrogen peroxide required (Fig. 1) are quite high. Significant catalatic activity is only observed at peroxide concentrations over 8 mM. The presence of 1 mg/ml BSA caused a significant increase in the initial rate of oxygen evolution, and at 10 mg/ml resulted in about 50% inhibition of the reaction. The ability to scavenge hydrogen peroxide by a catalase reaction, a complete detoxication of ROS, is an attractive reaction to consider

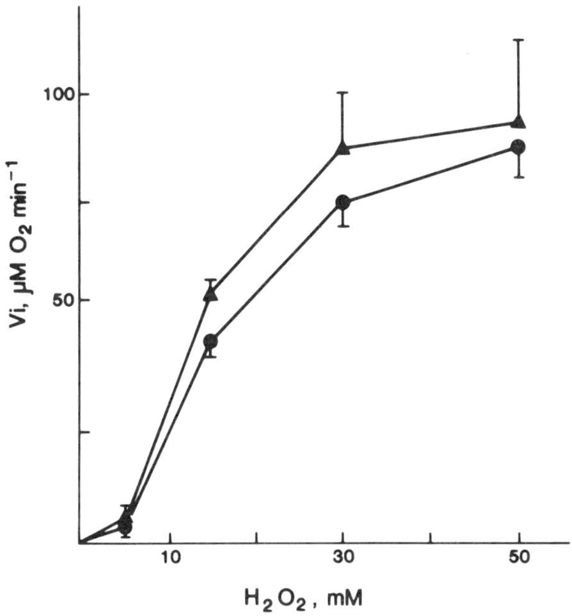

Figure 1. Catalase Activity of CuDIPS and $CuSO_4$ Concentration of CuDIPS (●) and $CuSO_4$ (▲) was 100 μM.

in regard to the biological activities of CuDIPS. The high concentrations of peroxide and catalyst required, however, raise serious doubts as to the occurrence of this reaction in an intact cell or tissue.

An alternative mechanism for the removal of peroxides is utilized by peroxidases:[21]

$$H_2O_2 + 2\ DH \longrightarrow 2\ H_2O + 2\ D^{\bullet} \qquad \text{Eq. 3}$$

Most peroxidases are nonspecific in their choice of donor molecules (DH) which supply the requisite electrons for peroxide reduction. These donations usually occur by sequential one-electron donations from different donor molecules, thus producing two radicals for each peroxide reduction cycle. The donor molecule employed in these studies, TMPD, is a well-characterized peroxidase substrate. Both CuDIPS and $CuSO_4$ catalyze the hydrogen-peroxide-dependent oxidation of TMPD (Table 2). In contrast to the catalase activity (Fig. 1), the TMPD

Table 2. TMPD Peroxidase Activity[a]

Catalyst	BSA	Vi
HRP, 100 ng/ml	0 mg/ml	96.2 μM min^{-1}
	1	97.7
	10	93.5
CuDIPS, 50 μM	0	85.1
	1	74.3
	10	2.8
$CuSO_4$, 50 μM	0	72.3
	1	67.7

[a] All reactions contained 80 μM TMPD and were initiated by the addition of 1 mM H_2O_2.

peroxidase activity is seen at relatively low concentrations of both catalyst and hydrogen peroxide. The effects of BSA resemble the effects on the SOD activity of CuDIPS and $CuSO_4$ rather than the effects noted on the catalase activity. Horseradish peroxidase (HRP), at a concentration which gives a comparable rate of TMPD oxidation, is unaffected by BSA at either 1 or 10 mg/ml. Finally, as has been the case in all assays, the activities of $CuSO_4$ and CuDIPS at a given concentration are nearly identical.

To summarize our results, we find that CuDIPS is not only an effective scavenger of superoxide, but that it also scavenges hydrogen peroxide in the same concentration range. This scavenging of hydrogen peroxide may occur by either a catalase reaction or a peroxidase reaction. The choice of reaction path appears to be determined by the concentration of hydrogen peroxide and by the availability of peroxidase electron donor substrates. In all assays performed, $CuSO_4$ is equipotent with CuDIPS. Finally, BSA at 1 mg/ml does not inhibit the scavenging activities of CuDIPS or $CuSO_4$ but at 10 mg/ml BSA substantial inhibition is seen.

If the biological activities of CuDIPS stem from the ability of the complex to scavenge ROS, these data point to several important considerations. The first is that it is not sufficient to consider only superoxide as a target for CuDIPS. The complex is routinely assumed to be a specific superoxide scavenger when it actually can scavenge a variety of ROS. Secondly, the tremendous increase in biological activity when copper is complexed, as in CuDIPS, does not appear to result from any change in the inherent reactivity of the complex. Rather, this more likely results from changes in the absorption and distribution of the lipophilic complex relative to that of $CuSO_4$. The last consideration involves the effects of protein on the ROS-scavenging activities. The concentrations of BSA employed are much lower than normal serum levels. As such, any application of CuDIPS in which systemic distribution is involved must consider the effects of albumin on the delivery and activity of the complex.

The biological activities of copper complexes are numerous and potentially powerful. These may prove important as both therapeutic and chemopreventive agents, and are currently of great value as tools for probing the fundamental basis of complex biological processes. Without a complete understanding of the mechanism by which these complexes exert their activities, however, the rational use of CuDIPS and related compounds is not possible.

ACKNOWLEDGMENTS

These studies were supported by a grant from the American Cancer Society (#IN-115) as administered by the University of Kansas Cancer Center. The authors also thank Ms. Maxine Floyd for her skillful preparation of this manuscript.

REFERENCES

1. Halliwell, B.; Gutteridge, J. M. C. Biochem. J. 1984, 219, 1.

2. DiGuiseppi, J.; Fridovich, I. CRC Crit. Rev. Toxicol. 1984, 12, 315.

3. Ames, B. N. Science 1983, 221, 1256.

4. Kensler, T. W.; Trush, M. A. Env. Mutagenesis 1984, 6, 593.

5. Troll, W.; Weisner, R. Ann. Rev. Pharmacol. Toxicol. 1985, 25, 509.

6. Sorenson, J. R. J. J. Med. Chem. 1976, 19, 135.

7. Solanki, V.; Yotti, L.; Logani, M. K.; Slaga, T. J. Carcinogenesis 1984, 5, 129.

8. Kensler, T. W.; Trush, M. A. Biochem. Pharmacol. 1983, 32, 3485.

9. Kensler, T. W.; Bush, D. M.; Kozumbo, W. J. Science 1983, 221, 75.

10. Nakamura, Y.; Colburn, N. H.; Gindhart, T. D. Carcinogenesis 1985, 6, 229.

11. Leuthauser, S. W. C.; Oberley, L. W.; Oberley, T. D.; Sorenson, J. R. J.; Ramakrishna, K. J. Natl. Cancer Inst. 1981, 66, 1077.

12. Oberley, L. W.; Rogers, K. L.; Schutt, L.; Oberley, T. D.; Leuthauser, S. W. C.; Sorenson, J. R. J. J. Natl. Cancer Inst. 1983, 71, 1089.

13. de Alvare, L. R.; Goda, K.; Kimura, T. Biochem. Biophys. Res. Commun. 1976, 69, 687.

14. Younes, M.; Lengfelder, E.; Richter, C.; Shubotz, L. M.; Weser, U. In: Chemical and Biochemical Aspects of Superoxide and Superoxide Dismutase (J. V. Bannister and H. A. O. Hill, eds); Elsevier; New York, 1980; p. 336.

15. Sigel, H.; Angew. Chem. Internat. Ed. 1969, 8, 167.

16. Flohe, L.; Otting, F. In: Methods in Enzymology, Vol. 105 (L. Packer, ed.); Academic Press; Orlando, 1984; p. 93.

17. Van der Ouderaa, F. J.; Buytenhek, M. In: Methods in Enzymology, Vol. 86 (W. E. M. Lands and W. C. Smith, eds.); Academic Press, New York, 1982; p. 60.

18. Betts, W. H.; Cleland, L. G., Whithehouse, M. W. In: Inflammatory Diseases and Copper (J. R. J. Sorenson, ed.); Humana Press; Clifton; 1982; p. 553.

19. Sorenson, J. R. J., personal communication.

20. Oberly, L. W.; Leuthauser, S. W. C.; Pasternack, R. F.; Oberley, T. D.; Schutt, L.; Sorenson, J. R. J. Agents and Actions 1984, 15, 535.

21. Dunford, H. B. In: Advances in Inorganic Biochemistry, Vol. 4 (G. L. Eichorn and L. G. Marzilli, eds.); Elsevier Biomedical; New York, 1982; p. 41.

DISCUSSION

SORENSON: What were the molar relationships between amounts of Cu(II)SO_4 or Cu(II)(3,5-dips)$_2$ and amounts of albumin put into those preparations?

REED: One mg/ml BSA is approximately 15 uM so 10 mg/ml is 150 uM. I would have expected that at low concentrations of Cu(II)SO_4 or Cu(II)(3,5-dips)$_2$, say 10 to 30 uM Cu, that the 15 uM BSA would have had some depressing effect on activity. That did not happen with the SOD-like activity. That was also not observed with TMPD oxidation by peroxidase. The high concentration of BSA, 150 uM, was inhibitory at all Cu concentrations for SOD-like activity and for peroxidase activity. The one observation that doesn't fit is the catalase activity which was stimulated somewhat by the low concentration of albumin and then inhibited, but only partially, by the high concentration of BSA. So the catalase activity responds to protein in a much different fashion then the peroxidase activity or the SOD activity.

SORENSON: What you have done is a real first with regard to showing that Cu(II)(3,5-dips)$_2$ has peroxidase-like activity even though that kind of activity is known for other Cu complexes and for inorganic Cu. It is important to be able to show disproportionation of hydrogen peroxide as has been shown for O_2^-. Cu complexes are also known to react with hydrogen atoms, high energy electrons, hydroxyl radical, and hydroperoxyl radical at rates of the order of $10^9 M^{-1}s^{-1}$. In addition, Cu complexes are probably capable of reducing excited singlet state oxygen to triplet state oxygen. Cu complexes then, would seem to be able to get rid of all oxygen radicals, hydrogen peroxide, and high energy electrons if they are allowed to react with them.

PICKART: You said that albumin negates SOD-like activity of small Cu complexes. These small Cu complexes have very similar biological actions in animals when compared to enzymatic SOD in conditions where you have trauma or where you have hypoxic tissues in reperfusion. If theories regarding SOD action are correct with regard to breaking down O_2^- then these small complexes are doing this in biological systems. The point I am making is that the albumin assay where you add albumin into the system may not be a valid measure of what's going to happen in the biological system. The albumin assay makes enzymatic SOD look like the only thing around, but in the biological system obviously the small Cu complexes are much more effective than Cu-Zn SOD.

REED: Obviously the only point that can be made from my data is that in these in vitro test systems, very high albumin concentrations do totally abolish the enzyme-like activity. Even the high concentration of albumin, 10 mg/ml, is substantially lower than albumin concentrations in plasma and so if we are talking about in vivo administration, this is a problem. If this complex with albumin is a very tight complex and doesn't dissociate then we can't consider SOD activity. I agree there is no correlation between activities, the pharmacological end points seen with Cu(II)(3,5-dips)$_2$ administration and with SOD administration. That would seem to imply that if albumin complexes are formed with these complexes they are not irreversibly formed.

FIELDS: Are the catalytic activities that you have seen specific for Cu or can other metals do the same thing?

REED: Other transition metal ions or complexes can serve as an electron shuttle to catalyze these dismutation reactions. There's a tremendous amount of work showing that inorganic Fe and, in particular, Fe complexes - Fe EDTA, Fe ADP, and Fe ascorbate can catalyze dismutation reactions. Catalase is Fe-dependent. Cu is not the only transition metal that can function in this way.

RISK: An immediate interpretation that I put on your data is that the Cu effect is probably not occurring extracellularly because of all the albumin and other proteins that could complex with or trap Cu. That, in conjunction with other work we've heard about such as the increased lipophilicity of some of these more effective molecules, would suggest that your work is substantiating the notion that Cu complexes may be acting in lipid matrices of the plasma membrane or endoplasmatic reticular regions.

REED: I would agree that the primary actions of things like Cu(II)(3,5-dips)$_2$ are intracellular. It has been established in the literature that the efficacy of Cu(II)(3,5-dips)$_2$ as compared to some other Cu complexes and Cu(II)SO_4 is in part based on its lipophilicity and its ability to cross membranes. That's definitely part of the activity in vivo.

HERNANDEZ: In our cat model of ischemia we can measure a decrease in Cu-Zn SOD and Cu-Zn SOD does protect against ischemia. Also, even though catalase is in the ischemic tissue, we find protection with catalase. Perhaps the endogenous catalase is inactivated by a pH change associated with ischemia so that exogenous catalase is beneficial. It also protects against reperfusion injury.

REED: In going through the literature and looking at these more complex systems it seems as though requirements for protection are highly dependent on which model system is being examined. In some cases SOD is seen as the key scanvenger that needs to be supplemented, in others catalase is the key scavenger, and in still other cases both SOD and catalase have to be supplied in order to see protection. It seems to be very system dependent.

SORENSON: It seems reasonable that if a Cu complex is either a binuclear or square planar complex and in that form either O_2^- or hydrogen peroxide can approach and bond to the free bonding sites they will be disproportionated and the complex will evidence enzyme-like activity. If albumin bonds to those free sites so that they're occupied by this peptide in such a way as to not allow O_2^- or hydrogen peroxide to approach those free sites then disproportionation will not occur and enzyme-like activity is absent. However, this tertiary complex might serve as a much more stable transport form of the particular complex in vivo. The tertiary complex may also be a form that is actively transported across cell membranes if receptors recognize albumin.

REED: The only caution on that is that effects of BSA on catalase activity do not seem to fit. With 1 mg/ml BSA we see stimulation of catalase activity by $Cu(II)(3,5\text{-dips})_2$. At 10 mg/ml BSA we do see inhibition but not nearly as profound as we see with peroxidase or SOD activities of $Cu(II)(3,5\text{-dips})_2$.

SORENSON: This might be interpreted as activation for catalase-like activity. The actual structures of these tertiary albumin complexes might be important in understanding stimulation for one type of activity versus inhibition of another.

Analgesic and Antimicrobial Activities of Copper Complexes

COPPER COMPLEXES HAVE POTENT ANALGESIC ACTIVITY AND THEY MAY ACTIVATE OPIOID RECEPTORS.

Shigeru Okuyama, Sanae Hashimoto, Hironaka Aihara, William M. Willingham*, and John R. J. Sorenson**;

Research Center, Taisho Pharmaceutical Co., Ltd., 1-403, Yoshio-cho, Ohmiya, Saitama 330, Japan; *Department of Natural Sciences, University of Arkansas at Pine Bluff, Pine Bluff, Arkansas 71601, U.S.A. and **Department of Biopharmaceutical Sciences, College of Pharmacy, University of Arkansas for Medical Sciences, Little Rock, Arkansas, 72205, U.S.A.

Introduction

Copper complexes of non-steroidal antiinflammatory agents are more effective antiinflammatory agents than their parent drugs[1,2]. This led to the suggestion that copper complexes were active metabolites of these drugs[1], which was supported by observations that copper complexes are effective in treating arthritic and other degenerative diseases of man[3].

Since it is well known that non-steroidal antiinflammatory agents have analgesic activity[4], the question remained as to whether or not copper complexes of these agents have analgesic activity. To answer this question three mononuclear copper complexes: Cu(II)(anthranilate)$_2$, Cu(II)(salicylate)$_2$, and Cu(II)(3,5-diisopropylsalicylate)$_2$ [Cu(II)(3,5-dips)$_2$], as well as three binuclear copper complexes: Cu(II)$_2$(aspirinate)$_4$, Cu(II)$_2$(niflumate)$_4$, and Cu(II)$_2$(indomethacin)$_4$ were studied in two pain models. Results obtained with the mononuclear complexes have been published elsewhere[5]. All of the studied complexes were more effective than their parent ligands and loosly bonded forms of copper, Cu(II)(chloride)$_2$ and Cu(II)$_2$(acetate)$_4$, in the Acetic Acid-Induced Writhing Mouse and Adjuvant-Induced Arthritic Rat pain models following oral and subcutaneous administration. Cu(II)$_2$ (indomethacin)$_4$ was found to be as effective as morphine in both pain models. These data and pertinent literature support the hypothesis that this analgesic activity is due to activation of copper-dependent opioid receptors.

Materials and Methods

Details concerning the source of chemicals used in this work, the use of the acetic acid writhing pain[6] and the adjuvant arthritic pain models[7,8,9], as well as treatment regimens have also been published elsewhere[5].

All drugs were suspended with 5% gum arabic in saline solution or wetted with enough propylene glycol to give a final concentration of 4% propylene glycol and the wetted solid suspended in 1.4% polyvinyl alcohol in saline.

The ED_{50} value and 95% confidence limits obtained for each compound studied in the writhing pain model were calculated from percent inhibition by the method of Litchfield and Wilcoxon[10]. Student's t-test was used to determine significance of differences between complex-treated and vehicle-treated groups. The ED_{50} value and 95% confidence limits for data obtained with the adjuvant arthritis pain model were calculated from the positive rate for non-treated controls, vehicle-treated adjuvant arthritic rats, for each compound studied by the method of Litchfield and Wilcoxin[10].

Results

All three binuclear complexes; Cu(II)$_2$(aspirinate)$_4$, Cu(II)$_2$(niflumate)$_4$, and Cu(II)$_2$(indomethacin)$_4$ were more effective than their parent ligands in the acetic acid-induced writhing pain model following oral administration in both 5% gum arabic and PG-PVA, and they were most effective following administration in PG-PVA (Table I). All three of these complexes were also more effective than their parent ligands following s.c. administration, with ED_{50} values ranging from 0.02 to 0.003 mmol/kg. Data obtained for the analgesic activity of Cu(II)(chloride)$_2$ and Cu(II)$_2$(acetate)$_4$ (Table I) show that these compounds were not as effective as the drug complexes with ED_{50} values of only 0.60 and 0.32 mmol/kg respectively.

Data in Table II compare analgesic activities of these binuclear complexes with their parent ligand in the adjuvant arthritis pain model. Cu(II)$_2$(aspirinate)$_4$ and Cu(II)$_2$-(niflumate)$_4$ were more effective than their parent ligand in this pain model following oral administration in 5% gum arabic while all three binuclear complexes were more effective than their parent ligand following oral adminstration in the PG-PVA vehicle. In addition, all three complexes were either as effective or more effective following s.c. administration than they were following oral administration. With ED_{50} values ranging from 0.06 to 0.002 mmol/kg in this pain model, these complexes appear to be as effective or slightly more effective in this pain model than in the writhing pain model. Both Cu(II)(chloride)$_2$ and Cu(II)$_2$(acetate)$_4$ were devoid of analgesic activity in this pain model at doses which produced statistically significant analgesic effects in the writhing pain model.

Another matter of interest is the comparison of the analgesic activity of these copper complexes with the activity of morphine. ED_{50} values for morphine in the writhing and adjuvant arthritis pain models were found to be 0.002 mmol/kg (Table III), the same value obtained for Cu(II)$_2$(indomethacin)$_4$ following oral and s.c. administration in both the writhing pain model (Table I) and the adjuvant arthritis pain model (Table II).

Discussion

It was generally found that analgesic activities of copper complexes were greater following oral treatment using the PG-PVA suspending system than they were with the gum arabic suspending system. A copper complex firmly bonded to gum arabic, or competition by gum arabic for bonding sites on copper and ligand exchange, may account for reduced activity associated with gum arabic suspension Neither propylene glycol nor polyvinyl alcohol have functional groups capable of causing this interference.

Table I. Analgesic effects of parent ligands and copper complexes on acetic acid induced writhing in mice

Compound	Oral (mmol/kg)[a]	Oral (mmol/kg)[b]	Subcutaneous (mmol/kg)[b]	% Inhibition	ED50 (mmol/kg) (95% C.L.)
Aspirin					
	0.56			28.8	1.30
	1.11			41.9*	(0.57 - 2.95)
	2.22			66.0**	
		0.21		32.1	0.48
		0.42		41.2*	(0.22 - 1.04)
		0.84		66.7***	
			0.28	28.0	0.52
			0.56	56.6**	(0.28 - 0.96)
			1.11	71.3***	
$Cu(II)_2(aspirinate)_4$					
	0.06			37.1	>0.24
	0.12			38.1*	
	0.24			33.0	
		0.09		31.6*	0.14
		0.18		58.8**	(0.09 - 0.25)
		0.36		79.7***	
			0.03	36.4	0.04
			0.06	59.0*	(0.03 - 0.08)
			0.24	81.5***	
Niflumic acid					
	0.18			52.3**	0.15
	0.35			64.5***	(0.05 - 0.49)
	0.71			74.6***	
		0.13		37.9*	0.21
		0.27		51.1**	(0.11 - 0.40)
		0.53		79.5***	
			0.04	21.0	0.10
			0.09	53.8**	(0.05 - 0.19)
			0.18	63.6**	
$Cu(II)_2(niflumate)_4$					
	0.04			37.4**	0.09
	0.08			47.7***	(0.03 - 0.31)
	0.16			59.5***	
		0.06		33.3*	0.10
		0.12		56.5***	(0.05 - 0.20)
		0.24		72.9***	
			0.01	37.8*	0.02
			0.02	48.1**	(0.01 - 0.08)
			0.04	58.4***	

TABLE I. (Continued)

Indomethacin					
	0.01			36.2*	0.04
	0.03			41.7**	(0.01 - 0.13)
	0.06			58.3**	
		0.007		26.8	0.01
		0.014		60.1**	(0.01 - 0.03)
		0.028		68.0***	
			0.007	18.8	0.02
			0.014	36.2	(0.01 - 0.04)
			0.056	60.1**	
$Cu(II)_2(indomethacin)_4$					
	0.006			33.3	0.012
	0.013			50.0**	(0.005- 0.028)
	0.026			73.3***	
		0.001		37.7	0.002
		0.002		52.6**	(0.001 - 0.003)
		0.004		75.4***	
			0.002	34.1	0.003
			0.003	53.2*	(0.001 - 0.006)
			0.006	69.9**	
$Cu(II)(chloride)_2$					
			0.56	47.6**	0.60
			1.12	68.6***	(0.30 - 1.2)
			2.24	82.9***	
$Cu(II)_2(acetate)_4$					
			0.21	41.5*	0.32
			0.41	58.5***	(0.08-1.30)
			0.83	61.0***	

[a]administered in 5% gum arabic in saline, [b]administered in 5% propylene glycol 1.4% polyvinyl alcohol in saline for s.c. administration.
*$p<0.05$, **$p<0.01$, ***$p<0.001$ versus vehicle treated group.

Dissociation of these copper complexes in the stomach as a result of gastric acidity can not be used to account for analgesic activities of these complexes since their ligands and more freely dissociable forms of copper, $Cu(II)(chloride)_2$ and $Cu(II)_2(acetate)_4$, have less activity than observed for these complexes. In addition, copper complexes have antisecretory activity and decrease gastric acidity[1,2]. Finally, irritant-induced antiinflammatory and concomitant stress-induced analgesic activity can not be used to explain these results since the most likely irritant, $Cu(II)(chloride)_2$, was the least effective compound.

TABLE II. Analgesic effects of parent ligands and their copper complexes on adjuvant-induced arthritic pain in rats.

Compound	Oral (mmol/kg)[a]	Oral (mmol/kg)[b]	Subcutaneous (mmol/kg)[b]	% Inhibition	ED50 (mmol/kg) (95% C.L.)
Aspirin					
	0.28			26.7	0.62
	0.56			43.3	(0.32 - 1.36)
	1.11			70.0	
		0.56		30.0	1.41
		1.11		40.0	(0.50 - 3.96)
		2.22		63.3	
$Cu(II)_2(aspirinate)_4$					
	0.06			6.7	0.29
	0.12			16.7	(0.13 - 0.65)
	0.24			43.3	
		0.03		13.3	0.09
		0.06		30.0	(0.04 - 0.21)
		0.12		60.0	
			0.03	23.3	0.06
			0.06	50.0	(0.03 - 0.12)
			0.12	73.3	
Niflumic acid					
	0.18			13.3	0.42
	0.35			26.7	(0.26 - 0.68)
	0.71			86.7	
		0.18		23.3	0.48
		0.35		43.3	(0.19 - 1.20)
		0.71		60.0	
$Cu(II)_2(niflumate)_4$					
	0.04			6.7	0.18
	0.08			13.3	(0.09 - 0.38)
	0.16			46.7	
		0.16		36.7	>0.16
			0.01	16.7	
			0.02	50.0	0.03
			0.04	60.0	(0.01 - 0.06)
Indomethacin					
	0.001			26.7	0.003
	0.003			43.3	(0.001 - 0.006)
	0.006			76.6	
		0.007		23.3	0.02
		0.014		43.3	(0.01 - 0.03)
		0.028		70.0	

Table II. (Continued)

$Cu(II)_2(indomethacin)_4$					
	0.003			26.7	0.007
	0.006			46.7	(0.004 - 0.15)
	0.013			70.0	
		0.001		20.0	0.002
		0.002		40.0	(0.001 - 0.003)
		0.003		86.7	
			0.001	16.7	0.002
			0.002	43.3	(0.001 - 0.005)
			0.003	60.0	
$Cu(II)(chloride)_2$		2.24		0.0	>2.24
			2.24	3.3	>2.24
$Cu(II)_2(acetate)_4$		0.83		0.0	>0.83
			0.83	3.3	>0.83

[a]administered in 5% gum arabic in saline, [b]administered in 5% propylene glycol and 1.4% polyvinyl alcohol in saline for s.c. administration.

Table III. Analgesic activity of morphine in the acetic acid induced writhing (W) and adjuvant arthritis (AA) pain models.

Dose (mmol/kg)[a]	% Inhibition	ED50(mmol/kg) (95% of C.L.) W	ED50(mmol/kg) (95% of C.L.) AA
0.001	35.2	0.002	
0.002	44.1*	(0.001-0.003)	
0.004	88.8***		
0.001	20.0		
0.002	40.0		0.002
0.004	60.0		(0.001-0.005)

[a]Administered subcutaneously as the hydrochloride salt in saline.

These data are consistent with the notion that nonsteroidal antiinflammatory agents form copper complexes in vivo[1] and support the possibility that copper complex formation in vivo accounts for the analgesic activity of these antiinflammatory agents as well. The observation that copper complexes are absorbed and mediate analgesia following tissue distribution suggests physiologic and biochemical roles for copper in analgesia. Existing literature also supports this conclusion: 1) DDC and PTT, Dopamine β Hydroxylase (DβH) inhibitors which form copper complexes in vivo[11], induce analgesia and potentiate antagonist reversible morphine analgesia[12,13]. That is, putative Cu-DDC and Cu-PPT complexes formed in vivo potentiate antagonist reversible morphine analgesia. 2) Both morphine and these DβH inhibitors, which potentiate morphine analgesia, increase CNS copper content[13-15]. 3) Levallorphan, morphine plus levallorphan, and morphine-tolerance cause decreases in CNS copper and decrease pain threshold[14,15]. 4) Severe pain is produced with penicillamine treatment and this effect of penicillamine is reversed with copper supplementation[15]. Penicillamine is also known to decrease β-endorphin analgesia. 5) Cu(II)(Chloride)$_2$ and Cu(II)SO_4 produce analgesia and potentiate antagonist reversible morphine analgesia[17,19-21]. Cu(II)(Chloride)$_2$ or Cu(II)SO_4 can yield complexes in vivo[22] and these complexes could mediate analgesia.

In 1954, Beckett and Casey[23] proposed an "anionic" site on the opiate-receptor to account for bonding of opiate agonists and antagonists which were recognized as having a cationic component at biological pH values as shown in the following equilibrium.

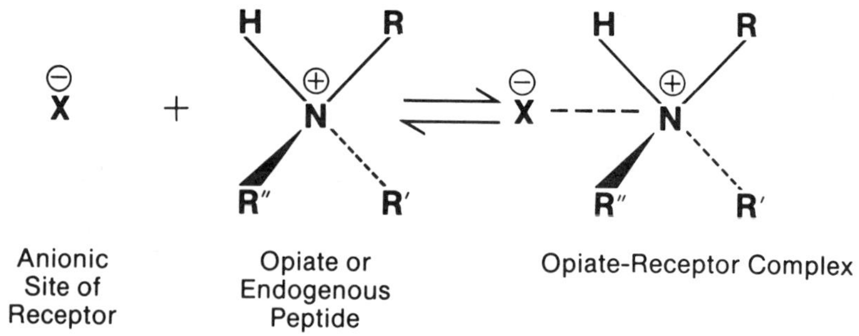

Results presented in this publication and existing literature support the hypothesis that the originally proposed "anionic" site may be better represented as cationic due to the presence of an atom of copper bonded to the proposed "essential" thiol group[24-26] of the opiate-receptor with an as yet unknown oxidation state, Cu(I) or Cu(II), and with as yet unknown additional coordinate-covalent bonding to the receptor. This hypothesis utilizes the "essential" thiol group on the "inactive" opiate-receptor and proposes activation of this inactive form, by bonding with copper. Subsequent bonding of the opiate to the active form of the receptor leads to receptor stimulation as shown in the following scheme.

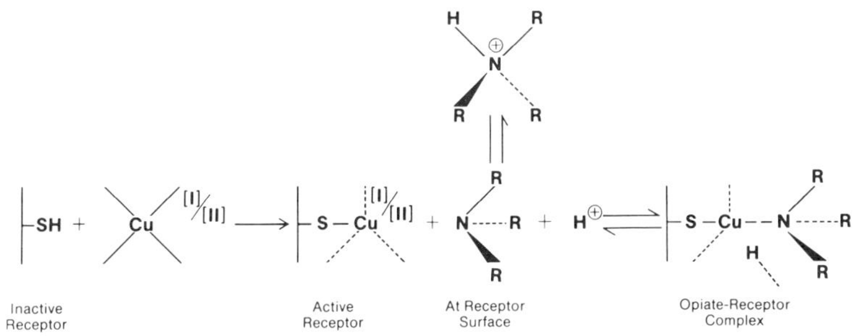

Activation of inactive receptors with copper is consistent with analgesia produced by copper complexes of DDC or PTT formed in vivo following treatment with DDC or PTT, analgesia produced by copper complexes formed in vivo following treatment with Cu(II)(chloride)$_2$ or Cu(II)SO$_4$ increased analgesia for Cu(II)(methionine-enkephalin) and Cu(II)(leucine-enkephalin)complexes when compared to the parent opioid peptides[27], and analgesia produced by treatment with copper complexes reported in this publication.

References

1. Sorenson JRJ. Med. Chem. 1976, 19, 135.
2. Sorenson JRJ.: In: Metal Ions in Biological Systems, Volume 14, (Ed H. Sigel). Marcel Dekker, New York, N.Y. 1982.
3. Sorenson JRJ; Hangarter W. Inflammation, 1977, 2, 217.

4. Jaffe JH; Martin WR. In: Goodman and Gillman's The Pharmacological Basis of Therapeutics, 7th Edition. (Eds A.G. Gillman, L.S. Goodman, and A. Gillman) Mac Millan Publishers Co. Inc. New York, N.Y. 1985, p 491.
5. Okuyama S; Hashimoto S; Aihara H; Willingham WM; Sorenson JRJ. Agents and Actions, In Press.
6. Koster R; Anderson M; De Beer EJ. Fed. Proc. 1959, 18, 412.
7. Kuzuna S; Kawai K. Chem. Pharm. Bull. 1975, 23, 1184.
8. Winter CA; Kling PJ; Tocco DJ; Tanabe K. J. Pharmacol. Exp. Ther. 1979, 211, 678.
9. Rosenthale ME; Capetola RJ. Fed. Proced. 1982, 41, 2577.
10. Litchfield JT; Wilcoxon F. J. Pharmacol. Exp. Ther. 1949, 96, 99.
11. Johansson B; Stankiewicz Z. Biochem. Pharmacol. 1985, 34, 2989.
12. Watanabe K; Matsui Y; Iwata H. Experientia 1969, 25, 950.
13. Bhargava HN; Way EL. J. Pharmacol. Exptl. Therap. 1974, 190, 165.
14. Iwata H; Watanabe K; Miichi H; Matsui Y. Pharmacol. Res. Comm. 1970, 2, 213.
15. Iwata H, Watanabe K; Matsui Y. Pharmacol. Res. Comm. 1971, 3, 147.
16. Keiser HR; Henkin RI; Kare M. Proc. Soc. Exp. Biol. 1968, 129, 516.
17. Iwata H; Watanabe K. Japan. J. Pharmacol. 1971, 21, 731.
18. Iwata H; Watanabe Y; Matsui Y. Europe. J. Pharmacol. 1970, 11, 298.
19. Stern P. Wein. Klin. Wochenschr. 1968, 80, 181.
20. Bhargava HN. Psychopharmcol. 1978, 57, 223.
21. Marzullo G; Hine B. Science 1980, 208, 1171.
22. Chutkow JG. Proc. Soc. Exp. Biol. Med. 1978, 158, 113.
23. Beckett AH; Casey AF. J. Pharm. Pharmacol. 1954, 6, 986.
24. Simon EJ; Hiller JM; Edelman I. Proc. Nat. Acad. Sci. 1973, 70, 1947.
25. Simon EJ; Groth J. Proc. Nat. Acad. Sci. 1975, 72, 2404.
26. Pastrnack GW; Wilson HA; Snyder SH. Mol. Pharmacol. 1975, 11, 340.
27. Sharrock P; Day R; Lemaire S; St.-Piere S; Mazarguil H; Gairin JE; Haran R. Inorg. Chim. Acta 1982, 66, L91.

DISCUSSION

ELLIOTT (Graham): I understand morphine acts upon mu and gamma receptors, mu receptors are responsible for analgesic effects of morphine and the other are responsible for excitatory effects of morphine. If you block mu receptors you allow expression of the excitory effects of morphine. Do you have any evidence that your Cu complexes will activate the gamma receptors if you block with naloxone? Do you have any evidence that your Cu compounds induce total addiction in these animals?

SORENSON: No, we've not done these studies and have not, as yet, collaborated with anyone to look at effects of those specific receptors. We have not done anything with regard to addiction either. I've just recently gotten some data from Dr. Okuyama showing that the analgesic effects of $Cu(II)_2(indomethacin)_4$ are reduced by naloxone.

ELLIOTT (Graham): Were any behavioral studies done with these animals? Do you see the full spectrum of morphine effects or just the analgesic effect.

SORENSON: Analgesia is all that has been measured.

BHATHENA: The more specific opiate, other than morphine, for mu receptors will be endorphin. Endorphin also activates epsilon receptors. Have you made any Cu complexes of β-endorphin?

SORENSON: No. We'd be willing to make some for you if you have some β-endorphin.

BALL: Ascorbate, at very high levels, is quite effective in treating both addiction and pain of people who are on opiates. Do you know whether the Cu is monovalent or divalent? Ascorbate might be reducing Cu(II) to Cu(I).

SORENSON: The ascorbate involvment is interesting. I don't know what the oxidation state is in the activated receptor.

BALL: You don't know whether it's in the bivalent or monovalent form?

SORENSON: That's correct, and there is no way I could get a clue as to which is the oxidation state. One is tempted to say that if there is an interaction with a thiol group then it might be Cu(I) but thiols that can not be oxidized can form Cu(II) complexes and the Cu(II) oxidation state could be maintained by other interacting ligands. It's an open question but it has a solution if we can get some receptor protein and study it. Membranes containing these receptors can be isolated and bonding studies as well as electron spin resonance and other spectrophotometric studies can certainly be done to determine whether it's Cu(I) or Cu(II).

BARNEA: Marzullo and Hine (Science 1980, 208, 1171.) suggested that Cu(II) glutathione disulfide inhibited opiate receptor bonding and the Cu(II) fully accounted for this inhibition. That's exactly opposite what you are telling us.

SORENSON: Yes, they found inhibition of receptor bonding using uM concentrations of $Cu(II)Cl_2$ with in vitro preparations of brain membranes. They also found that nanomolar concentrations of Cu(II) chloride injected intracerebroventricularly had potent opioid analgesic activity which was reversed by naloxone and dithiothreitol. Their interpretation of their results was that analgesia was being mediated by some process other than Cu-dependent opioid receptor activation. This notion that Cu activates the opioid receptor is different from what they were suggesting and it is consistent with all of their results. The apparent inconsistency in their results may be due to the very large difference in concentration of Cu(II) needed to show bonding inhibition in vitro and analgesia in vivo.

BARNEA: It's also possible that when you inject Cu(II) chloride into the brain you get release of opioids that account for analgesia. Analgesia may be due to released opioids and not due to Cu. When they did their studies in vitro with membranes, there was direct interaction with the added Cu. Related to this direct interaction with the receptor, opioid receptors are very strongly effected by all kinds of cations like Na and K. Could it be that some of these effects, if it is really indeed a Cu effect, which has to be shown, are similar to effects found with those cations even though it's a completely different species of metal? Could it just mimic other cations that produce some kind of confirmational change(s) which would not be specific for Cu but, perhaps, more easily observed with Cu?

SORENSON: Could be. That means I don't know and nobody else does either. I am collaborating with Prof. Garvil Pasternak to answer these questions. He participated in the work showing that the opioid receptor contained an "essential thiol" group while he was at John Hopkins, along with Candice Pert and Soloman Snyder who showed that a high concentration of NaCl caused allosteric conformational changes which affect opioid bonding. The suggested requirement of Cu for opioid receptor activation is something I offer as a rationale, and it's open for challenge.

PEPTIDES AND OPIATES IN COPPER DEFICIENCY

Sam J. Bhathena and Lillian Recant*
Carbohydrate Nutrition Laboratory
BHNRC, ARS, USDA
Beltsville, Maryland 20705

* Diabetes Research Laboratory
V.A. Medical Center
Washington, D.C. 20422

A link between copper metabolism and endogenous opiates or their receptors and peptides has been suggested by the work from several laboratories (1-4). In humans, Kolb and Sadee proposed a relationship between copper metabolism, endorphins and schizophrenia (5). Early work in animals have shown that central nervous system disorders such as ataxia, motor incoordination and brain and spinal cord degeneration occur in experimental copper deficiency (6-9). Though not specifically mentioned, it is possible that some alteration in opiates (besides alteration in neurotransmitters) could have occurred. A critical role of copper in the brain development has been suggested by Zimmerman et al. (10). O'Dell et al. have reported decreased concentration of dopamine and norepinephrine in brain in experimental copper deficiency (11-13) and that repletion with copper restores the level of norepinephrine (13). It is known that dopamine-β -monooxygenase (14,15) and tyrosinase (16,17), enzymes involved in catecholamine synthesis are copper bound and the third enzyme tyrosine-3-monooxygenase, an iron bound enzyme is controlled by copper ions in that it is decreased to the same extent as catecholamine levels in the brain of copper deficient rats (13). Similarly, the enzyme involved in opiate processing namely peptidyl-glycine-α-amidating monooxygenase is a copper bound enzyme (18) and hence alterations in opiates in copper deficiency can be expected. Copper is also involved in glucagon metabolism (19). In genetic copper-deficiency such as Menke's disease, defect in neural functions have been reported (20). Conversely, peptide as well as steroid hormones also affect copper metabolism (21). Recently, Formicka-Kozlowska et al. (22) reported formation of copper complexes of met-enkephalin. Sharrock et al. (23) reported potentiation of biological activity of met- enkephalin when complexed with copper, and that copper complexes of enkephalin do not affect binding of either leu- or met-enkephalin with the receptor. However, copper decreases the binding of gonadotropin releasing hormone to pituitary membrane preparations (24).

Several studies have demonstrated colocalization and in some cases cosecretion of opiates and peptides, or peptides and catecholamines in the same cells in several tissues. Thus, enkephalins and norepinephrine coexist in chromaffin granules of adrenal medulla (25), enkephalins and insulin in B cells of the pancreas (26,27), β-endorphin and glucagon (28), vasopressin with β-endorphin and dynorphin in luteal cells of the ovary (29), arginine-vasopressin and dynorphin in hypothalamic neurosecretory magnocellular neurons (30), vasopressin and dynorphin in neurohypophysis (31), neuropeptide Y(NPY) and norepinephrine in perivascular and cardiac sympathetic nerves (32), NPY and epinephrine in adrenal medulla (33) and enkephalin with arginine-vasopressin and oxytocin in nerve terminals of rat neurohypophysis (34). Cosecretion of opiate peptides and catecholamines has also been reported in humans (31).

Recently, Meister et al. (35) reported coexistence of growth hormone releasing factor with tyrosine hydroxylase (tyrosine monooxygenase) in specialized cells of hypothalamus and Crouch et al. (36) reported the presence of superoxide dismutase in B cells of pancreas with insulin. Copper has been shown to stimulate the release of luteinizing hormone from pituitary (24) and luteining hormone releasing hormone from pituitary (37) as well as hypothalamus (38) in vitro. In vivo, Allen et al. (39), reported decreased basal thyrotropin releasing hormone stimulated thyroxine release in copper deficient rats. No effect was observed, however, on the level of thyroid-stimulating hormone. It is therefore possible that copper deficiency can alter content as well as secretion of peptides, opiates and neurotransmitters in several tissues simultaneously. Significant hypo- or hypertrophy of several endocrine tissues such as pancreas, adrenal, heart and brain have been observed in copper deficient animals (40).

Fields et al. (40,41) in rats and Reiser et al. (42) in humans have demonstrated that copper deficiency symptoms are aggravated by dietary fructose compared to starch. Thus, opiate and peptide changes observed in copper deficiency could also be magnified by dietary fructose. The present report summarizes our recent findings on opiates and peptides in copper deficiency in humans and rats fed either starch or fructose.

Human Study

Normal male volunteers were fed diets containing 1 mg/day or 0.35 mg/1000 kcal. of copper for 7 weeks with 20% of calories either as fructose or cornstarch. The subjects were then crossed over to diets containing the other carbohydrate source for 4 weeks. All subjects were then fed starch diet supplemented with 3 mg/day of copper for 3 weeks. The average level of copper in pretest meals was 0.57 mg/1000 kcal. The design of the study and other details have been described elsewhere (42,43). Met- and leu-enkephalin like-immunoreactivity and β-endorphin immunoreactivity were measured from plasma (43). Plasma β-endorphin levels tended to increase during the feeding of low copper diets, but the rise was not significant. The level fell towards normal on copper repletion. Leu- as well as met-enkephalin levels decreased significantly during the feeding of low copper diets and returned to normal on copper supplementation. The changes were impressive as 19 out of 20 subjects showed decrease in leu-enkephalin and all subjects showed a decrease in met-enkephalin. It thus appears that plasma enkephalin levels may be better markers of copper status (nutriture) in humans when copper deficiency is only marginal.

Studies in Rats

Changes in catecholamines in brain in copper deficient rats along with decreased brain copper levels have been described by several laboratories (11-13). As stated earlier, since catecholamine, opiates and/or peptides coexist and possibly cosecrete, and since enkephalin levels in plasma are decreased in humans fed low copper diet, we looked for opiate and peptide changes in endocrine tissues of the rats namely pancreas, pituitary, hypothalamus, heart and adrenals.

Rats were fed copper adequate (6.0 μg Cu/g diet) or copper deficient (0.6 μg Cu/g diet) diets along with 62% carbohydrate as either starch or fructose for 7 weeks. The presence of copper deficiency was ascertained by low weight gain, atrophy of the pancreas, decreased levels of copper, ceruloplasmin and superoxide dismutase and significant increases in triglyceride and cholesterol levels in plasma. The details of the rat study are described elsewhere (27,45).

In plasma, significant decrease was observed in insulin level in rats fed copper deficient diet. Even greater decrease in plasma insulin was observed in rats fed copper deficient diet with fructose as compared to starch. Plasma glucagon levels were not significantly altered by either the copper levels in the diet or the source of carbohydrates. The response of plasma somatostatin was variable and no clear conclusion could be drawn.

Pancreas. Significant increase in insulin content was seen in rats fed copper deficient diets whether expressed as total pancreatic content or per wet weight of the pancreas or per protein content, regardless of the source of the dietary carbohydrate. When pancreatic insulin extracts were subjected to gel filtration, no increase in large molecular weight species, such as proinsulin was observed, indicating that conversion of proinsulin to insulin was not affected. Pancreatic glucagon content was higher in rats fed copper deficient diet only when expressed per wet weight of the pancreas or per unit of protein content. The effect was significant in fructose fed rats. Similar results were observed for pancreatic somatostatin, namely increased level when expressed per weight of the pancreas or per protein content. Fructose had greater effect on somatostatin content in rats fed copper deficient diet.

Increased pancreatic insulin content and decreased plasma insulin level indicates possible defect in insulin secretion. Pancreata were therefore perfused in separate groups of animals with low and high glucose (46). Preliminary data show significantly reduced response in copper deficient rats. This observation differs from that of Weaver et al. (47), who observed no change in insulin secretion in response to glucose in copper deficient rats treated with penicillamine.

Total pancreatic β-endorphin levels tended to decrease in rats fed copper deficient diets. However, when expressed per protein content, there was significant increase in rats fed fructose along with low copper. Leu- as well as met-enkephalins were significantly higher in copper deficient rats when expressed on protein basis. Fructose feeding compounded the effect. Effect of copper deficiency was more pronounced for leu-enkephalin as compared to met-enkephalin.

Since enkephalins are derived from proenkephalin A and prodynorphin (proenkephalin B), they may exist in large molecular weight forms. It is possible that in copper deficiency enkephalin precursors are not processed to pentapeptides. We therefore, measured total leu- and met-enkephalins. Aliquots of pancreas extract were digested successively first with trypsin and then with carboxypeptidase B. This converts precursor peptides to enkephalins. Similar differences in total (precursor) enkephalins persisted, in that, in rats fed copper deficient diets, total enkephalins were higher. Further, the ratio of free to total leu-enkephalin was significantly lower compared to that of met-enkephalins indicating that there may be a selective defect in the precursor conversion to enkephalin for leu-enkephalin. Though prodynorphin is a major source of leu-enkephalin, it is not clear whether it contributes to leu-enkephalin. Leu-enkephalin also comes from proenkephalin A but there is more met-enkephalin compared to leu-enkephalin derived from proenkephalin A. Since there is a selective defect in leu-enkephalin formation, at least in pancreas, prodynorphin pathway may be more selectively affected in copper deficiency.

As stated earlier, enkephalins are colocalized with insulin and therefore possibly could be cosecreted. Since both insulin and enkephalins in pancreas are altered in copper deficiency, we studied the relationship between insulin and free as well as total leu- and met-enkephalins. Positive correlationship was observed between insulin and total leu- as well as met-enkephalins while a negative correlationship was observed between insulin and free enkephalin levels. These changes indicate possible impairment in the conversion of precursor to small enkephalins. Since alterations in enkephalins in pancreas precedes changes in insulin, it is possible that changes in enkephalins are primary which then affects insulin secretion. Several studies have shown that in vivo (48) as well as in vitro (49,50) administration of enkephalins alter insulin secretion. It is also possible that changes in enkephalins and insulin brought about by copper deficiency are independent events. More data are required to resolve this. However, since enkephalins and insulin are colocalized in the pancreas, they are therefore more likely to be cosecreted as well.

In order to understand decreased secretion of insulin from the pancreas, we studied pancreatic morphology at the electron microscopic level. Significant changes include: 1) contraction of the nuclei with folding of the nuclear membrane, 2) increase in cytoplasmic filopodia and pinocytotic vesicles, 3) loss of electron density and disorganization in the capillary basement membranes and 4) the presence of redundant membrane material surrounding lucent spaces at the juncture of the cell membrane with the basal lamina (51). The latter may have been derived from defective recycling of the membranes. These changes would lead to impaired endocytosis and emiocytosis in the pancreatic B cell resulting in decreased insulin (and possibly enkephalins) secretion causing increased pancreatic and decreased plasma levels (51). Weaver et al. (47), however, did not observe any alteration of endocrine pancreas in copper deficient penicillamine treated rats.

Pituitary and Hypothalamus. No significant changes in either β-endorphin or the enkephalins, free as well as precursor, were observed in the anterior or the posterior pituitaries of rats fed copper deficient diets. The only significant change was a decrease in leu-enkephalin precursor in the the posterior pituitary of rats fed copper deficient diet with fructose. Posterior pituitary of copper deficient rats fed fructose also showed significant increase in somatostatin. No other changes in pituitary somatostatin were observed. Somatostatin was also measured from hypothalamus and no effect of copper deficiency or the dietary carbohydrate was observed.

Adrenals. No change in free or precursor met-enkephalin were observed in adrenals.. Free leu-enkephalin was very low in adrenals to measure accurately. No copper effect was observed on precursor leu-enkephalin. However, adrenals from fructose fed rats showed a significant decrease in precursor leu-enkephalin content compared with starch fed rats.

Heart. Changes in cardiac abnormalities including death due to the rupture of the heart, in severe experimental copper deficiency have been reported by Klevay (52) and Fields (53). Heart, especially atria, have been now recognized as an endocrine tissue. They contain and secrete natriuretic peptides. Significant increases in plasma atrial natriuretic peptide (ANP) have been reported in patients with congestive heart failure (54,55) and hypertension (55). We therefore measured ANP in plasma of copper deficient rats. Preliminary study showed significant increase in ANP in copper deficient rats fed fructose (56). No change was observed in copper deficient rats fed starch, showing a protective effect of starch. It is important to note that only copper deficient rats fed fructose and not starch die of cardiac rupture. Further, catecholamines (57,58), enkephalins (57-60), and very recently β-endorphin (61) have been reported to be involved in cardiac regulation. It is therefore important to study the precise role and relative importance of β-endorphin, enkephalins, catecholamines and atrial peptides in cardiac function in severe copper deficiency.

It is thus clear that opiates and peptide changes in copper deficiency are not generalized to all endocrine tissues, but occur specifically in pancreas and possibly in heart. Secondly, copper deficiency syndrome also involves changes in opiates and peptides in plasma besides decreased copper levels, ceruloplasmin, superoxide dismutase and a host of other parameters previously described. In fact, decreased plasma enkephalins when confirmed in other studies could be a more sensitive marker of copper deficiency, at least in humans.

REFERENCES

1. Marzullo, G; Hine, B. Science 1980, 208, 1173.
2. Chapman, D.B; Way, E.L. Ann. Rev. Pharmacol. Toxicol. 1980, 20, 553.
3. Sadee, W; Pfeiffer, A; Herz, A; J. Neurochem. 1982, 39, 659.
4. Marangos, P.J; Patel, J., Clark-Rosenberg, R; Martino, A.M. J. Neurochem. 1982, 39, 184.
5. Kolb, V.M; Sadee, W. Res. Commun. Psychol. Psychiatry Behav. 1983, 8, 343.
6. Bennetts, M.W. Aust. Vet. J. 1932, 8, 137 and 183.
7. Everson, G.J; Tsai, H.C; Wang, T.I. J. Nutr. 1967, 93, 533.
8. DiPaolo, R.W; Kanfer, J.N; Newberne, P.M. J. Neuropath. Exp. Neurol. 1974, 33, 226.
9. Carlton, W.W; Kelly, W.A. J. Nutr. 1969, 97, 42.
10. Zimmerman, A.W; Matthiew, J; Quarles, R.H; Roscoe, B.O; Jing, M.H. Arch. Neurol. 1976, 33, 111.
11. O'Dell, B.L. Med. Clin. North Am. 1976, 60, 687.
12. Morgan, R.F; O'Dell, B.L. J. Neurochem. 1977, 28, 207.
13. Feller, D.J; O'Dell, B.L. J. Neurochem. 1980, 34, 1259.
14. Friedman, S; Kaufman, S. J. Biol. Chem. 1965, 240, 4763.
15. Ljones, T; Skotland, T. In: Copper Proteins and Copper Enzymes, Vol. 2; Lontie, R., Ed.; CRC Press, Inc; Boca Raton, FL; 1984; p. 131.
16. Vaneste, W.H; Zuberbuhler, A. In: Molecular Mechanisms of Oxygen Activation; Hayashi, O., Ed.; Academic Press; New York; 1974, p. 371.
17. Robb, D.A. In: Copper Proteins and Copper Enzymes, Vol. 2; Lontie, R., Ed.; CRC Press, Inc; Boca Raton, FL; 1984; p. 207.
18. Mains, R.E; Eipper, B.A. Endocrinol. 1984, 115, 1683.
19. Alexander Ross, J.B; Rousslang, K.W; DeHaen, C; Lavis, V.R; Deranleau, D.A. Biochim. Biophys. Acta 1979, 576, 372.
20. Danks, D.M. In: The Metabolic Basis of Inherited Disease, 5th Ed.; Stanbury, J.B; Wyngaarden, J.B., Fredrickson, D.S., Goldstein, J.L., Brown, M.S., Eds.; McGraw-Hill; New York; 1983, p. 1251.
21. Ettinger, M.J. In: Copper Proteins and Copper Enzymes, Vol. 2; Lontie, R., Ed.; CRC Press, Inc; Boca Raton, FL; 1984; p. 175.

22. Formicka-Kozlowska, G; Pettit, L.D., Steel, I., Livera, C.E., Kupryszewski, G., Rolka, K. J. Inorg. Biochem. 1985, 24, 299.
23. Sharrock, P; Day, R; Lemaire S; St-Pierre, S; Mazarguil, H; Gairin, J.E; Haran, R. Inorg. Chim. Acta 1982, 66, L91.
24. Hazum, E. Biochem. Biophys. Res. Commun. 1983, 112, 306.
25. Viveros, H; Diliberto, E.J., Jr; Hazum, E; Chang, K.J. In: Neural Peptides and Neuronal Communications; Costa, E., Trabucchi, M., Eds.; Raven Press, New York, 1980, p. 191.
26. Timmers, K; Voyles, N.R; King, C; Wells, M; Fairtile, R; Recant, L. In: Neural and Endocrine Peptides and Receptors; Moody, T., Ed., Plenum Press, 1986 (In Press).
27. Recant, L; Voyles, N.R; Timmers, K; Zdenski, C; Fields, M; Bhathena, S.J. Peptides (In Press).
28. Grube, D; Voight, K.H; Weber, E. Histochemistry 1978, 59, 75.
29. Lolait, S.J; Autelitano, D.J; Marwick, A.J; Toh, B.H; Funder, J.W. Peptides 1986, 7, 267.
30. Watson, S.J; Akil, H; Fischli, W; Goldstein, A; Zimmerman, E; Nilaver, G; Van Wimersma Greidamus, T.B. Science 1982, 216, 85.
31. Anhut, H; Knepel, W. Neurosci. Lett. 1982, 31, 159.
32. Lundberg, J.M; Martinsson, A; Hemsen, A; Theodorsson-Norheim, E; Svedenhag, J; Ekblom, B; Hjemdahl, P. Biochem. Biophys. Res. Commun. 1985, 133, 30.
33. Lundberg, J.M; Hokfelt, T; Hemsen, A; Theodorsson-Norheim, E; Pernow, J; Hamberger, B; Goldstein, M. Regulatory Peptides 1986, 13, 169.
34. Martin, R; Voigt, K.H. Nature 1981, 289, 502.
35. Meister, B; Hokfelt, T; Vale, W.W; Sawchenko, P.E; Swanson, L; Goldstein, M. Neuroendocrinol. 1986, 42, 237.
36. Crouch, R.K; Gandy, S; Patrick, J; Reynolds, S; Buse, M.G; Simson, J.A.V. Exp. Mol. Pathol. 1984, 41, 377.
37. Tsou, R.C; Dailey, R.A; McLanahan, C.S; Parent, A.D; Tindall, G.T; Neill, J.D. Endocrinol. 1977, 101, 534.
38. Burrows, G.H; Barnea, A. Endocrinol. 1982, 110, 1456.
39. Allen, D.K; Hassell, C.A; Lei, K.Y. J. Nutr. 1982, 112, 2043.
40. Fields, M; Ferretti, R.J; Smith, J.C; Reiser, S. Am. J. Clin. Nutr. 1984, 39, 289.

41. Fields, M; Feretti, R.J; Reiser, S; Smith, J.C; Jr. Proc. Soc. Exp. Biol. Med. 1984, 175, 530.
42. Reiser, S; Smith, J.C; Jr; Mertz, W; Holbrook, J.T; Scholfield, D.J; Powell, A.S; Canfield, W.K; Canary, J.J. Am. J. Clin. Nutr. 1985, 42, 242.
43. Bhathena, S.J; Recant, L; Voyles, N.R; Timmers, K.I; Reiser, S; Smith, J.C; Jr; Powell, A.S. Am. J. Clin. Nutr. 1986, 43, 42.
44. Bhathena, S.J; Recant, L; Voyles, N.R; Reiser, S; Smith, J.C; Jr. "unpublished results"
45. Bhathena, S.J; Voyles, N.R; Timmers, K; Fields, M; Kennedy, B.W; Recant, L. Fed. Proc. 1986, 45, 235.
46. Bhathena, S.J; Perrino, P.V; Voyles, N.R; Smith, S.S; Wilkins, S.D; Coy, D.H; Schally, A.V; Recant, L. Diabetes 1976, 25, 1031.
47. Weaver, Fr. C; Sorenson, R.L; Kaung, H.L.C. Diabetes 1986, 35, 13.
48. Ipp, E; Dhorajiwala, J; Pugh, W; Moosa, A.R; Rubenstein, A.H. Endocrinol. 1982, 111, 2110.
49. Green, I.C; Perrin, D; Pedley, K.C; Leslie, R.D.G; Pyke, D.A. Diabetologia 1980, 19, 158.
50. Green, I.C; Ray, K; Perrin, D. Horm. Metabol. Res. 1983, 5, 124.
51. Hummel, G.M; Bhathena, S.J; Voyles, N.R; Fields, M; Timmers, K; Recant, L. "unpublished results"
52. Klevay, L.M. In: Metabolism of Trace Metals in Men: Developmental Biology and Genetic Implications, Vol. 1; Rennert, O.M; Chan, W-Y; Eds.; CRC Press, Boca Raton, Fl.; 1984; p. 128.
53. Fields, M. Int. Med. 1985, 6, 91.
54. Shenker, Y; Sider, R.S; Ostafin, E.A; Grekin, R.J. J. Clin. Invest. 1985, 76, 1684.
55. Tang, J; Song, D.L; Suen, M.Z; Xie, C.W; Chang, D; Chang, J.K. Peptides 1986, 7, 33.
56. Bhathena, S.J; Zamir, N; Kennedy, B.W; Fields, M. Recant, L. "unpublished results"
57. Eiden, L.E; Ruth, J.A. Peptides 1982, 3, 475.
58. Ruth, J.A; Eiden, L.E. Neuopeptides 1984, 4, 101.
59. Holaday, J.W. Ann. Rev. Pharmacol. Toxicol. 1983, 23, 541.
60. Greenberg, M.J; Painter, S.O; Price, D.A. Neuropeptides 1981, 1, 309.
61. Haddad, G.G; Jeng, H.J; Lai, T.L. Am. J. Physiol. 1986, 250, H796.

DISCUSSION

ELLIOTT (Graham): Did you notice any behavioral changes in these people who showed the decrease in enkephalin levels?

BHATHENA: We stopped the study because some subjects had cardiac problems. Two subjects had tachycardia and one subject had a myocardial infarction.

ELLIOTT (Graham): The fructose fed Cu-deficient rats were highly stressed. Any change you saw could be just a result of a change in hormone balance due to stress. Did you have a specific effect on enkephylin production or breakdown?

BHATHENA: Dr. Fields sequentially studied rats fed fructose from 1 to 7 weeks and she found decreases in ceruloplasmin and Cu-Zn SOD. These opiate peptides were not measured earlier so we don't know the specificity of these effects.

HARRIS: The insulin increase in association with degeneration of the pancreatic endocrine membrane, found with Cu deficiency, was explained as a lack of insulin release. Is part of the problem associated with enkephalins? You also found a very high elevation of somatostatin which, as I understand it, is going to be a factor that will affect the release of insulin. Is it possible that there is a direct effect by that delta cell hormone to cause a feedback block of insulin release?

BHATHENA: Yes, absolutely possible.

HARRIS: Changes found in Cu deficiency are similar to changes found in vitamin E deficiency. Are these changes specific for Cu deficiency?

FIELDS: The diet was supplemented with all other minerals and vitamins, including vitamin E.

KEEN: Were you suggesting that there might have been oxygen damage?

HARRIS: Yes.

KEEN: We've studied pancreatic function with Mn deficiency and found less insulin, which is opposite to the accumula tion found with Cu deficiency and there was free radical damage.

CROOK: Were changes in atrial natriuretic peptide (ANP) associated with feeding fructose negative diet and no Cu?

BHATHENA: Yes.

CROOK: How long was the experiment?

BHATHENA: Five weeks on the Cu-deficient diet.

CROOK: Were these male rats?

BHATHENA: Yes, only male rats were studied.

CROOK: ANP has been identified in lungs and some other non-atrial tissues.

BHATHENA: Yes, it is present in pituitary also. The reason why we measured ANP is that it is increased in hypertension and myocardial infarction. It was an obvious extension for us to measure ANP in these particular rats because they die of cardiac rupture with significant hypolipemia.

PICKART: Do people who ingest alot of fructose have different enkephalin or endorphin levels or pain thresholds?

BHATHENA: I can only speculate. These data may be significant in the sense that the American Diabetic Association suggests that diabetics should not consume too much fructose. Levels we use are not out of the ordinary in the sense that the normal American diet does contain 20% fructose and that's what we fed these humans. With rats we went up to 62% to evaluate pathological changes.

RISK: Fructose could be encouraged in the diet rather than sucrose if we made sure we took Cu supplements. Perhaps because of the AMA recommendation, I'm a great believer in vitamin and trace mineral supplementation as a routine procedure because we can never be sure that we are getting the proper amounts in our diet.

The electron microscopic photograph that you showed for plasma membrane changes in the pancreas has tremendous significance. I was thinking specifically of the mechanism proposed quite a few years ago by Lacey for insulin secretion. Seems to me that if there are degenerative changes they affect the rate of fusion of these granules and alter exocytosis. Have you considered that as a fundamental mechanism involved?

BHATHENA: Oh sure. That's the reason we presented the electron micrograph showing clearly that these membranes are completely distorted and that may be why we see these changes. Others could not find this pancreatic change in their electron micrographs. The only difference between their study and our study is that we produced Cu deficiency by feeding low Cu diet and fructose, while they used a Cu-deficient diet plus penicillamine (300 mg/kg).

MYLROIE: How old were these rats?

BHATHENA: They used adult rats. We used weanling rats. We produced Cu deficiency by dietary means in young growing animals. Their animals were adults, 300 grams. They may have had adequate stores of Cu in their tissues. I have no idea as to how penicillamine chelates Cu and causes its removal.

SORENSON: Do you have a citation for that paper?

BHATHENA: Weaver, Fr.C., Sorenson, R.L., Kaung, H.-L.C., Diabetes 1986, 35, 13. They used 300 gram rats and they used penicillamine to exacerbate the Cu deficiency. They found atrophy of the acinar pancreas but no significant change in the endocrine pancreas. Prof. Mylroie pointed out another paper (Smith, P.A., Sunter, J.P., Case, R.M. Digestion 1982, 23, 16.) clearly showing the same thing, that the acinar pancreas is atrophic but not the endocrine pancreas.

MYLROIE: The sulfhydryl group in penicillamine might be protective.

SORENSON: In answer to the question about how penicillamine chelates Cu to facilitate its removal, penicillamine can complex with Cu(I) to form a light yellow 1:1 Cu(I) complex with the thiol and carboxyl or amino group. This complex is polymeric. If penicillamine is oxidized to the disulfide this disulfide can form a blue 2:2 complex with Cu(II). The Cu complex of penicillamine that is excreted in the urine is a purple mixed valance complex containing 6 atoms of Cu(II), 8 atoms of Cu(I), 12 molecules of penicillamine and one chloride. This complex is a penta sodium salt and very water soluble.

SORENSON: Those were enkephalins that decreased so markedly and then rose so quickly when Cu was supplemented. Are enkephlins a normal plasma component?

BHATHENA: Yes they are. We can measure them in plasma.

SORENSON: Is Cu ever associated with enkephalins in plasma?

BHATHENA: No. They do form Cu complexes. Another group, which you have cited in your presentation, has reported that Cu complexes of enkephalins are more analgesic and their receptor binding is not at all altered. These complexes do decrease Lipotropin Releasing Hormone binding to pituitary cells in vitro.

SORENSON: I was wondering whether or not the enkephlins might contain Cu and function as a CNS transport form as an acute phase responder to pain to produce analgesia.

FIELDS: Did your radioimmunoassay for insulin detect pro-insulin?

BHATHENA: Yes. Actually we did measure pro-insulin. We found no change. The conversion of pro-insulin to insulin was not at all affected, as was the conversion of precursors to enkephalins.

MECHANISM OF THE ANTIMYCOPLASMAL ACTIVITY OF 2,2'-BIPYRIDYL COPPER COMPLEXES

H. van der Goot, H. Smit, H.-D. Gaisser and H.Timmerman
Department of Pharmacochemistry,
Vrije Universiteit,
De Boelelaan 1083,
1081 HV Amsterdam, The Netherlands

INTRODUCTION

Some years ago we discovered the remarkable antimycoplasmal activity of the easily accesible 1-amino-3-(2-pyridyl)isoquinolines[1,2]. In these studies *Mycoplasma gallisepticum*, the causative agent of a respiratory disease in poultry, was used as the testorganism. It appeared that for antimycoplasmal activity in the isoquinoline series the presence of a 2,2'-bipyridyl system is a prerequisite together with ortho substitution towards the nitrogen atoms, a possibility of a cis coplanar conformation of the bipyridyl system and a certain degree of lipophilicity[3,4]. Furthermore the presence of a trace of copper is necessary; without the addition of copper no activity could be established, while the copper itself at the concentrations used showed no inhibitory effect in the MIC-tests[5,6]. From these results it was concluded that the combination of copper and the 2,2'-bipyridyl compound, i.e. a copper complex, causes the growth-inhibitory effect. A quantitative structure-activity relation study[7] revealed an important contribution of the electronic character of the ortho substituents as well as their steric properties. An optimal lipophilicity log $P_{oct./water}$ of 3.8 was found, although a small variation had only a slight effect on the antimycoplasmal activity. Nevertheless a certain extent of lipophilicity is necessary for antimycoplasmal activity. Remarkably this quantitative structure-activity relationship has been derived from physico-chemical properties of the ligands themselves. For an understanding of the mechanism of action the fact that copper is needed for activity has to be included. Therefore we have studied the mode of action of the copper/2,2'-bipyridyl system and investigated the interactions with physiological processes which eventually lead to growth-inhibition or celldeath.

MODE OF ACTION

In general the mode of action via which a drug exerts its growth-inhibitory effect on micro-organisms can be inhibition of macromolecular synthesis (DNA, RNA or proteins), inhibition of the energy yielding metabolism, inhibition of cellwall synthesis and disturbance of the cell membrane integrity. In our studies we used bis(2,9-dimethyl-1,10-phenanthroline) copper[I] nitrate ($Cu(DMP)_2NO_3$) as a testsubstance unless otherwise indicated.

Fig.1. Influence of $Cu(DMP)_2NO_3$ on macromolecular synthesis of *Myc.gallisepticum*

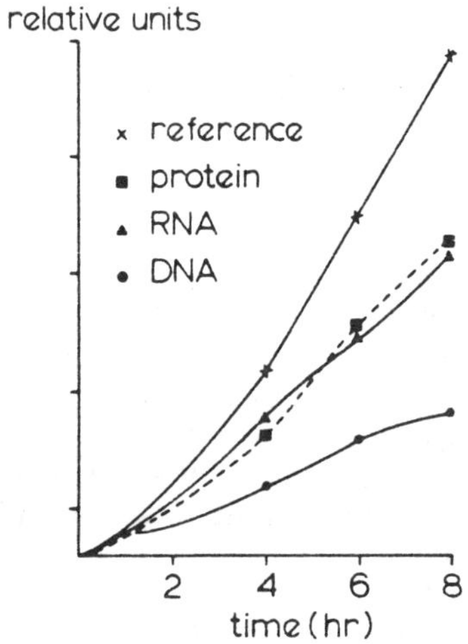

Fig. 1. shows the influence on the macromolecular synthesis established from incorporation of ^{14}C-labelled precursors. The control for all these three experiments could be transferred to one experimental curve. It can be seen that the DNA synthesis is somewhat more inhibited than the RNA and protein synthesis after addition of $Cu(DMP)_2NO_3$. It might indicate that the DNA synthesis is primary inhibited slightly by the complex, whereas the inhibition of the RNA and protein synthesis is rather a consequence of the general growth-inhibition[9].

As mycoplasmas lack a cellwall, the possibility of interfering with cellwall synthesis can be ruled out. Furthermore a strong influence of $Cu(DMP)_2NO_3$ on the oxygen consumption of a growing culture of *Mycoplasma gallisepticum* was found, viz. immediately after its addition a strong decrease was found. However, addition of yeast extract restored the original oxygen consumption rate. As yeast extract contains copper coordinating compounds, it indicates that decomposition of the complex abolished the activity. Nevertheless, these experiments point to an interference of the copper/DMP complex with the respiratory chain of the mycoplasma. To study the influence on the metabolic pathway a pH auxostat culture of *Mycoplasma gallisepticum* was developed from which the glucose consumption rate and the lactate, pyruvate and acetate production rate was established. Addition of $Cu(DMP)_2NO_3$ yielded instantaneously a strong decrease of glucose consumption and of lactate and acetate production, whereas the pyruvate production showed in first instance an increase (fig. 2).

Fig. 2. Influence of $Cu(DMP)_2NO_3$ on specific rate of glucose consumption and lactate, pyruvate and acetate production of *Myc.gallisepticum.*

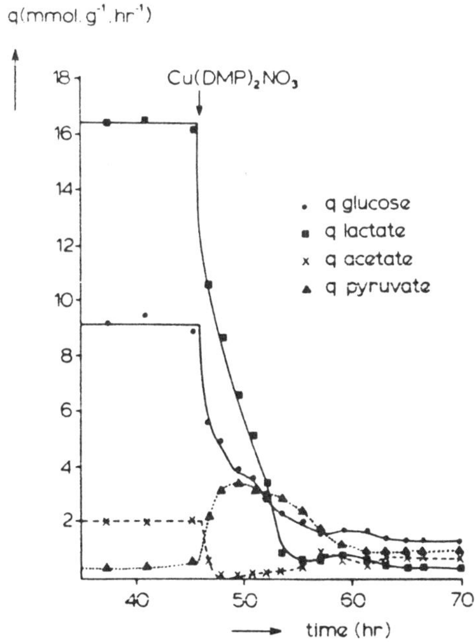

These results indicate that enzymes involved in the energy yielding metabolism, e.g. LDH and NADH-oxidase are inhibited by the copper complex[9]. To verify the influence of $Cu(DMP)_2NO_3$, on the activity of these enzymes, inhibition experiments with these enzymes, isolated as crude cell extracts of *Mycoplasma gallisepticum*, were performed. It was found that $Cu(DMP)_2NO_3$ indeed inhibits LDH and NADH oxidase. In the same experiment DMP appeared to be inactive, but $CuSO_4$ caused a much stronger inhibition than $Cu(DMP)_2NO_3$[9]. This rather surprising result indicates that copper itself is more potent to inhibit these enzymes than the complex.

MECHANISM OF ACTION

As from our MIC tests is clear that $CuSO_4$ - in contrast to $CuSO_4$ and ligands together - has no antimycoplasmal activity it is worthwhile to establish the enzyme inhibiting effect of a series of copper complexes and to relate them to the MIC-values. For these experiments a series of 1-substituted 3-(2-pyridyl)isoquinolines was used. The results of in vitro experiments measuring the inhibition of LDH and NADH oxidase are expressed as IC_{50}-values (table 1).

Table 1. Inhibition of LDH and NADH-oxidase by copper[II] complexes.

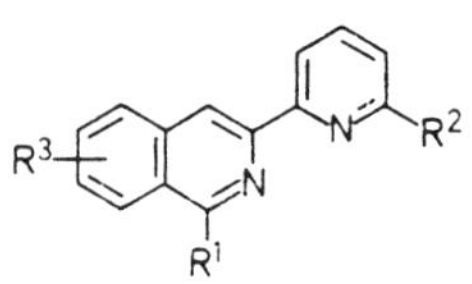

R^1	R^2	R^3	LDH IC_{50}(μM)	NADH-oxidase IC_{50}(μM)	MIC(μM)
H	H	H	21.40	6.74	25.0
NH_2	H	H	11.07	-	1.6
MeO	H	H	8.32	7.82	0.8
Cl	H	H	3.53	2.09	12.0
NH_2	CH_3	H	6.24	2.75	0.1
NH_2	H	6,7-DiMeO	2.93	18.78	39.0
DMP			7.26	7.35	0.1

However, no correlation between the MIC-values and the IC_{50}-values could be established. It even appeared that some ligands having a low IC_{50}-value can have a high MIC-value. Therefore we conclude that a direct inhibition of these enzymes by the complexes without other barriers cannot be the simple mechanism of action.

Principally, there are at least three different copper/2,2'-bipyridyl complexes, viz. Cu[II] 1:1; Cu[II] 1:2 and Cu[I] 1:2 complex. Regarding DMP as the ligand all three complexes showed the same MIC-value, indicating an equipotency of the complexes or more probable that two of them are converted into the other, being the active species[5]. The amount of a certain complex in the growth-medium depends on several features e.g. stability constants, concentration and redox potential of the system[11]. We have measured the latter in a growing culture of *Mycoplasma gallisepticum*. It was found that after inoculation of the growth-medium the redox potential decreased very rapidly from 250 mV to 50 mV, while at the same time not enough acid has been produced to give a positive result in the MIC test[10]. This result indicates that indeed before lowering of pH in the MIC test is observed, the redox potential of the growth-medium can reach a low level. For DMP it is known that at this level the concentration of the Cu[I] $(DMP)_2$ complex is higher than the concentration of the Cu[II] $(DMP))_2$ complex[12]. In general for ortho substituted 2,2'-bipyridyl ligands the Cu[I] complex is favoured over the Cu[II] complex for sterical reasons (tetrahedral vs square planar surrounding) leading to a high redox potential. On the other hand one should realize that also ortho substitution is necessary for a high antimycoplasmal activity.

All these experiments make it probable that DMP favour the copper to interfere with the cell. Indeed copper uptake experiments with ^{67}Cu alone showed a small copper uptake, possibly mostly caused by absorption to the cell envelope. However, when with the same copper concentration also two equivalents of DMP are present, much more copper has been taken up, or at least bound by the cell[13]. A more detailed study (fig. 3) shows the ^{64}Cu uptake vs time in Adler medium and phosphate buffered saline (PBS) whether or not in the presence of

Fig. 3. ^{64}Cu-uptake by *Myc. gallisepticum*

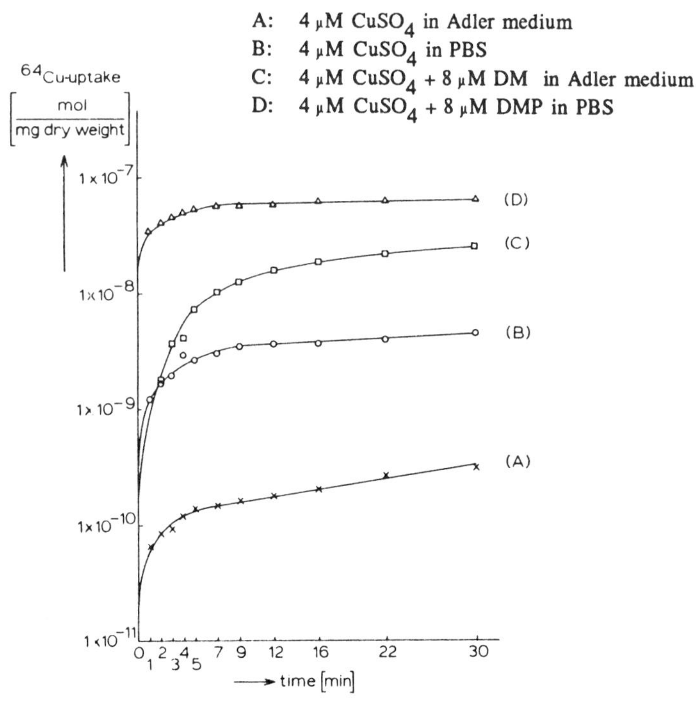

DMP[10]. It appeared that in both cases the copper uptake is very fast (completed within about 10 min.). Furthermore again copper uptake in Adler medium in the presence of DMP is much higher than without DMP. Copper uptake in PBS is somewhat higher because in Adler medium a part of the copper is coordinated by medium components. The analogous experiments with radiolabelled DMP yielded comparable results. Also the uptake of DMP is a fast process and now the presence of copper increases the DMP uptake. Because both the copper and DMP uptake are increased, it is likely that a complex is absorbed. However, experiments with both labelled ^{64}Cu and ^{14}C-DMP revealed that these molecules are after absorption not present in the cell in a 1:2 ratio. Remarkably much more copper is found than would be expected[13]. This result can be explained if the process of adsorption of the complex is followed by dissociation after which DMP leaves the cell.

We have tried to verify these propositions using a small series of ligands with electronically and sterically different substituents. If the above mentioned ideas are correct, the lipophilicity of the copper[I] complexes may correlate with the MIC-values. Therefore the octanol/water partition of the copper[I] complexes have been measured. Because some of the complexes appear to be extremely susceptible for oxydation by air, we made use of a filter probe technique performing the whole procedure under oxygen free conditions[14]. It was found that all copper[I] complexes, notwithstanding their ionic structure, have positive log $P_{oct./water}$ values (table 2).

Table 2. Log $P_{oct./water}$ of copper[I] complexes and their ligands

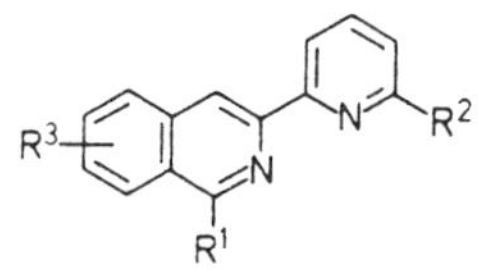

R^1	R^2	R^3	log P (ligand)	log P (complex)
NH_2	H	H	2.74	1.33
NH_2	CH_3	H	3.18	2.60
NH_2	H	6-CH_3	3.04	1.35
NH_2	H	8-CH_3	3.05	0.85
H	H	H	3.05	0.68
Cl	H	H		1.22
DMP			2.46	1.50

The corresponding copper[II] complexes (not all measured) have in general highly negative log P values. These results make it probable that the copper[I] complexes pass the lipophilic mycoplasma membrane. However, comparing the MIC-values with the log $P_{oct/water}$ of these complexes showed that there is no good correlation between these two (r=0.62). Therefore other processes must involve in the growth-inhibition too. Additionally it is noteworthy that the lipophilicity of the complexes do not correlate with the lipophilicity of the ligands.
For a further evaluation the stability constants of the copper[I] complexes were determined according to the procedure of James and Williams[12], including the measurement of the redox potential at various pH's in the presence of a 1:1 ratio of total copper[I] and total copper[II] and an excess of ligand. This experiment typically yields the ratio of β_2^{I} and β_2^{II}. To obtain β_2^{I}, β_2^{II} was determined by standard procedures (table 3). It appeared that there is a reasonable correlation between the stability of the copper[I] complexes and the MIC-value (r=0.87), indicating that within this series active ligands give stable copper[I] complexes[15].

Table 3. Stability constants of copper[I] and copper[II] complexes

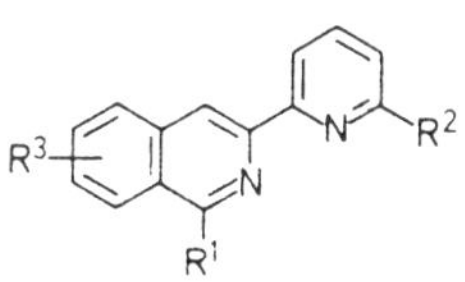

R^1	R^2	R^3	log β_1^{II}	log β_2^{II}	log β_2^{I}	MIC µmol/1
NH_2	H	H	6.29	13.38	14.77	1.6
NH_2	CH_3	H	6.14	10.95	16.67	0.1
NH_2	H	6-CH_3	7.17	15.60	17.57	1.5
NH_2	H	8-CH_3	8.55	14.99	16.22	0.2
H	H	H	4.91	7.60	8.01	25.0
Cl	H	H	3.51	6.73	12.56	12.0
NH_2	H	6,7-diMeO	7.72	13.72	14.45	39.0
DMP			5.30	10.78	18.55	0.1

The significance of copper in the growth-inhibition of *Mycoplasma gallisepticum* by copper complexes could be demonstrated clearly by establishing the copper uptake in the presence of various ligands and at various concentrations[16]. These experiments showed that even with ligands with very different MIC-values always at the MIC the same amount of copper has been taken up (fig. 4).

Fig. 4. Copper uptake by *Myc.gallisepticum* in the presence of different ligands at various concentrations.

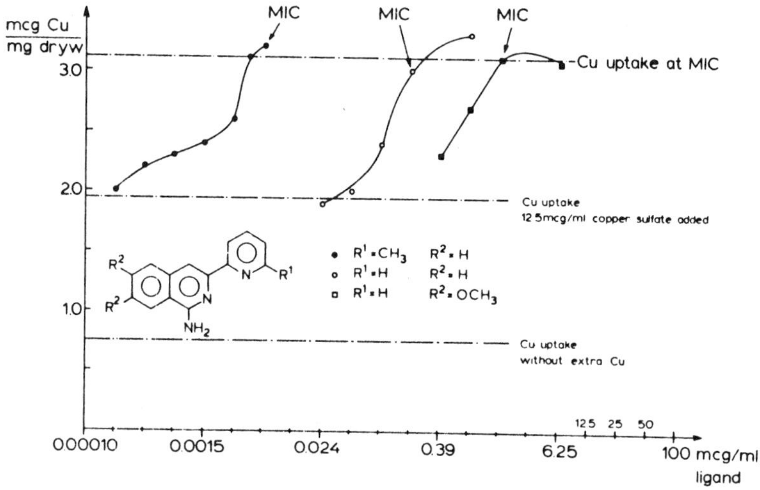

Therefore we conclude that the copper is the ultimate toxic species and that the ligand only serves as a carrier. The efficiency of the ligand is predominantly governed by lipophilicity and stability of the copper[I] complex.

Summarizing it may be said that the mechanism of the copper penetration into the cell consists of: copper[II] complex formation in the medium, reduction to the lipohilic copper[I]complex, penetration of the lipohilic cell membrane, entering the cytosol followed by dissociation, returning of the ligand to the medium and causing enzyme inhibition by copper.

REFERENCES

1. Van der Goot, H; Oostendorp, JG; Nauta, WTh. Eur. J. Med. Chem., 1975, 10, 603.
2. Pijper, PJ; Van der Goot, H; Timmerman, H; Nauta, WTh. Eur. J. Med. Chem., 1984, 19, 389.
3. Pijper, PJ; Van der Goot, H; Timmerman, H; Nauta, WTh. Eur. J. Med. Chem., 1984, 19, 393.
4. Pijper, PJ; Van der Goot, H; Timmerman, H; Nauta, WTh. Eur. J. Med. Chem., 1984, 19, 399.
5. Antic, BM; Van der Goot, H; Nauta, WTh; Balt, S; De Bolster, MWG; Stouthamer, AH; Verheul, H; Vis, RD. Eur. J. Med. Chem., 1977, 12, 573.
6. Antic, BM; Van der Goot, H; Nauta, WTh; Pijper, PJ; Balt, S; De Bolster, MWG; Stouthamer, AH; Verheul, H; Vis, RD. Eur. J. Med. Chem., 1978, 13, 565.
7. Pijper, PJ. PhD Thesis, Vrije Universiteit, Amsterdam, 1980.
8. Hahn, FE. In: Topics in Current Chemistry, Vol. 72; Springer Verlag; Heidelberg, 1977; p 1.
9. Smit, H; Van der Goot, H; Nauta, WTh; Timmerman, H; De Bolster, MWG; Jochemsen, AG; Stouthamer, AH; Vis, RD. Antimicrob. Agents Chemother., 1981, 20, 455.
10. Gaisser, HD. PhD Thesis, Vrije Universiteit, Amsterdam, 1985.
11. Eriks, JCh; Gaisser, HD; Van der Goot, H; Timmerman, H. Pharm. Weekbl. Sci. Ed., 1983, 5, 319.
12. James, BR; Williams, RJP. J. Chem. Soc., 1961, 2007.
13. Smit, H; Van der Goot, H; Nauta, WTh; Timmerman, H; De Bolster, MWG; Stouthamer, AH; Vis, RD. Antimicrob. Agents Chemother. 1982, 21, 881.
14. Gaisser, HD; Van der Goot, H; Timmerman, H. Pharm. Res., 1985, 181.
15. Gaisser, HD; Van der Goot, H; Timmerman, H. Eur. J. Med. Chem., 1986, 21, 285.
16. Gaisser, HD; Van der Goot, G; Timmerman, H. Eur. J. Med. Chem., 1985, 20, 513.

DISCUSSION

FIELDS: What happens to the host given this compound?

van der GOOT: When we treated mice infected with Trichomonas we found that these compounds were very toxic and they produced very large necrosis at the injection site. On the other hand, Prof. deZwart's compounds turned out to be much less toxic and we are quite optimistic about the usefulness of his compounds.

PICKART: We did some very similar work years ago looking for Cu complexes that would inhibit tumor growth. We found one factor, pyridinecarboxaldehyde pyridylhydrazone. It inhibited cultured cell growth at concentrations of 10^{-9} to 10^{-12} M. It also killed tumor cells but we had a lot of necrosis in the surrounding tissue. It was so toxic I didn't want to work with it. It was finally sold as a color reagent for Cu analysis.

van der GOOT: We know about this literature. We also sent our compounds to NIH in Bethesda and they found no anti-cancer activity at all for neither these ligands nor their Cu complexes.

ELLIOTT (Graham): Is the main message of your presentation that the dissociation of the complex within the cell is the rate limiting step in determining their activity?

van der GOOT: Yes. Bipyridyl compounds transport Cu(I) into cells because they are lipophilic.

ELLIOTT (Graham): Depending on the sort of complex that you have, you could have different mechanisms of uptake. One could be passive based upon lipophilic character and the other could be active transport couldn't it?

van der GOOT: You are asking whether it is possible that other complexes not giving the Cu(I) complexes but staying in the divalent state could be active. We have non-ortho substituted phenanthrolines that are well known as forming divalent Cu complexes that are not easily reduced and these compounds turned out to be inactive. In addition, bipyridyl and substituted bipyridyl Cu(II) complexes were also inactive. We also tested sulfonic acids substituted dimethylphenanthroline Cu(I) complexes, which are hydrophilic and inactive. So that indicates we need lipophilicity to get Cu into cells.

BARNEA: You indicated that Cu(II)sulfate inhibited these enzymes in vitro whereas Cu complexes inhibited them when you had the whole cell. In the beginning I thought you told us that the mechanism of action involved the complex but then I got the impression that it's dissociated Cu that is mechanistically important so I'm confused.

van der GOOT: The completely dissociated complex would not cause enzyme inhibition. Copper(II)sulfate itself is ionic Cu(II) and it is more active in inhibiting certain enzymes but as I showed you in a previous slide Cu complexes themselves inhibit. When the Cu complex is inside the cell exchange occurs to form complexes inside the cell. The Cu complex itself can inhibit these enzymes together with ionic Cu compounds. On the other hand, cells treated with di-labeled DMP complex contained only small quantities of DMP so most of the DMP had left these cells and the Cu remained coordinated with peptides or other cellular components to cause inhibition.

BARNEA: Then maybe the lipophilic Cu complexes that you are working with might be followed if you double labeled one of your complexes by labeling the ligand and Cu. You may be able to quantitate dissociation and measure leakage of the ligand from cells.

WILLINGHAM: It's my understanding that the Cu(II) phenanthroline complexes are ionic. How do you account for the fact that the complex would be more lipid soluble than the ligand itself?

van der GOOT: That complex is not more lipid soluble than the ligand itself. I showed in my slide of lipophilicity that these ligands had values of about 2.7, 3.0, and so on. The Cu(II) complexes had a lipophilicity value of about 1.0.

WILLINGHAM: I misunderstood you. I thought you said the Cu(II) complex was more lipid soluble than the individual ligand.

van der GOOT: No. What I meant was that the Cu(I) complex is more lipid soluble than the Cu(II) complex.

SORENSON: In answer to Prof. Barnea's question, it is plausible that ligand exchange and induction of some Cu-dependent enzymes or facilitating some Cu-dependent process, as you suggest, are likely fates of Cu in our complexes. Since we don't have any evidence for that as yet, the other option that it is the chemical reactivity of the complex itself that is important is open. It is likely that both of these are equally plausible. If you were to ask what my preference is, it's ligand exchange and induction of Cu-dependent enzyme activity such as Cu-Zn SOD or lysyl oxidase activity. I think part of the plasma Cu increase in response to disease, provides Cu in a form that allows for activation of Cu-Zn SOD and other Cu-dependent enzymes which may exist in tissues, in part, as the apoenzyme. One of the questions in the SOD world is why are there two identical subunits. Understanding Zn metabolism might provide a clue. Zn deficiencies result in death very fast because Zn-dependent processes are vital and they are much more important with regard to maintaining life of cells and cell division. The Zn in Cu-Zn SOD may be incorporated in the protein along with apoenzyme synthesis and only one of the subunits is fully activated with Cu. The other subunit acquires Cu in a physiologic process in response to some burst of O_2^-. That may be why one sees increases in Cu-Zn SOD activity in response to different diseases. Yes, these Cu complexes we work with may be forms of Cu that are useful in facilitating the full activation of Cu-Zn SOD. My question for Prof. van der Goot is: Do you get the MIC values that you showed in that second to last slide when you give the Cu complex of those ligands?

van der GOOT: No. Normally we add a fixed amount of Cu to the growth medium and then we add the ligand. Complex formation is always a very fast process and it is much more work to prepare these Cu complexes first. We prepared the Cu complex of one of our most active compounds but that Cu complex was very insoluble and that lack of solubility makes them difficult to work with, so we didn't use them.

PICKART: I mentioned previously we had worked with similar complexes and we tried to find out what they were doing in the cell to inhibit division. We found no binding to DNA. ESR studies suggested that these complexes were reacting with a sulfhydryl ligand somewhere in the cell.

van der GOOT: We also found a decrease in SH groups in these cells.

BALL: It was suggested at our last symposium that mycoplasma might well be an etiologic agent in producing rheumatoid arthritis. Have there been any additional developments in that understanding with regard to rheumatoid arthritis?

van der GOOT: I haven't seen any. Indeed, mycoplasma could be responsible for some types of arthritis.

SORENSON: When you responded to Dr. Fields' question concerning toxicity, was that for these Cu complexes?

van der GOOT: I think that was a general remark. If you ask me whether we prepared Cu complexes of all of these ligands, the answer is no. We always used the ligand and added the calculated amount of Cu to it.

SORENSON: If the toxicity of the ligand is due to removal of Cu from some site, wouldn't the Cu complex be less toxic than the ligand?

van der GOOT: Cu was necessary for activity. We thought at first that maybe removal of Cu or some other essential metal could be the mechanism of action but later on we found out that if there was no Cu added to the growth medium we couldn't find any activity at all. So that could not be, in this case, the reason for toxicity.

SORENSON: How is it known that Cu attaches to the outer surface of mycoplasma to cause toxicity?

van der GOOT: I've shown that part of the Cu we added to the growth medium will bind to phosphate or carboxylate groups on the mycoplasmal cell surface. Then after that, when we add these ligands we see a dramatic increase in Cu content of these cells. That's why these ligands are present.

ELLIOTT (Graham): Have you ever done any fractionation studies to look at the distribution of membrane and non-membrane attached Cu?

van der GOOT: These studies were not reproducible because a lot of radioactivity is lost on washing these cells.

ELLIOTT (Graham): So Cu binding to the membrane was a pretty loose attachment?

van der GOOT: I think so, but I can't be sure whether it was actually bound to membrane or dissolved in fluid between membranes?

ELLIOTT (Gary): You've postulated that there is dissociation of the chelate in the cell with subsequent diffusion of the ligand out of the cell. I'm curious as to how that might occur. We have a lipophilic ligand in a fairly lipophilic intracellular environment and yet we're postulating that there is movement into the external environment again. If these are truly synthetic non-endogenous compounds, there are only two ways that I could explain that. One would be an active transport process which at first thought doesn't seem very plausible although that could be the case. Secondly, we're changing the aqueous solubility of these compounds intracellularly. These compounds should be able to bind Fe for instance. When I see the DNA inhibition results that you've generated, I'm wondering whether we're dealing with a system that has more than one mechanism of action. Activity due to Cu addition and then activity due to subsequent removal of Fe and inhibition of ribonucleotide reductase.

van der GOOT: In that case you have to recognize that our active bipyridyl compounds have ortho substituents. For example, the dimethylorthophenanthroline (DMP) is very specific for Cu because Cu(I) has tetrahedral configuration and it can coordinate to 2 phenanthroline molecules in the tetrahedral configuration. Coordination with Fe(III) requires another ligand which is sterically hindered and not favored in complex formation with Fe(III). It's known that Fe(III)DMP complexes are rather unstable because of steric hindrance. Ortho substitution of these ligands make them specific for Cu(I).

COPPER(I)-NICOTINIC ACID COMPLEX: AN IMMUNOPOTENTIATOR IN CHICKENS VACCINATED AGAINST NEWCASTLE DISEASE

S.A. Musa,[1] A.H. Hafez,[2] A.Y. Nassar[3] and M.A. Gohar[4]
[1]Department of Poultry Diseases & [2]Animal Hygiene, Faculty of Veterinary Medicine, Assiut University; [3]Department of Biochemistry, Faculty of Medicine, Assuit University and [4]Department of Chemistry, Faculty of Science Alexandria University

ABSTRACT

Copper-nicotinic acid complex was fed as a food additive to baby chicks and their immune status estimated after vaccination with Hitchner B_1 and inactivated ND vaccines. This work was done in order to determine the immunostimulant effect of copper(I)-nicotinic acid complex in chickens. Immunity was estimated by haemagglutination inhibition, agar-gel precipitation, and challenge tests. Results showed much higher immunity status and post vaccination protection in birds supplemented with the complex. These data can be used in developing and realizing veterinary preventive programs or passive immunity in childhood.

INTRODUCTION

Immunopotentiation has received considerable attention with regard to factors that stimulate antibody production.[1] Also the role of vitamins, especially those of the B complex, in biochemical pathways including protein synthesis has been studied in immunopotentiation.[2]

The field of essential metalloelement complexes has attracted many authors in this last decade. Biochemical mechanisms have been postulated concerning the mode of action of such vitamin metal complexes on enzyme reactivities.[3,4]

This work has been conducted in order to study the effect of dietary copper (I)-niacin complex on immunoresistance to infection. Chickens were vaccinated against Newcastle disease and supplemented with the complex. The treatment schedule was based upon normal biological vitamin B_3 or niacin requirements.

MATERIAL AND METHODS

Experimental Chicks

In this study 100 one-day-old Hubbard chicks in five groups (A,B,C,D and E) were kept under hygenic scheme of isolation. All birds were checked at 30 days of age for freedomfrom specific haemagglutination inhibiting antibodies to Newcastle disease virus, and thereafter they were vaccinated. All birds received a normal balanced diet.

Chicken Embryos

Nine-day-old embryonated chicken eggs were obtained from the government operated Beni-mur Poultry Farm for the propagation of the virus, titration, and reisolation.

Newcastle Disease Vaccines Used

1- Hitchner B_1 vaccine strain in 1000 dose vials ($10^{8.5}$ EID_{50}) produced by TAD Pharmazeutishes Werk GMBH, West Germany was administered in the drinking water.

2- Inactivated oil adjuvant vaccine produced by IVaz Lab., Padowa, Italy was administered by intramuscular injection.

Copper(I)-Nicotinic Acid Complex

Copper(I)-nicotinic acid complex (Fig. 1) is a bright red pure crystalline compound [CuCl$(HNA)_2$]. The complex has been synthesized by the method of Gohar and Dratovisky.[5]

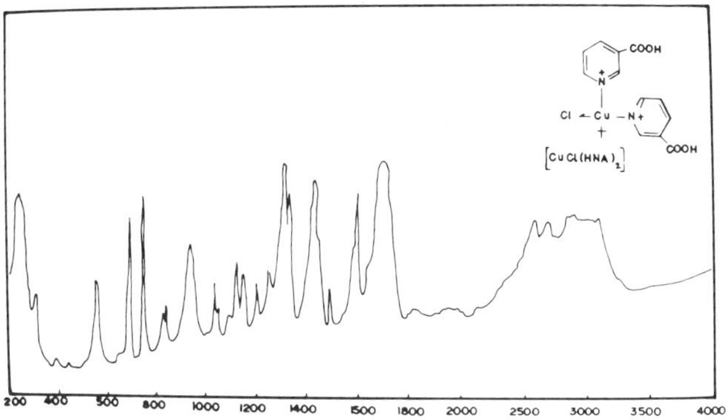

Figure 1: Infrared spectra of the Cu^{+}-nicotinic acid complex.

Two hundred mg of the complex was dissolved in 100ml of saline solution containing 0.5g of L-ascorbic acid.

Immunological Tests

1- Challenge Test: A locally virulent velogenic viscerotropic isolate of ND virus (Newcastle Disease) was used for intramuscular challenge in doses of 10^6 EID_{50} (100 CLD_{50}) per bird. Challenged birds were observed for two weeks and mortality was taken as a criterion of infection. Virus was reisolated from internal organs of dead birds.

2- Haemagglutination-Inhibition (HI) Test: The micro-technique using the beta procedure was employed. A stock virus suspension of the Komarov strain of ND virus was prepared and 4 HA units were used per serum dilution.

3- Virus Titration: Titrations of ND viruses were done with ten fold virus dilutions. Chicken embryos were inoculated via the allantoic sac (5 eggs/dilution) and the EID_{50} was calculated.

4- Agar Gel-Precipitation (AGP): Two fold dilutions of sera were quantitatively tested for specific precipitating antibodies using the Nansi microtechnique.[6]

Experimental Design

Two groups of chicks (A and B) were given 1.5 mg of the copper complex/100g of feed for seven successive days prior to vaccination. Birds in group A were vaccinated with Hitchner B1 vaccine while those of group B were vaccinated with inactivated vaccine when they were 30 days old.

Groups C and D did not receive the complex and birds of group C were vaccinated by the Hitchner B1 vaccine while birds of group D were vaccinated with inactivated vaccine when they were 30 days old.

Group E chicks were kept unvaccinated and they were not treated with the complex during the study.

After 3 and 6 weeks post vaccination 10 serum samples were taken from each group of chicks and subjected to HI and AGP tests (10 birds were challenged at each time of sampling).

RESULTS

Geometric means of HI and precipitin titers are recorded in Tables 1 and 2. Results obtained 3 weeks post-vaccination showed that protection percentages were 100, 100,80,70 and 10% in chicks of groups A,B,C,D and E respectively. Moreover, the protection percentages revealed 6 weeks postvaccination were 90,100,60,70 and zero% in chicks of groups A,B,C,D and E respectively.

Table 1: Geometric mean (GM) titres (Log 2) of HI antibodies, 3 and 6 weeks (W) postvaccination.

Individual Birds	Group A		Group B		Group C		Group D		Group E	
	3W	6W	3W	6W	3W	6W	3W	6W	3W	6W
1	6	5	8	7	3	4	8	5	0	0
2	6	5	9	7	4	3	6	4	0	0
3	5	6	9	8	5	3	7	5	0	0
4	5	5	8	7	5	4	7	5	0	0
5	7	5	9	7	4	3	7	4	0	0
6	6	5	9	8	5	4	6	5	0	0
7	6	4	7	6	6	4	6	6	0	0
8	6	6	9	6	5	4	8	5	0	0
9	6	5	8	6	5	3	6	6	0	0
10	7	5	9	5	5	3	8	5	0	0
GM	6	5.1	8	6.7	4.7	3.5	6.9	5	0	0

Table 2: Geometric mean titres (Log 2) of Precipitin antibodies, 3 and 6 weeks postvaccination.

Individual Birds	Group A		Group B		Group C		Group D		Group E	
	3W	6W	3W	6W	3W	6W	3W	6W	3W	6W
1	4	0	4	1	4	1	4	1	0	0
2	4	2	4	1	4	1	4	1	0	0
3	3	1	4	0	3	1	3	1	0	0
4	4	0	4	2	4	0	4	0	0	0
5	4	1	3	1	3	1	4	1	0	0
6	2	0	4	0	3	0	3	1	0	0
7	4	0	4	1	4	0	3	1	0	0
8	4	0	5	1	3	0	4	1	0	0
9	3	1	4	0	3	0	3	0	0	0
10	4	0	4	1	4	1	4	0	0	0
GM	3.6	0.5	4	0.8	3.5	0.5	3.6	0.7	0	0

DISCUSSION

It is of interest and value to find that all birds given the copper(I)-nicotinic acid complex showed a significant high protection rate after vaccination by 3 and 6 weeks. Also, there was no clear correlation between precipitin in sera of vaccinated chicks administered the complex and those that did not receive it.

From the biochemical point of view, it is suggested that most of the absorbed complex was biologically utilized as such. This suggestion is based upon the structural stability of the complex with regard to competing ligands.[5] Moreover, the most probable modification of the ligand in this copper complex is reduction to the pentadienyl derivatives[7] shown in Figure 2.

Hence, this complex could be conjugated in hepatocytes with phosphoribose resembling that of nicotinic acid.[8] Accordingly, NAD and NADP-like structures could be biosynthesized. These structures are suggested to be more reactive as hydrogen carriers. This is due to the electronegativity of the chloride ion. The lone pair of electrons on nitrogen in turn are attracted toward the copper atom as shown in Figure 2.

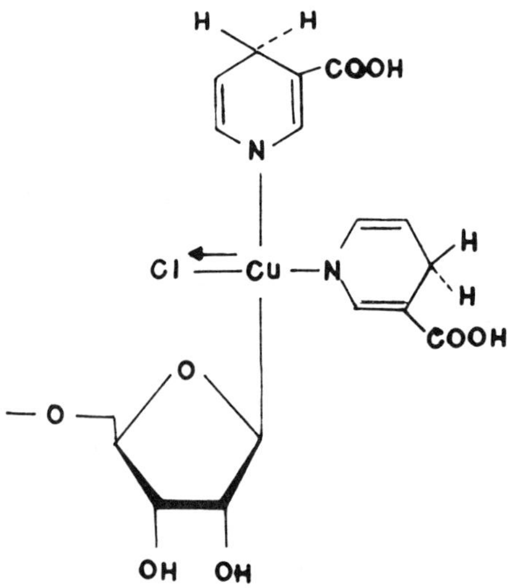

Figure 2: Probable Cu^{+} conjugation to ribose.

This electron shift enhances stability of the complex and the reactivity at C_4 in the two pyridine nuclei as shown in Figure 3. The predicted higher reactivity of NAD and NADP-like structures enhances the activity of oxido-reductases which catalyze anaerobic and aerobic oxidations that result in accumulation of ATP, GTP, and UTP. These nucleotides are essential compounds for the promotion of initiating factors that control protein biosynthesis and immunoglobulin production.[9]

Further studies utilizing radioisotopic techniques are needed to determine the biochemical fate of this copper complex. For the time being, the complex of Cu^+-niacin could be successfully utilized in preventive veterinary programs and in producing passive immunity in children.

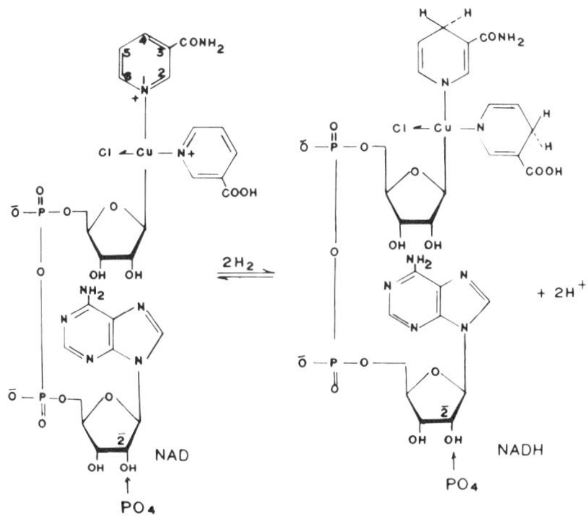

Figure 3: Predicted NAD and NADP-like structures derived from the Cu^+-nicotinic acid complex.

REFERENCES

1. Good, R; Jose, D; Cooper, W; Fernandes, G; Klamer, T; Yaines, E. In: Malnutrition and the Immune Response, Raven Press; N.Y., 1977.
2. Karran, S; Alberti, M. In: Practical Nutritional Support, Pitman Medical Publishing Co. LTD; England, 1980.

3. Franklin, K; Richardson, M. Inorganica Chemica Acta 1980, 46, 191.
4. Gray, A; Adeyemo, A. Inorganica Chemica Acta 1981, 55, 93.
5. Gohar, M; Dratovsky, M. Coll. Cerch. Chem. Commun. 1975, 40, 26.
6. Nansi, W. Nature 1958, 18, 1289.
7. Cotton, F; Takats, J. J.An. Chem. Soc. 1970, 92, 2353.
8. Petrack, B; Greengard, A; Craston, H; Kalinsky, H. Biochem. Biophys. Res. Commun. 1963, 13, 472.
9. Haselkorn, R; Rothman-Denes, L. Annu. Rev. Biochem. 1973, 42, 397.

(Editor's note: There was no discussion of this paper. None of these authors were able to attend and present this paper.)

GENERAL DISCUSSION

SORENSON: We are reporting immuno-incompetency in Cu-deficient animals and that incompetency was exacerbated by high sucrose diet. We also noticed that the usual impairment in Fe utilization associated with Cu deficiency was also made worse as a result of feeding a high sucrose diet. There was less Fe in tissues of animals fed a semi-purified diet and given adequate Cu in comparison to animals fed Certified Purina Rodent Chow. We used a Cetified Purina Rodent Chow fed control group, which we think has some merit in these kinds of studies, as a standard diet fed group and then compared results obtained with this group with results obtained for animals fed the semi-purified diet and one of 3 levels of Cu; adequate, marginally deficient, and deficient. Those animals given adequate dietary Cu and the semipurified diet containing 30% sucrose had impaired immunocompetency and less Fe in their tissues then the Certified Rodent Chow fed animals. Animals fed the same semipurified diet and marginally deficient or deficient dietary Cu had further compromised immune responses and reduced tissue Fe contents.

BARNEA: I wonder if anyone has some thoughts or ideas on what the role of metallothioneine (MT) is in the cell. Is it only a depot for getting rid of extra Cu or could it be a source of Cu as it is needed homeostatically. If that's the case, why is it so that we only find a lot of MT in liver, kidney, and intestine? Those who have been looking for evidence of mRNA for MT in the brain cannot demonstrate its presence. A lot of data was presented showing that the brain is second to the liver in Cu content, so where is all this Cu located in the brain? Axonal terminals or secretory granules can not be used to account for all of this Cu. What is holding the Cu there so tightly? Why is it so difficult to alter brain content of Cu with Cu-deficient diets for example? Rats kept 3 or 4 months on Cu-deficient diets lose very little, 20% to 30% brain Cu, whereas there is a 90% reduction in blood and liver Cu content. What in the brain is doing such a good job of holding on to Cu?

LOVEN: Could MT possibly be a detoxifying agent? It may be that toxic levels of metals cause induction of this protein in tissues and concomitant storage. Hilderbrand and Toby have shown that if they induce these MTs in CHO cells, they become resistent to certain anticancer drugs which bind to the sulfhydryl rich sites. It just detoxifies metals by binding them and its primary purpose may not be storage but detoxification of anything that will bind to sulfhydryls.

KEEN: There is a lot of evidence to support what Dean (Loven) is suggesting: that it acts, in part, as a detoxifier. I would strongly support it as a reservoir of trace elements, at least during the prenatal period. I'm willing to accept that perhaps it doesn't serve as an essential reservoir in adults but consistently we find, in virtually every species I'm aware of, that prenates have high concentrations of liver MT binding both Cu and Zn that appear to be rapidly mobilized in the postnatal period. Some will argue that this is still detoxifying during the fetal period because you don't have bile for Cu and Zn excretion. Perhaps you would just get too much build-up and that also would cause damage. We recently got some data that argues against that. What we have discovered is that we can give a transitory Zn-deficiency diet to pregnant mice during the last two-thirds of gestation, this is a 5 ppm Zn diet, so dams are not severely Zn deficient. The offspring had more MT which was not too suprising. If we challenge those offspring with Zn, we get an amplification of MT synthesis. That does not sound to me like it's a detoxifying protein, but actually stores every bit of Zn that it can store during the fetal period. This is where data supports it being a reservoir of Zn, not just a detoxifier.

HARRIS: I tend to agree with both comments. We're dealing with what appears to be a major storage protein. Carl's (Keen) suggestion is very important, a temporary reservoir of important metals. You find MT in practically all tissues and it seems that if there is going to be a certain reserve of these important metals like Cu and Zn, this protein fits that bill. Regarding its role in intercellular metabolism, Cu does make its way into the liver, which seems to be a sorting organ that transfers Cu into functional components such as ceruloplasmin, intracellular enzymes, or into an excretory component such as found in bile. There is another pathway involving MT which normally does not contain a great deal of the newly absorbed Cu. Van den Hamer has measured this amount of Cu under nontoxic and non-loading conditions and found that only about 5% to 10% of the total Cu in the liver is present as MT. That is a shocking figure because that suggests that MT is not an indispensible component for the sorting mechanism involved in redistributing Cu to other extracellular and cellular components. On the other hand, Cu found in bile may originate in MT. Under those conditions then, you do see Cu committed to a specific step in the pathway when it becomes bound to MT. Another thought that I want to leave with you is a point that Petering has been making very strongly, and that is that MT may actually serve in some instances as a doner of Cu for intracellular apoenzymes. This point is a very debatable point. A lot of people have quite a bit of reserve as to its correctness, but it's still something for us to think about. What Petering is arguing is that in the liver discrete forms of Zn are amassed in one portion, while discrete forms of Cu are amassed in another portion. MT becomes an equalizer and allows a homogeneity of the metal mix.

RISK: This is just a speculative answer to Prof. Barnea's question. Frequently metals bind to reactive membranes, especially at the synapse, which are high in unsaturated lipids. Is there a possibility that Cu or Cu complexes might bind to some of those unsaturated regions?

CROOK: Prof. Barnea and I had discussed this question regarding Cu in the brain. We have unsuccessfully attempted to measure transport of ^{64}Cu(II)-citrate into dog brain. We did find some strange pools which don't really correspond to any neuroanatomical structures that I'm aware of. At that time Prof. Barnea explained that animals made deficient had brain Cu pools which were only reduced about 30% whereas other organ Cu stores were markedly reduced, so typically the brain is always the last frontier. We would like to understand what is happening in the basal ganglia and other brain structures involved in movement disorders.

SORENSON: Joe Prohoska told me that 25% of the brain Cu is Cu-Zn SOD. The need to prevent lipid oxidation in the brain helps me understand why the brain has so much Cu. In addition, catecholamine syntheses in the brain may require two forms of DBH and Extremely Acidic Copper Containing Protein. Also, pigmentations requiring tyrosinase accounts for some of the Cu in brain.

Jim (Crook), did you say you used Cu(II)-citrate to examine brain transport? Cu(II)-citrate would probably clear pretty fast through the kidney because it's so water soluble. Chutkow (Proc. Soc. Exp. Biol. Med. 1978, 158, 113) measured the rate of Cu uptake in the brain following subclavian injection and within 15 seconds there was uptake of Cu which was markedly increased if they coadministered amino acids. It was not a passive uptake since it was a saturable uptake process.

With regard to MT, telelogically, it can't be a protein that just stores something that doesn't turn over in any tissue without causing some degeneration. Accumulation of Cu in the liver and brain results in liver and CNS degenerative disease in Wilson's patients. It was Gary Evans who reported that MT from Wilson's patients had a rather high stability constant in comparison with MT from normal humans. Perhaps Wilson's disease is associated with marked accumulation, without turnover, to such an extent that it causes loss of tissue function.

FIELDS: We fed radioactive Cu with sucrose, fructose, or starch diets to Cu-deficient and supplemented animals and killed them at 8, 24, 48, and 96 hours and we were surprised to find that the level of radioactivity in the brain was so low that we didn't even include it in our results. It was less than 1% of the administered dose. Most of the activity was either in feces, intestine, or liver. Secondly, is it not possible that Cu is present in glial cells in addition to neurons since some of the neuropeptides bind to glial cells and not to neurons? As I recall, Cu is important for binding of various neuropeptides to glial cells.

SORENSON: What is the interpretation of those results showing no or very little radiolabeled Cu in the brain?

FIELDS: We offered no interpretation because it didn't seem to be significant at the time.

SORENSON: Those results may indicate that the turnover in brain is very slow compared to other tissues.

FIELDS: I agree but the study was performed in order to see if animals fed fructose absorbed less than those fed starch. Therefore this 1% did not make a difference in an absorption study. Now when I look back, it's a very interesting observation but not for the study which involved absorption and excretion.

MELETHIL: Following oral administration and absorption the liver may remove it before it can get to the brain. If you had given it IV, for instance, perhaps you might have seen a different distribution.

KEEN: If you do the same experiment with a 12 or 14 day old rat, the brain takes up Cu if it is given orally and instead of being 1% or 2% as in an adult you'll get 3% or 4%. It doesn't sound like a lot more but there's definitely more there. If you catch it during the window when brain Cu is increasing, which is between day 10 and day 20 in the rat, you'll see a nice accumulation and it has a very slow turnover. I think the fact that you're getting 1% is really consistent with this observation. We very often get almost the same numbers and this is telling us something about turnover. It's hard to really amplify that turnover even if you give D-penicillamine for long periods of time. We'll get tissue Cu down in virtually all organs except brain. It might come down 5% when other tissues decrease 40% or 50%.

FIELDS: When you refer to the brain, are you referring to whole brain, cerebrum plus cerebellum, or just the cerebrum?

KEEN: The whole brain.

BARNEA: Now I'm confused because Prof. Sorenson brought up the point that much of the Cu is associated with neurotransmitter synthesis and storage in noradrenergic granules. These subcellular organelles are in very rapid turnover due to constant release of their contents in brain function. They cannot retrieve the Cu. For example, peptidergic granules are not reused for that function because there is unidirectional movement of granules that are formed in the cell body and accumulate in axonal terminals. This constant loss of Cu, and peptide after amidation for example, from neurons may create a demand for Cu. Noradrenergic granules are in a similar situation although there is retrieval and reuse of the granule membrane to make storage vesicles but I am not so sure that those vesicles can make norepinephrine. They can take up norepinephrine but I am not sure that they have the full capacity to make norepinephrine and if they do they have to get Cu from somewhere. It's hard to imagine that the entire brain content is only very slowly turning over when some compartments loose their Cu constantly. Also, if MT is a detoxifying agent in some tissues, what is the detoxifying agent in the brain?

PICKART: On this general idea of another MT-like factor, I can tell you about our search for an endogenous liver proliferation inhibitor. When part of a liver is cut off it grows out to its original size and then stops. We have isolated an ethanol soluble fraction that has this inhibitory activity. Our most active fraction is a peptide of 1700 Daltons which is 35% cysteine, but it is not part of bovine liver MT according to amino acid sequence. The first 3 amino acids were glutamic acid. This compound would appear to be a tremendous chelator of transition metals. There may be some other chelating agents in neurons that retain copper.

HARRIS: You will see MT induction when there appears to be a rather abnormal amount of Cu reaching a tissue that can respond by synthesizing this protein. In the case of the brain there is some protection by the blood brain barrier which may protect against rapid crossing and uptake into the brain. This might be explained as a buffering effect. Wilson's disease is a rather slow insidious increase of Cu in the brain, which eventually leads to psychotic disease. Where is that Cu being stored? Why is that Cu penetrating into the brain, in spite of the blood brain barrier, to actually cause tissue damage?

FIELDS: Didn't we all learn in anatomy that there are billions of neurons and that we may lose ten thousand a day without replacement? Within a year we lose millions, right? So another question is: If Cu is being deposited in neurons and neurons die, where is the Cu? What is happening to the Cu of the neuron? Regarding Wilson's disease, the biochemical explanation was that Cu causes damage because it is not being carried by ceruloplasmin and its form is one that allows deposition not only in the brain but also in other tissues. Cu without the protein is toxic. The whole idea of a protein transporting mechanism is to provide a form of Cu which is nontoxic.

HARRIS: That's an interesting hypothesis. However, Wilson's disease is not simply Cu deposition in the liver or brain. Large deposits of Cu have been found in the liver without any degeneration. In fact, I remember one story told by Irmin Sternlieb about a swan that was found dead on the front lawn of a New York pathologist. The animal had so much Cu in its liver that the liver was pitch black and it did not die of liver toxicity, it died of starvation. This points up the fact that it's not necessarily just getting Cu into a tissue, there must be sensitive areas in tissues. In the case of Wilson's disease there's almost total absence of any biliary excretion of Cu which may allow it to accumulate in a more sensitive area which produces pathology.

BHATHENA: Prof. Barnea suggested that Cu is released along with peptides. To this date, I don't think anybody has clearly demonstrated that released peptides, opiates, or neurotransmitters are associated with Cu. Cu is definitely associated with enzymes which are involved in the synthesis of neurotransmitters and hormonal or opioid peptides. It is likely that Cu is retained by these enzymes when these neurotransmitters and peptides are released from neurons? If it is retained there may not be a requirement for high concentrations of Cu by the brain.

FIELDS: The report by Icker and Manes showed co-release of beta endorophin and alpha amidating enzyme and it's known that DβH is co-released with catecholamines. DβH is found in blood. These enzymes are found in the extracellular space upon release of neurotransmitters or peptides. It may not be that the entire compliment of enzyme is released. Co-release of Cu is as likely as ascorbate release and ascorbate is co-released.

PICKART: Most of those early results that suggested that the brain shrunk with age are now being called into question. Stature measured in Europe and the United States has been increasing by about 1.2 mm/year during the past sixty years. There has been an increase in size of the average person, which is probably due to better nutrition. Correction for the increase in actual head size among healthy individuals does not show shrinkage with age. Earlier studies compared young people who were physically bigger with their grandparents. As a result, the number of brain cells lost per day may actually be one thousand cells a day rather than ten thousand a day.

Anticancer Activities of Copper Complexes and Iron Chelators

ANTICANCER EFFECTS OF CU(II)(3,5-DIISOPROPYLSALICYLATE)$_2$ IN MICE INOCULATED INTRAMUSCULARLY WITH EHRLICH ASCITES CARCINOMA CELLS

Scott L. Kasemeier, Hamid Salari, and John R. J. Sorenson;

Department of Biopharmaceutical Sciences, College of Pharmacy, University of Arkansas for Medical Sciences, Little Rock, Arkansas, 72205, U.S.A.

Introduction

In 1979, Oberley and Buettner published a review[1] which suggested that all neoplastic cells were deficient in total superoxide dismutase (SOD) activity. To determine whether or not Cu-Zn SOD had an anticancer effect, Oberley and Leuthauser injected mice with a single dose of this enzyme one hour after they implanted Sarcoma 180 tumor cells in the hind limb of CF1 mice and measured average leg diameter versus time after implantation. In animals treated with saline, tumor growth was unimpeded.[2] Animals treated with Cu-Zn SOD showed a reduction in tumor growth. Animals treated with Cu-Zn SOD also showed a slight increase in survival in comparison to those treated only with saline-vehicle.

It was then reported that copper complexes of nonsteroidal antiinflammatory drugs, including Cu(II) (salicylate)$_2$, had SOD-mimetic activity.[3,4] The question thus arose as to whether small molecular weight copper complexes might have anticancer activity. Cu(II)(3,5-diisopropylsalicylate)$_2$ [Cu(II)(3,5-DIPS)$_2$] was compared with several other copper complexes and found to be a very effective anticancer agent in CBA/J mice implanted in the hind limb with Ehrlich carcinoma cells.[5-7] Mice were treated with 0.5 to 1.0 mg of one of three different copper salicylates for 5 days after tumor cell implantation. Tumor growth was assessed by measuring the diameter of the tumor bearing leg. It was found that treatment with Cu(II)(3,5-DIPS)$_2$ produced a marked depression in tumor growth.

Survival was also markedly increased in those animals treated with copper complexes. Deaths began to occur early in the vehicle-treated animals, and then more rapidly after day 24. All vehicle-treated animals were dead by day 58. In the Cu(II)(3,5-DIPS)$_2$-treated group, one death occurred very early, but all other animals survived to day 46 when they began to die. After all of the vehicle-treated animals had died, 45% of the Cu(II)(3,5-DIPS)$_2$-treated animals were yet surviving.

Leuthauser found cell differentiation within Ehrlich tumors produced by implantation of undifferentiated transformed Ehrlich cells.[5] Differentiated epithelial cells forming ducts containing what appeared to be mucoid material were found in sections of these tumors. Leuthauser and Oberley suggested[6] that this could have resulted if some of the Ehrlich cells, which are derived from mouse mammary tissue, had differentiated to normal cells.

Sahu subsequently conducted cell culture experiments in which he grew neuroblastoma cells in medium containing Cu(II)(3,5-DIPS)$_2$.[8] Results of his study showed a reduction in cell number and an increase in differentiation as evidenced by the development of axons and dendrites on neuronal cell bodies. Further reduction in undifferentiated cell number and further increase in percent differentiation occurred with increasing concentrations of Cu(II)(3,5-DIPS)$_2$. Reduction in cell number was shown not to be due to cell killing, by trypan blue exclusion.

Subsequently, Kensler et al.[9] demonstrated an anticarcinogenic effect for Cu(II)(3,5-DIPS)$_2$ and Solanki et al.[10] demonstrated antimutagenic activity for Cu(II)(3,5-DIPS)$_2$. Attention was increasingly focused upon $Cu(II)(3,5\text{-}DIPS)_2$ when Wattenberg[11] and Cerutti[12] suggested that this drug represented a class of compounds that might be used to prevent cancer and carcinogenesis and Copeland's most recent suggestion that $Cu(II)(3,5\text{-}DIPS)_2$ be studied in greater detail.[13] These suggestions along with the observed anticancer, anticarcinogenic, and antimutagenic activities have led to attempts to develop a prolonged treatment paradigm using $Cu(II)(3,5\text{-}DIPS)_2$ to depress tumor growth and increase survival of mice bearing solid Ehrlich cell tumors.

Materials and Methods

In the first study, there were four treatment groups composed of ten C57BL/6 female adult mice. Mice in three of these groups were inoculated intramuscularly (im) in the left rear limb with Ehrlich ascites carcinoma cells, 5×10^5 cells in 0.1 ml sterile saline. The fourth group of ten mice were not inoculated with Ehrlich ascites carcinoma cells. Mice of this fourth group as well as mice of one of the inoculated groups were injected subcutaneously (sc) every twelve hours with $Cu(II)(3,5\text{-}DIPS)_2$. Mice of another of the inoculated groups were injected sc every twenty-four hours with this complex. Mice of the remaining inoculated group were injected sc every twenty-four hours with a vehicle containing 1.4% (w/v) polyvinyl alcohol and 4% (v/v) propylene glycol in sterile saline. In the first study, mice were treated for 64 days and then treatment was interrupted for 3 days, after which time treatment was resumed at half dosage until day 76.

In the second study, there were two treatment groups composed of ten C57BL/6 female adult mice and two treatment groups composed of ten Viable Yellow Black (VYB) female adult mice. All mice were injected im in the left rear limb with Ehrlich ascites carcinoma cells, 1×10^6 cells in 0.1 ml sterile saline. For each of the two strains, there was one group which was injected sc every twelve hours with $Cu(II)(3,5\text{-}DIPS)_2$ and another group which was injected sc every twelve hours with vehicle. In the second study, mice were treated for 48 days. In both studies, mice treated with $Cu(II)(3,5\text{-}DIPS)_2$ were given the dosage of 25 mg/kg body weight at each injection.

Tumor growth was assessed approximately every 4 days by measuring the thickness (lateral-medial dimension) and the width (anterior-posterior dimension) of the implanted, tumor-bearing leg. The product of these two dimensions for each individual mouse was calculated on each occasion of measurement, and the average of these individual values was determined for each treatment group. These data were further analyzed by analysis of variance and regression. Survival was compared between the various treatment groups.

Results

In the first study, repeated measurements of tumor size were not statistically different ($p \leq 0.05$) in tumor-bearing mice, whether they were injected with vehicle or with Cu(II)(3,5-DIPS)$_2$ once every twenty-four hours (once daily) as shown in Figure 1. On the other hand, repeated measurements of tumor size amongst mice injected

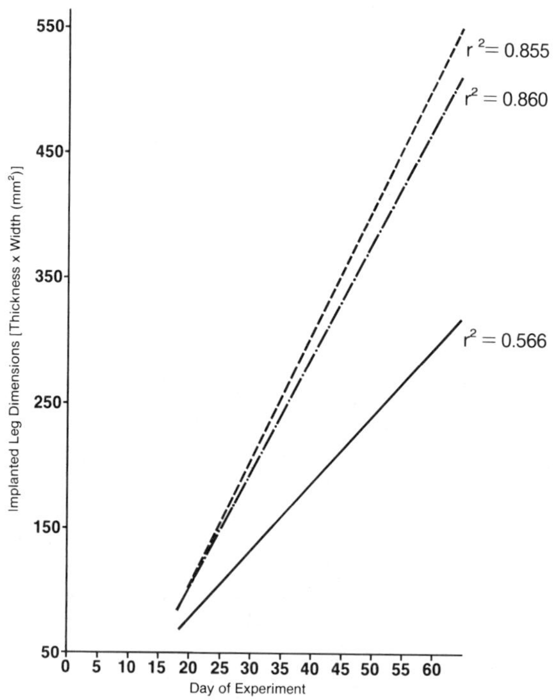

Figure 1. Regression for tumor growth in C57BL/6 mice treated with vehicle (– –), 25 mg/kg Cu (II) (3,5-DIPS)$_2$ once daily (–·–), or 25 mg/kg Cu (II) (3,5-DIPS)$_2$ twice daily (——).

every twelve hours with Cu(II)(3,5-DIPS)$_2$ were statistically ($p \leq 0.05$) smaller than mice injected with either vehicle or Cu(II)(3,5-DIPS)$_2$ every twenty-four hours. Correlation analysis of growth records for individual mice demonstrates that tumor growth was quite linear with time. For mice treated with vehicle, the slope of the least squares fitted plot was 10.150; for mice treated with Cu(II)(3,5-DIPS)$_2$ every twenty-four hours, the slope was 9.323; and, for mice treated with Cu(II)(3,5-DIPS)$_2$ every twelve hours, the slope was 5.453. Each slope significantly ($p \leq 0.001$) differed from all others.

As graphed in Figure 2, survival was substantially shortened in those mice which were tumor-bearing and injected with Cu(II)(3,5-DIPS)$_2$ in comparison to mice which either were not tumor-bearing or were injected with vehicle. Survival was similar in tumor-bearing animals,

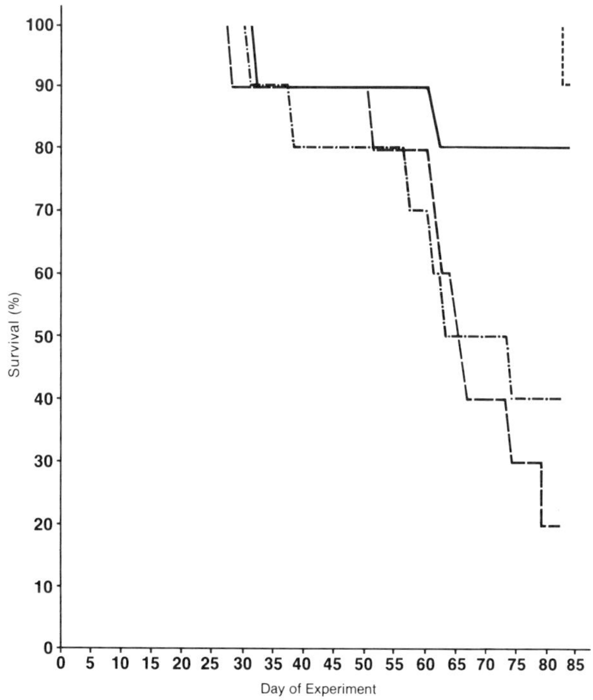

Figure 2. Survival of tumor bearing C57BL/6 mice treated with vehicle (----), 25 mg/kg Cu (II) (3,5-DIPS)$_2$ once daily (–·–) or twice daily (– –), and non-tumor bearing C57BL/6 mice treated with 25 mg/kg Cu (II) (3,5-DIPS)$_2$ twice daily (——).

whether injected once or twice daily.

In the second study, correlation analysis for tumor growth of individual mice showed that growth was again quite linear with time and treatment with Cu(II)(3,5-DIPS)$_2$ twice daily significantly decreased tumor growth, as shown in Figure 3. The slope for Cu(II)(3,5-DIPS)$_2$-treated VYB mice (4.706) was higher than for Cu(II)(3,5-DIPS)$_2$-treated C57BL/6 mice (3.853) whereas the slope for vehicle-treated C57BL/6 mice (9.033) was higher than for vehicle-treated VYB mice (6.849). Thus, the slope was lower for the Cu(II)(3,5-DIPS)$_2$-treated group than for the respective vehicle-treated group for each mouse strain. All comparisons of slope met or exceeded significance levels of $p \leq 0.05$. As shown in Figure 4, Cu(II)(3,5-DIPS)$_2$-treated VYB mice exhibited a higher mortality rate than did the Cu(II)(3,5-DIPS)$_2$-treated C57BL/6 mice. Vehicle-treated

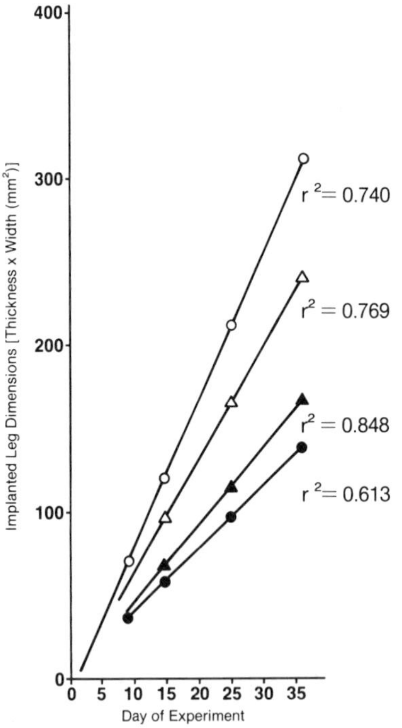

Figure 3. Regression for tumor growth in C57BL/6 mice treated with vehicle (○) or 25 mg/kg Cu (II) (3,5-DIPS)$_2$ (●) twice daily and VYB mice treated with vehicle (△) or 25 mg/kg Cu (II) (3,5-DIPS)$_2$ (▲) twice daily.

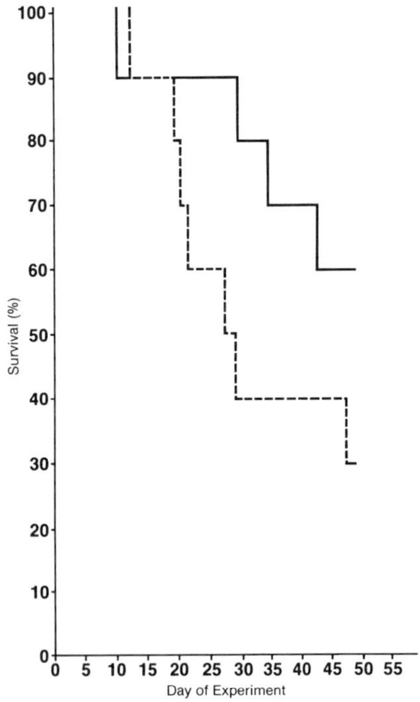

Figure 4. Survival of C57BL/6 mice (——) and VYB mice (– –) treated with 25 mg/kg Cu (II) (3,5-DIPS)$_2$ twice daily.

mice of either strain exhibited no mortality during the entire experimental period.

All repeated measurements in the C57BL/6 mice showed that tumors were significantly ($p \leq 0.02$) larger in mice treated with vehicle. However, only one out of six repeated measurements of tumors in the VYB mice showed a significant ($p < 0.05$) difference between vehicle-treated and Cu(II)(3,5-DIPS)$_2$-treated mice. On this occasion, tumors were significantly ($p \leq 0.02$) larger in the vehicle-treated mice. Only the initial measurement showed that tumors of vehicle-treated C57BL/6 mice were significantly ($p \leq 0.002$) larger than tumors of vehicle-treated VYB mice, all other such comparisons not being significantly ($p \leq 0.05$) different. On this single occasion, tumor size was larger for the C57BL/6 mice. On no occasion were measurements of tumor size significantly ($p < 0.05$) different upon comparison of Cu(II)(3,5-DIPS)$_2$-treated C57BL/6 mice and Cu(II)(3,5-DIPS)$_2$-treated VYB mice.

Discussion

These investigations demonstrate different rates of growth for the Ehrlich ascites tumor amongst different mouse strains as well as a difference in the anti-carcinogenic potential for Cu(II)(3,5-DIPS)$_2$ amongst both strains studied. It should be noted that when Cu(II)(3,5-DIPS)$_2$ is administered twice daily the rate of tumor growth is further reduced and that the variance of growth from linearity in regression is greater. One factor to which this may be attributed is that Cu(II)(3,5-DIPS)$_2$, administered at the proper dosage, may completely inhibit neoplasia. Indeed, there was one mouse which did not show tumor growth.

In our current ongoing study (data not reported), three of ten mice inoculated with tumor cells, treated initially with 25 mg/kg Cu(II)(3,5-DIPS)$_2$ for two weeks and thereafter with 6.25 mg/kg, have survived longer than all mice inoculated with tumor but not injected with Cu(II)(3,5-DIPS)$_2$. These three surviving mice show no evidence of continued presence of tumor beyond the fifth week of the experiment. Thus the possibility is posed that a treatment regimen using Cu(II)(3,5-DIPS)$_2$ can be developed to effectively treat Ehrlich cell tumor in mice.

Acknowledgments

We are indebted to the Arthur Armbrust Cancer Research Fund, International Copper Research Association, Max and Victoria Dreyfus Foundation, Elsa U. Pardee Foundation, and Denver Roller Corporation for financial support and the National Center for Toxicological Research for the mice used in these studies. We are also indebted to Professor Larry Oberley for supplying us with cells and his comments on this manuscript.

References

1. Oberley, LW; Buettner, GR. Cancer Res. 1979, 39, 1141.
2. Oberley, LW; Leuthauser, SWHC; Buettner, GR; Sorenson, JRJ; Oberley, TD; Bize, IB. In: Pathology of Oxygen, Academic Press; New York NY, 1982; p 207.
3. DeAlvare, LR; Goda, K; Kimura T. Biochem. Biophys. Res. Comm. 1976, 69, 687.
4. Brigelius, R; Spottl, R; Bors, W; Lengfelder, E; Saran, M; Weser, U. Febs Lett. 1974, 47, 72.
5. Leuthauser, SWHC. Antitumor activities of superoxide dismutase and copper coordination compounds (Dissertation #8012387); University Microfilms International; Ann Arbor MI, 1979.
6. Leuthauser, SWC; Oberley, LW; Oberley, TD; Sorenson, JRJ; Ramakrishna, K. J. Nat. Cancer Inst. 1981, 66, 1077.
7. Sorenson, JRJ; Oberley, LW; Oberley, TD; Leuthauser, SWC; Ramakrishna, K; Vernino, L; Kishore, V. In: Trace Subst. Environ. Health--XVI; University of Missouri; Columbia MO, 1982; p 362.
8. Sahu, SK. Effects of dexamethasone on neuroblastoma cell differentiation, superoxide dismutase activity, and the possible role of negative oxygen ion on cell differentiation (Dissertation #8012417); University Microfilms International; Ann Arbor MI, 1979.
9. Kensler, TW; Trush, MA. Biochem. Pharmacol. 1983, 32, 3485.
10. Solanki, V; Yotti, L; Logani, MK; Slaga, TJ. Carcinogenesis 1984, 5, 129.
11. Wattenberg, LW. Cancer Res. 1985, 45, 1.
12. Cerutti, PA. Science 1985, 227, 375.
13. Copeland, ES. Cancer Res. 1986, 46, 5451.

DISCUSSION

JOHNSON: Are you talking about remission or just reduction of tumor size and growth in that time period?

KASEMEIER: We are talking about reduction in tumor growth.

JOHNSON: Not remission of the tumor itself?

KASEMEIER: No, not yet.

JOHNSON: Would these legs of the Cu(II)(3,5-dips)$_2$ treated mice sluff off if you had let them go on in time? Did you just slow down the process?

KASEMEIER: Apparently the degenerative loss of the limb was prevented with treatment.

CROOK: You didn't say anything about metastasis. Did you happen to notice if that was a phenomenon in the Cu(II)(3,5-dips)$_2$ treated rats?

KASEMEIER: Yes, metastasis is observed and that will be covered in the histopathological presentation by Ms Torregrosa which is the next presentation.

ELLIOTT (Graham): Do these tumors produce prostaglandin E_2 and does your compound inhibit this production? It could account for your decreased vascularization.

KASEMEIER: That is quite likely. We did not measure PGE_2.

ELLIOTT (Graham): This reaction could explain your decreased tumor growth.

KASEMEIER: Yes it could. These studies did not cover that topic. At this point we are simply attempting to determine whether or not we can effectively inhibit tumor growth.

PICKART: The differentiation of cells grown in culture looks alot like what has been reported for many cancer cell lines using dibutyryl cyclic AMP. Do you see the comparison there?

KASEMEIER: Yes, there may be some interaction leading to stimulation of cyclic AMP synthesis.

EGNER: Have you measured myeloperoxidase or any other indicator of inflammation, or just leg size?

KASEMEIER: Just tumor leg diameter and mortality at this point.

EGNER: With topical applications of Cu(II)(3,5-dips)$_2$ we have been able to show that it is a real strong biometic SOD but I have never personally been able to get it to inhibit inflammation across the board. Concentrations we use seem to reduce myeloperoxidase about half the time. Secondly, I'm not sure why you picked C57 mice and I have no idea what a Viable Yellow Black mouse is. Could you explain your mouse models? For topical applications of tumor promoters and carcinogens that we use, the C57 mouse is one of the least sensitive strains. That's why I was surprised you used it. It might be very sensitive in your system.

KASEMEIER: It is true that Prof. Oberley originally used CBA mice and not C57 Black mice. These were preliminary studies aimed at determining whether or not we could develop a treatment protocol to reduce tumor growth. We used the C57 Black and Viable Yellow Black mice because they are free. We do not have to buy them.

CRISPENS: Why did you decide to use such aggressive therapy? Why did you go after this with such high doses, twice daily, and get the toxicity that you got?

KASEMEIER: Susan Leuthauser and Prof. Oberley originally used the 25 mg/kg dose in their evaluations of Cu salicylates as antineoplastic agents. To begin our work, intended to develop a treatment protocol, we used that same dose and gave it twice a day to achieve beneficial effects. We did not intend to produce any toxicity. In our current study, that is ongoing at the moment, we have reduced the dose and we are achieving beneficial effects with much less toxicity.

CRISPENS: I understand that, but every 12 hours?

KASEMEIER: Yes, that is a lot of injecting but that should allow us to achieve a more constant blood level which may produce more effective reduction in tumor size.

EFFECTS OF $Cu(II)(3,5\text{-}DIPS)_2$ ON SOLID EHRLICH CELL TUMOR IN MICE: A PATHOLOGICAL STUDY

Doris Torregrosa[1], Scott L. Kasemeier[2], John R.J. Sorenson[2], and Louis W. Chang[1]

[1]Department of Pathology, College of Medicine and [2]Department of Biopharmaceutical Sciences, College of Pharmacy, University of Arkansas for Medical Sciences, Little Rock, Arkansas 72205, U.S.A.

INTRODUCTION

The first evidence indicating that Cu (II) (3,5-diisopropylsalicylate)$_2$ (CuDIPS) possessed antitumor activity was published in 1981 by Leuthauser et al.[1]. This group of investigators showed that when Ehrlich ascites cells were injected into thigh muscle of adult mice, solid tumors developed. Subsequent administration of CuDIPS directly into the tumor site yielded both retardation and reduction of tumor growth with an increased survival rate when compared to the saline treated controls. Using light microscopy, these investigators found that the metastatic activity of the tumor cells was also reduced and delayed in CuDIPS treated mice. Small nodules of metastatic tumor cells were found in the lungs of the saline controls as early as 10 days after implantation whereas, metastasis was not observed until day 45 in CuDIPS treated mice.

Recently published data also suggests that CuDIPS had anticarcinogenic activities both *in vitro*[2,3] and *in vivo*[4-6].

The present investigation represents a preliminary study designed to provide some histological evidence that will clarify the role of CuDIPS as an antitumor and/or anticarcinogenic agent. Our findings should clarify:

1) the effects of systemically administered CuDIPS on the solid Ehrlich cell tumor at the cellular and tissue levels, 2) the effect of treatment on the metastatic activity of the tumor cells, and 3) any CuDIPS related toxicity.

EXPERIMENTAL DESIGN

Sixty young adult (20 gram) female mice (C57BL/6N) were used in this experiment. They were divided at random into four different treatment groups:

VT: vehicle treated, tumor bearing
1 Cu T: CuDIPS treated once a day, tumor bearing
2 Cu T: CuDIPS treated twice a day, tumor bearing
2 Cu: CuDIPS treated twice a day

Solid tumors were induced by implantation of 5×10^6 ascitic cells of the Ehrlich carcinoma cell line (in 0.1 ml saline) intramuscularly into the thigh muscle. Following tumor implantation, mice were injected subcutaneously on their backs with either vehicle (composed of 4% propylene glycol and 1.4% polyvinyl alcohol in saline) or CuDIPS in vehicle. Injections consisted of 25 mg/kg b.w. of CuDIPS once or twice a day, every 12 hours.

Two-dimensional measurements of the thighs were made every week in order to estimate tumor growth.

After 2.5 months (76 days), treatments were terminated. Surviving mice (25 animals) were kept in their cages until they were close to death. Tissues from tumor masses and various organs (lung, liver and kidneys) were excised, processed by standard histological techniques and stained with hematoxylin and eosin for pathological examination. Data was statistically evaluated by analysis of variance and Duncan's multiple range test.

RESULTS AND DISCUSSION

Survival of Animals and tumor growth:

The following table summarizes data obtained from survival post-treatment from each group and also the average leg measurements of mice prior to death:

Treatment	Survival Post-Treatment	Leg Measurements (Average size mm^2, $\pm$ S.E.M.)
VT	3 weeks	578.2 $\pm$ 76.1
1 Cu T	1 week	750.5 $\pm$ 42.7
2 Cu T	2 weeks	225.6 $\pm$ 44.4*

* significant change over control, $p < 0.05$

Our findings clearly demonstrated that CuDIPS, in a dose-related fashion, reduced growth of the primary tumor. The shorter survival time of CuDIPS treated animals may be related to the dose employed or to metastatic consequences following CuDIPS treatment (see following sections).

Histopathology of tumor:

Examination of primary tumors revealed distinct differences between vehicle and CuDIPS treated animals. While vehicle treated mice showed much larger tumors with massive central necrosis and cellular invasion, tumors from CuDIPS treated animals displayed much less necrotic and invasive activities.

Closer examination of these tumors also revealed a striking difference in the number of observable mitotic figures (mitotic index) in vehicle and CuDIPS treated tumors. The mitotic index appeared to be higher in CuDIPS treated animals than in vehicle treated mice. Since most of these mitotic figures were in the same mitotic phase (metaphase) (Figure 1,2) and CuDIPS treated animals had smaller tumors, we suggest that this phenomenon reflected an "arrest" of cellular division at metaphase, resulting in an "increased" detectable mitotic figures at metaphase with a decreased tumor growth in the CuDIPS treated animals.

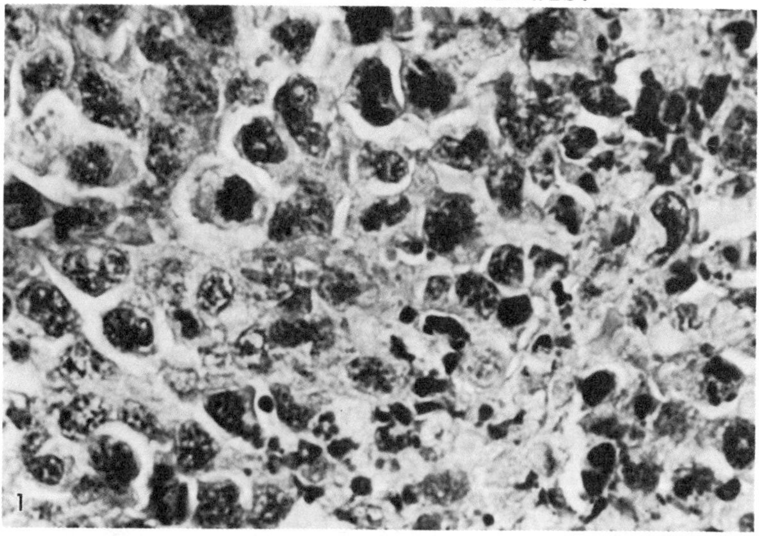

Fig. 1. Primary tumor, VT. The tumor consists of large cells with hyperchromatic nuclei and few mitotic figures. 40 X

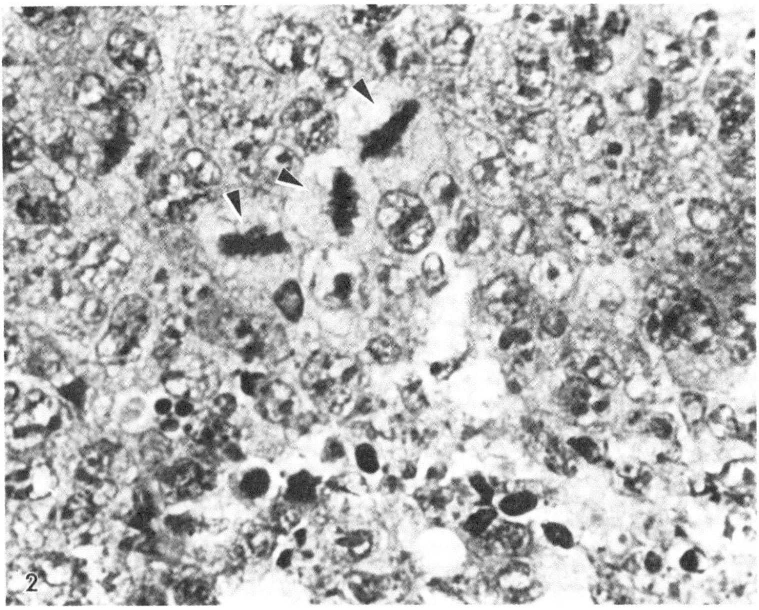

Fig. 2. Primary tumor, 1 Cu T. Note the increased number of mitotic figures at metaphase (►). 40 X

Our present investigation, also for the first time, demonstrates antitumor properties of CuDIPS even when the compound is administered systemically, at a distance from the tumor mass.

Metastasis

Among the animals examined, vehicle treated tumor bearing animals (VT) showed both renal and pulmonary tumor metastases with 40% and 20% incidence in the kidney and lung respectively. Based on these data, we suggest that the primary or major site of metastasis was the kidney under normal conditions.

With CuDIPS treatment, the metastatic pattern seemed to be altered showing higher incidence of lung metastasis than renal metastasis. The overall results are summarized in the following table:

Incidences of Tumor Metastasis in Animals Examined

Treatment	Kidney	Lung
VT	40%	20%
1 Cu T	20%	75%
2 Cu T	0%	50%

Although both CuDIPS treated groups showed higher incidences of pulmonary metastasis than the control (VT) animals, the higher dose group (2 Cu T) demonstrated a reduction of lung metastasis compared to the lower dose group (1 Cu T). Thus, while low levels of administration yielded more pulmonary metastasis, high concentration of CuDIPS might suppress such occurrence. Indeed, by injecting CuDIPS directly into the tumor site (therefore introducing a tremendously high concentration of CuDIPS into the tumor mass), Leuthauser et al.[1] reported a delayed and lower incidence of pulmonary metastasis than the vehicle-treated counterparts.

These results also indicate that CuDIPS reduced the potential of Ehrlich cell tumor metastasis to its primary metastatic site (kidney) in a dose-related fashion, i.e. VT > 1 Cu T > 2 Cu T. In fact, higher doses of CuDIPS totally eliminated or blocked renal metastasis. Histologically, metastatic tumors resembled those of the primary tumor. Sizes of renal metastatic tumors were also much larger in vehicle treated animals than in CuDIPS treated mice (Figures 3,4).

The larger dose of CuDIPS (2 Cu T) also showed a lower incidence of pulmonary metastasis than those in lower dose treatment group (1 Cu T). Histological examination also revealed that tumors in lungs of 2 Cu T animals were much smaller than those in 1 Cu T treated animals

(Figures 5,6).

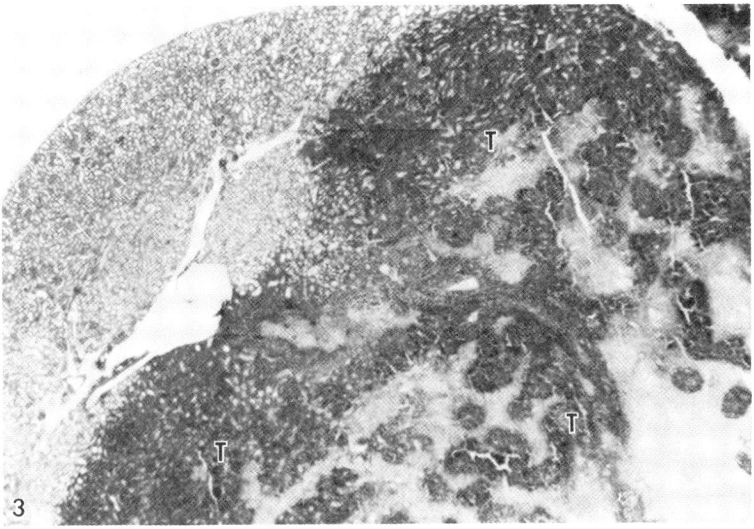

Fig. 3. Renal metastasis, VT. Large metastatic tumor (T) in the kidney. 2.5 X

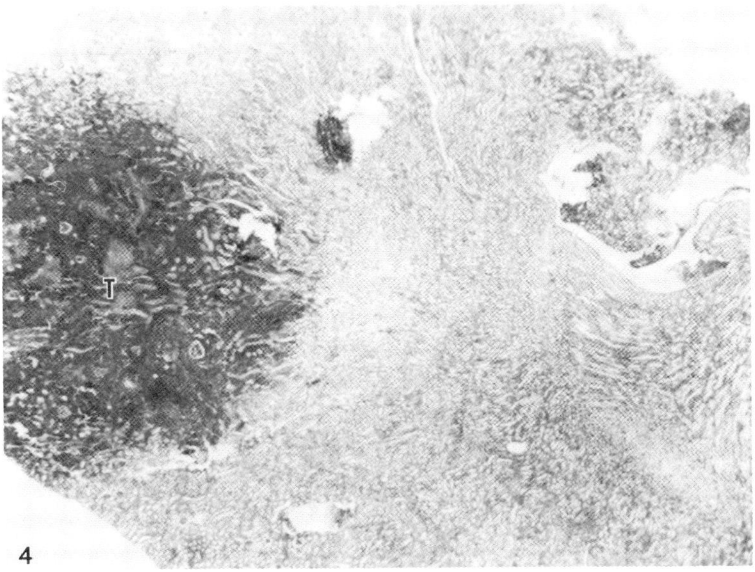

Fig. 4. Renal metastasis, 1 Cu T. A small metastatic tumor nodule (T) in the kidney is demonstrated. 2.5 X

The lower incidence of pulmonary metastasis and the shift of metastatic pattern with CuDIPS treatment was indeed intriguing. In the vehicle treated mice, the primary site of metastasis appeared to be the kidney with lungs being the secondary site. Under the influence of CuDIPS, an increase in the secondary route of metastasis (lung) occurred when the primary route (kidney) was reduced or blocked. The basis for this change of metastatic preference after CuDIPS treatment still awaits further investigation.

Since the mortality rate was found to be directly correlated with the incidence of pulmonary metastasis (1 Cu T > 2 Cu T > VT), these data suggest that the cause of death in these mice was closely related to or associated with pulmonary tumor development. This finding is also in agreement with observations reported by Leuthauser et al.[1]

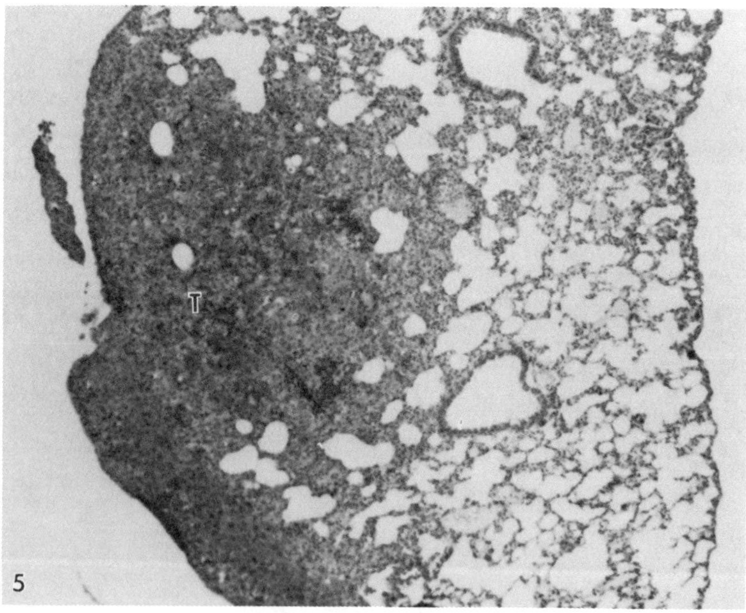

Fig. 5. Pulmonary metastasis, 1 Cu T. Large area of the lung was occupied by metastatic tumor (T). 2.5 X

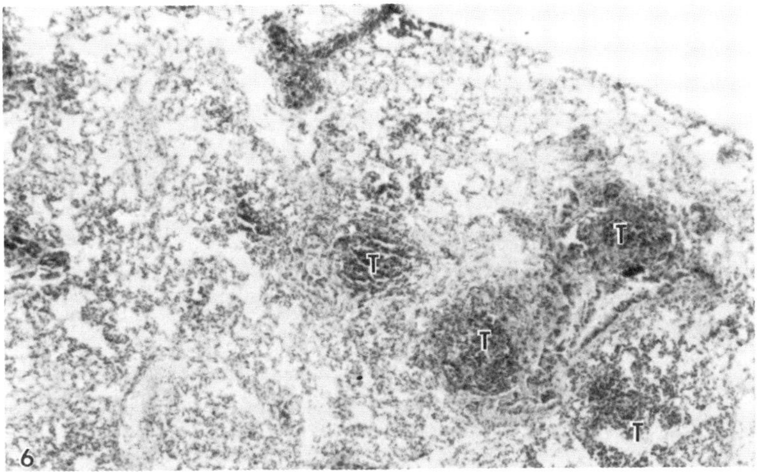

Fig. 6. Pulmonary metastasis, 2 Cu T. Only microscopic nodules of metastatic tumor (T) were observed. 2.5 X

CuDIPS induced hepatic changes:

Histological examination of the livers revealed massive accumulations of inflammatory cells (PMN) around hepatic sinusoids in vehicle treated mice (Figure 7). A reduction of PMN's occurred with CuDIPS treatment which may be related to the antiinflammatory actions of CuDIPS.[7,8] A dose related proliferation and hypertrophy of Kupffer cells was also observed in CuDIPS treated animals. This feature was most prominent in non-tumor bearing CuDIPS treated animals (Figure 8). A similar macrophage reaction was observed in the liver of rats treated with twenty times the effective antiinflammatory dose of Cu $(II)_2$ $(aspirinate)_4$ in a chronic toxicity study, which remitted when treatment was stopped. The precise mechanism of this interaction and phenomenon is still unclear at this time. However, one may postulate that the Kupffer cells (macrophages) were stimulated and activated for the removal of increased amounts of Cu-thioneine stored in the liver.

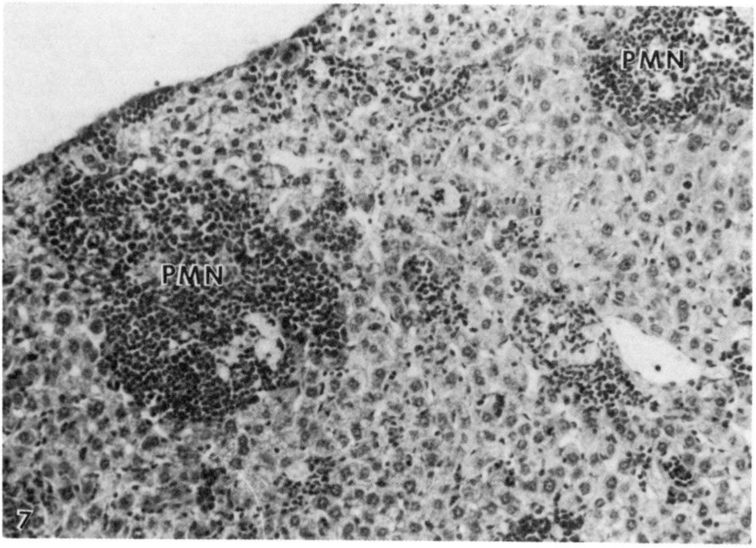

Fig. 7. Liver, VT. Aggregates of infiltration of inflammatory cells (PMN) were observed. 10 X

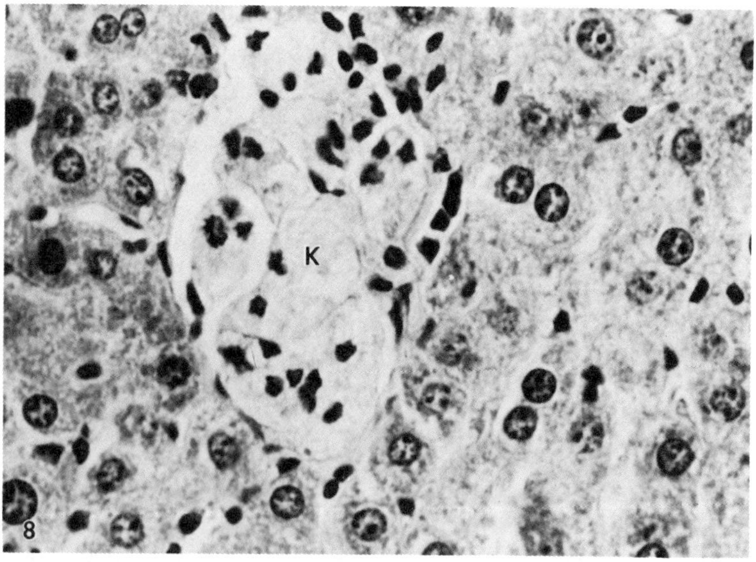

Fig. 8. Liver, 2 Cu. Hypertrophy and aggregation of Kupffer cells (K) were observed in the sinusoidal space. 40 X

CONCLUSIONS

Based upon this preliminary investigation of CuDIPS as an antitumor agent, the following conclusions may be drawn:

1. CuDIPS retarded tumor growth when administered systemically to mice.

2. CuDIPS reduces the growth of the primary tumor and also alters the histological characteristics of the tumor (i.e. less invasive, less necrotic, mitotic arrest).

3. The primary metastatic site of Ehrlich cell tumor is the kidney with lung as the secondary site. CuDIPS reduces or blocks renal metastasis. Higher dose of CuDIPS also induced less pulmonary metastasis than lower dose.

4. The dose of CuDIPS used in this study displayed a stimulation of Kupffer cell hypertrophy and activity in liver.

5. Under our present experimental conditions, CuDIPS treated mice, had a shorter survival time than vehicle treated animals. The cause of death in these mice is possibly related to pulmonary metastasis. Future studies using more appropriate doses and treatment schedules may decrease toxicity, reduce pulmonary metastasis, and increase survival time without a loss of antitumor effects.

ACKNOWLEDGMENT

We would like to thank Ms. Terri Bryant for her excellent secretarial help.

REFERENCES

1. Leuthauser, S.W.C.; Oberley, L.W.; Oberley, T.D.; Sorenson, J.R.J.; Ramakrishna, K. J. Natl. Ca. Inst. 1981, 66, 1077.

2. Nakamura, Y.; Colburn, N.H.; Gindhart, T.D. Carcinogenesis 1985, 6, 229.

3. Kennedy, A.R.; Troll, W.; Little, J.B. Carcinogenesis 1984, 5, 1213.

4. Solanki, V.; Yotti, L.; Logani, M.K.; Slaga, T.J. Carcinogenesis 1984, 5, 129.

5. Kensler, T.W.; Bush, D.M.; Kozumbo, W.J. Science 1983, 221, 75.

6. Egner, P.A.; Kensler, T.W. Carcinogenesis 1985, 6, 1167.

7. Weser, U.; Richter, C.; Wendel, A.; Younes, M Bioinorg. Chem. 1978, 8, 201.

8. Sorenson, J.R.J.; Oberley, L.W.; Crouch, R.K.; Kensler, T.W.; Kishore, V.; Leuthauser, S.W.C.; Oberley, T.D.; Pezeshk, A. Biol. Trace Elem. Res. 1983, 5, 257.

9. Sorenson, J.R.J.; Rolniak, T.M.; Chang, L.W. Inorg. Chim. Acta. 1984, 91, L31.

DISCUSSION

LOVEN: Another possible explanation for your change in metastatic pattern, as Gloria Happener and several others have shown, is that as the tumor progresses it develops specific cell types that have a propensity to metastasize to certain organs and your treatment may actually be killing off a specific metastatic population so that the complex may be actually acting on the tumor and not on the target organ.

TORREGROSA: Yes, that could be.

CROOK: Do you then think that $Cu(II)(3,5\text{-dips})_2$ is or is not going to be an effecive anticancer agent?

TORREGROSA: Indeed we definitely saw some antitumor effects in terms of growth of the primary tumor and inhibition of metastasis to kidneys. This was just a preliminary study. We still have much more work to do concerning morphologic changes and also concerning different doses of this compound to determine whether or not it can or cannot be used as an anticancer agent.

CROOK: In one of your slides you compared survival time and treatment. The longest survival, if I understood your table, was for the vehicle treated group.

TORREGROSA: The longest survival time was found for the $Cu(II)(3,5\text{-dips})_2$-treated animals that were non-tumor bearing.

CROOK: You had 76 days for each one of them?

TORREGROSA: Yes, for the most part the study lasted 76 days and those groups that survived longest were $Cu(II)(3,5\text{-dips})_2$-treated animals, that were non-tumor bearing.

CROOK: Of those 3 groups that had tumors, the vehicle-treated tumor bearing animals survived the longest and the other 2 had somewhat shorter survival times. Based on that, you are in essence saying that the vehicle was the best agent and that $Cu(II)(3,5\text{-dips})_2$ given once a day or twice a day was less effective.

TORREGROSA: Recall that all treatments of these animals were stopped. Metastasis, which was probably the ultimate cause of death, could have developed post treatment. Tumor bearing animals that were vehicle-treated actually had higher rates of kidney metastasis. So the kidney was the target for metastasis in these vehicle-treated animals. We still have to play with the dose to see if lowering the dose will make any difference in terms of metastasis and survival of these animals. You are right, it does seem as though the vehicle-treated animals do better because they survive longer but they have much more kidney metastasis. So kidney metastasis is killing these animals.

CROOK: If you showed me those 3 tissues, I would have to say they are wildly growing tumors, in the presence of Cu(II)(3,5-dips)$_2$.

TORREGROSA: But we see the absence of tumors in kidneys of tumor bearing mice treated with Cu(II)(3,5-dips)$_2$ twice a day.

CROOK: It's immaterial if you die of pulmonary metastasis or renal metastasis. No one dies of a primary tumor unless it happens to occur in an outflow tract of the heart or something like that. You've created a killer tumor with Cu(II)(3,5-dips)$_2$. With what you have shown it would be difficult for me to cast a vote in favor of Cu(II)(3,5-dips)$_2$ as an antineoplastic agent.

SORENSON: Would you agree that this is a good first attempt to evaluate chronic use of anything wherein the evaluation also includes histopathology? The important point of this study is to teach that we have begun to look at the possible use of this compound with regard to treating tumor. We picked one of the most ferocious tumor model systems known to man to begin this work. We already know we can effectively treat liquid tumors.

CROOK: I think there are some other qualifying things that you might do to cast it in a better light if you will. For example, the number of tumor cells being injected could be decreased. We know this from experience using nude mice in our monoclonal and antibody work that that's really critical. As far as being able to cast a favorable ballot, I'm not sure which way you would propose going with the dose, up or down, to get around the longer survival being associated with the vehicle. I think I would go other ways and look at the type of tumor, the particular animal model being used, and maybe a different animal model.

CHANG: I think we are all jumping the gun to make a conclusive ballot as you put it. Obviously this is a very very early investigation regarding the usefulness of this compound. We cannot just use mortality or toxicity as a basis of whether a certain compound has useful anti-tumor effects. All anti-tumor compounds are toxic, right? Radiation is toxic. I am sure that those who first suggested the use of radiation met with comments like; my God, you're going to irradiate this patient, he's going to die from radiation. An appropriate response would have been; we must determine what type of tumor might be sensitive to radiation. What dose of radiation can we use? We are trying to determine whether or not Cu(II)(3,5-dips)$_2$ affects characteristics of the Ehrlich solid tumor. This is step one. Another question would be, do all tumors behave the same way? What kind of dose can we use? What kind of animal model could we use to evaluate the anti-tumor potential of Cu(II)(3,5-dips)$_2$ and other compounds like it? I think we are moving in the right direction at this time. If we are able to treat animals bearing the solid Ehrlich cell tumor and change the characteristics of its growth and its metastatic pattern we can play with this anti-tumor activity, and perhaps, find useful applications. We can modify the dose and frequency of treatment. If we can block kidney metastasis, we must find out why we can block it. If we can find out why maybe we can block lung metastasis. At this point we should not say these animals all died so there is no value in studying it further. I see alot of value in further study and I think we are moving in the right direction. We cannot ask anyone, now , to cast a ballot to say whether or not Cu(II)(3,5-dips)$_2$ can be used as a therapeutic drug to treat cancer. No one can say that. I don't think you would deny the fact that indeed we are changing the characteristics of this tumor. We have favorably influenced the characteristics of this tumor in some tissues but the situation is not quite what we want in other tissues yet. That's why we need money for further studies.

CROOK: Certainly all radiation is not toxic. Obviously if it were, none of us would travel in airplanes where we are exposed to substantial amounts of cosmic radiation. If you desire money, you are going to have to show that the drug is more effective than the vehicle, or you will consume a large amount of time and effort for no useful purpose. The drug is essentially a good one. One should look for a better use for it and the wound healing area may be a better field to plow at this time than the antineoplastic field.

BARNEA: How important is it to have the Cu complexed to 3,5-dips. Are there any studies of treatment of this type of cancer with Cu(II) bleomycin or other Cu binding glycoproteins? Is this a better complex then those, or are they acting differently? Did you do anything with these types of complexes? They are quite effective in some cancers.

SORENSON: Bleomycin, the anticancer drug, has caused death in patients because of its cardiac toxicity. Cu bleomycin is supposedly less toxic. We've never seen any cardiac toxicity with $Cu(II)(3,5\text{-dips})_2$. Based upon its profile of pharmacologic effects it is an outstanding candidate among all anticancer drugs for these kinds of studies.

COPPER AND COPPER ENZYMES IN THE DEVELOPMENT AND TREATMENT OF CANCER

Dean P. Loven, Ph.D.

East Carolina University

Greenville, North Carolina 27858-4354

INTRODUCTION

Recent work on copper and cancer has centered on the use of low molecular weight copper complexes with superoxide dismutase (SOD) activity for cancer therapy. The rationale for using such compounds is based on extensive studies of SOD activity in normal and malignant tissues. As the Cu-Zn SOD is a copper containing protein, changes in copper metabolism, should they occur in cancer, could alter levels of this enzyme. However, changes in copper metabolism, which are directly linked to cancer have been reported for relatively few malignancies. Copper complexes have also been used to block tumor promotion, suggesting superoxide is involved at this step of carcinogenisis. I will attempt to summarize these various findings in this paper, making reference to various reviews for a more extensive discussion of these areas.

COPPER AND COPPER ENZYMES IN TUMOR DEVELOPMENT

Copper supplementation decreased the carcinogenic effects of many carcinogens, possibly by promoting their metabolism (reviewed in [1]). While copper is carcinogenic *in vitro*, and high copper in the diet has been correlated to stomach cancer, it is felt copper is not a carcinogen, although copper can interfere with the uptake of the anticarcinogen selenium (reviewed in[1]).

Treatment with the tumor promoter 12-0-tetradeconylphorbol-13-acetate (TPA) depressed total SOD activity in FF1 fibroblasts, mouse epidermis, and human leukocytes, but not V79 fibroblasts (reviewed in [2]). TPA treatment of isolated mitochondria did not affect Mn SOD activity, indicating Cu-Zn SOD is probably affected. TPA caused generation of superoxide by macrophages, fibroblasts, and epidermal cells[2,3]. It is possible the depression of Cu-Zn SOD occurred by generation of excess hydrogen peroxide, the product of the dismutation reaction, which can inactivate the enzyme[4]. The promoting effects of TPA can be blocked by exogenous SOD, or by copper II (3,5-diisopropylsalicylate)$_2$, a lipophilic low molecular weight compound with[2] SOD-like activity[2,5,6,7,8].

COPPER ENZYMES IN CANCER

Various reports indicated certain copper containing enzymes were altered in specific cancers, however, no consistant pattern of change has emerged. Serum and tumor levels of tyrosinase were elevated in some cancers[9,19]. It was proposed this enzyme could hydroxylate tyrosine residues of proteins such as antibodies and interferons, and so disrupt the response of the host to the tumor[20]. However, it is not known how common the change is in tyrosinase levels.

Cu-Zn SOD activity was variably changed in different human and animal cancers, while Mn SOD was almost universally lower or absent[21]. It is important to have proper controls for comparison of SOD activity, as SOD activity can vary with tissue type, and even between cell types within a tissue. Further, less differentiated cells, or cells undergoing division, have different SOD activities when compared to differentiated or quiescent cells from the same population[21]. As changes in both Mn SOD and Cu-Zn SOD may occur, it is also necessary to differentiate the activity of the two enzymes. Cu-Zn SOD was depressed in neuroblastoma, chemically induced large bowel carcinoma (compared to crypt), and Morris[21] and Yoshida hepatoma[22], and human thyroid tumors, lymphomas, intermediate and squamous cell epithelioma, childhood acute lymphocytic, T and B cell leukemia (reviewed in[21]). Cu-Zn SOD activity was elevated in L1210 leukemia and human large bowel carcinoma (reviewed in[21]). The enzyme was unchanged in basal cell carcinoma and P388 lymphocytic leukemia. In addition, Cu-Zn SOD is not as readily induced in malignant cells by oxygen[26], thyroxine[27] or superoxide generated by paraquat[28].

Various studies also measured SOD activity in organs of tumor bearing animals. Hepatic Cu-Zn SOD activity was elevated at all times after implantation of the Ehrlich ascites carcinoma tumor (EATC) into mice, while splenic Cu-Zn SOD was depressed[29]. Renal and skeletal muscle Cu-Zn SOD was transiently depressed, then elevated following implantation, while the activity in the lung was transiently elevated after implantation. These effects were different from the changes observed in the same study for Mn SOD activity[29]. Total SOD activity in liver and lung did not change but erythrocyte SOD activity was elevated at twelve days and depressed at 15 days after implantation of the Lewis lung carcinoma into mice[30]. Of several organs studied only hepatic Cu-Zn SOD activity decreased one to five weeks after implantation of the Vx-2 carcinoma into rabbits. The Cu-Zn SOD activity was also depressed in the liver of animals bearing the Yoshida ascites hepatoma[22]. Erythrocyte SOD activity was not altered in patients with various visceral cancers, but was elevated in acute myelocytic leukemia, chronic lymphocytic leukemia, lymphosarcoma, and Hodgkins lymphoma[32,33]. Elevated SOD activity did not correlate with inflammatory syndrome, stage, or dissemination of the disease, but highest levels were seen prior to or shortly after the initiation of therapy. The authors speculated the SOD gene was in some way altered in the hematopoietic stem cells in leukemia and lymphoma. Some patients on Endoxan Alkaran Oncovin therapy had lowered erythrocyte SOD levels which correlated with intravascular hemolysis[34]. Removal from this therapy restored SOD levels to normal and eliminated hemolysis.

COPPER AND COPPER ENZYMES IN CANCER THERAPY

Copper sulfate injection did not effect the growth of the EATC line in mice[35]. The copper complexes of substituted thiosemicarbazones and 3,4,7,8-tetramethyl-1,10-phenanthroline do have antitumor effects in animal models (Reviewed in[9]). Whether these compounds have SOD-like activity, as discussed below, is not known.

In addition to altered SOD activity, catalase levels were low in most malignancies[21], and at least some tumors were defective in glutathione metabolism[36]. Oberley and colleagues have used these observations to modify tumor antioxidant enzyme activity in order to increase the efficacy of the superoxide generating drug adriamycin[37]. Treatments combining adriamycin with diethyldithiocarbamate (DDC) to inhibit Cu-Zn SOD and 1,3-bis(-2-chloroethyl)-nitrosurea (BCNU) to inhibit glutathione reductase significantly increased survival of mice bearing the EATC or L1210 leukemia[21,35,37,38]. It was not clear if the effect of diethyldithiocarbamate DDC was due to inhibition of Cu-Zn SOD or other copper containing enzymes by copper chelation, or due to sulfhydryl binding activity which caused inhibition of enzymes such as aldehyde dehydrogenase[39]. It was possible that Cu-DDC complexes have SOD-like activity[37]. Disulfiram, which is metabolized to DDC, is toxic when given with adriamycin and BCNU[37]. The copper chelating agent penacillamine had some antitumor effect, but was not as effective as DDC[37].

Oberley and colleagues proposed tumor cells might be sensitive to hydrogen peroxide due to lowered catalase activity, and thus be sensitive to hydrogen peroxide generated by increased SOD activity[38,40]. CuDiPS[35,38] and copper-II glycylglycylhistidine plus ascorbate (which has not been tested for SOD activity)[41] had antitumor activity in the EATC model. In contrast, Cu-Zn SOD slightly increased growth of the ascites tumor Perugia (mammary carcinoma) in vitro, possibly by protecting from hydroxyl radical formation[42]. It is not known how CuDiPS inhibits tumor growth. If tumor cells are sensitive to hydrogen peroxide then addition of CuDIPS could increase hydrogen peroxide generation from superoxide and kill the cell. The protective effect of glutathione and sensitizing effect of BCNU when given with CuDiPS supports this idea[35,38]. Tumor cells generate superoxide which is detoxified via either remaining SOD activity in the cell or by reaction with other intracellular components such as glutathione or sulfhydryls[21,38]. A second possibility is superoxide, or a product of the reaction of superoxide with a cellular component such as sulfhydryls, is needed for tumor growth[42]. Cancer patients and tumor bearing animals have elevated levels of serum aldehydes, possibly generated by lipid peroxidation. Aldehydes inhibited immune cell function by altering membrane fluidity or surface receptors[44]. Nonimmune cells might also be affected by such products[45]. Addition of a lipophilic compound with SOD-like activity could prevent lipid peroxidation and protect cells from aldehydes, thus allowing the host immune system to respond to the tumor. It is also possible that removal of aldehydes could prevent damage of receptors on the tumor cell and restore some aspects of normal cell function. A third possibility is these SOD mimetic compounds are involved in the induction of cellular response to interferons. Cu-Zn SOD may be necessary for the initiation of interferon action[46]. Replacement of Cu-Zn SOD in normal or malignant cells with low molecular weight compounds may be adequate for inducing the antiviral state.

COPPER AND COPPER ENZYMES IN RESISTANCE TO THERAPY

Cells which survive radiation had elevated SOD activity, however,it is not clear if this is due to induction of the enzyme or due to endogenous differences in enzyme activity[47]. Exogenous SOD or CuDiPS protected animals and cells from the toxic effect of radiation[48,49]. As SOD may be important in protecting cells from radiation toxicity, induction of this enzyme by other treatments may protect the cells from combination therapy involving radiation.

Copper induced the synthesis of selected heat shock protein in various fibroblast cell lines[49,50]. Copper may induce the synthesis of these proteins by binding to cellular sulfhydryl groups[51]. Cu-Zn SOD activity was elevated in Chinese hamster ovary and ovarian carcinoma cells following hyperthermia and during the time period when thermotolerance was maximal[52]. In contrast, the cytotoxic action of polyamines at normal and hyperthermic temperatures was due to hydrogen peroxide generation, and excaberated by glutathione depletion, suggesting heated cells were more sensitive to hydrogen peroxide[53].

Adriamycin resistant P388 leukemia cells had normal levels of SOD, however most adriamycin resistant cells have decreased drug uptake[54]. CHO cells with elevated metallothionein levels were resistant to the alkylating agent melphalan[55]. An L-cell line variant high in metallothionein, a copper inducible protein, was also resistant to cis-dichloroamineplatinum[56]. It is possible elevated levels of intracellular copper could decrease the efficacy of these and possibly other drugs.

CONCLUSIONS

Superoxide dismutase is one of the few enzymes that is almost universally changed in cancer cells when compared to the appropriate normal cell type. This finding has implications for the development of cancer, as well as maintenance of the malignant state. Efforts to restore SOD activity or to inhibit the activity further in tumor cells has met with success in animal models of cancer therapy. Clearly these findings have potential for translation into the clinic.

ACKNOWLEDGEMENTS

Supported by Grant CA-42833 from the National Cancer Institute of the National Institutes of Health, DHHS, Bethesda, Maryland, and a grant from the Burroughs Wellcome Foundation.

REFERENCES

1 Owen, C. In: Copper Deficiency and Toxicity; Acquired and Inherited, in Plants, Animals, and Man; Noyes Publications; Park Ridge NJ, 1981; p.148

2 Troll, W;, Wiesner, R. Ann. Rev. Pharm. Tox. 1985, 25, 509.

3 Nakamura, Y; Colburn, N; Gindhart, T. Carcinogenesis 1985, 6, 229.

4 Loven D. In: Superoxide Dismutase, Vol. III; CRC Press; Boca Raton, FL, 1985; p 151.

5 Armato, U; Andreis, P; Romano, F. Carcinogenesis 1984, 5, 129.

6 Solanki, V; Yotti, L; Logani, M; Slaga, T. Carcinogenesis 1984, 5, 129.

7 Egner, P; Kensler, T. Carcinogenesis 1985, 6, 1167.

8 Schwarz, M; Peres, G; Kunz, W; Furstenberger, G; Kittstein, W; Marks, F. Carcinogenesis 1984, 5, 1663.

9 Owen, C. In: Physiological Aspects of Copper; Copper in Organs and Systems; Noyes Publications; Park Ridge NJ, 1982; p.230.

10 Margalioth, E; Schenker, J; Chevion, M. Cancer 1983, 52, 868.

11 Chakravarty, P; Ghosh, A; Chowdhury, J. J. Comp. Path. 1984, 94, 607.

12 Rubin, D; Rubin, E. Med. Hypothesis 1983, 10, 469.

13 Oberley, L. In: Superoxide Dismutase Vol. II; CRC Press; Boca Raton FL, 1982; p 127.

14 Vanella, A; Geremia, E; Pinturo, R; Pantane, G; Scarvaglieri, E; Cappellani, A; Garrotto, A; Bianchi, A; Zanghi, M. Pharmacol. Res. Comm. 1983, 15, 641.

15 Yoshimitsu, K; Yohnosuke, K; Usui, T. Acta Paediatr. Scand. 1984, 73, 92.

16 Hoffman, C; Webster, N; Wiggins, P; Chisholm, E; Giles, G; Leveson, S. Br. J. Ca. 1985, 51, 127.

17 Borrello, S; Seccia, A; Galeotti, T; Bartoli, G; Farallo, E;, Serri, F . Arch. Dermatol. Res. 1984, 276, 338.

18 Simon, L; Robin, E; Theodore, J. J. Cell. Physiol. 1981, 108, 393.

19 Loven, D; Guernsey D; Oberley L. Int. J. Ca. 1984, 33, 783.

20 Fernandez-Pol, J; Hamilton, P;, Klos D. Ca. Res. 1982, 42, 609.

21 Leuthauser, S; Oberley, L; Oberley, T; Loven, D. J. Natl. Ca. Inst. 1984, 72, 1065.

22 Capel, I; Thornley, A. Eur. J. Ca. Clin. Oncol. 1982, 18, 507.

23 Takada, Y; Noguchi, T; Okabe, T; Kajiyama, M. Ca. Res. 1982, 42, 4233.

24 Gonzales, R; Auclair, C; Voisin, E; Gautero, H; Dhermy, D; Boivin, P. Ca. Res. 1984, 44, 4137.

25 Saito, T; Kurasaki, M; Kaji, H; Saito, K. Ca. Letter 1984, 24, 141.

26 Auclair, C; Dhermy, D; Boivin, P. Cancer Chemother. Pharmocal. 1980, 4, 281.

27 Leuthauser, S; Oberley, L; Oberley, T; Sorenson, J; Ramakrishna, K. J. Natl. Ca. Inst. 1981 66, 1077.

28 Meister, A. Fed. Proc. 1984, 43, 3031.

29 Oberley, L; Schreiber, A; Rogers, K; Schutt, L; Loven, D; Oberley, T; Pasternak, R. In: Oxy Radicals and their Scavanger Systems Vol II: Cellular and Medical Aspects; Elsevier; New York, 1983; p 242.

30 Oberley, L; Rogers, K; Schutt, L; Oberley, T; Leuthauser, S; Sorenson, J. J. Natl. Ca. Inst. 1983, 71, 1089.

31 Eneanya, D; Bianchine, J; Duran D; Andresen, B. Ann. Rev. Pharmacol. Toxicol. 1981, 21, 575.

32 Nathan, C; Arrick, B; Murray, H; DeSantis, N; Cohn, Z. J. Exp. Med. 1980, 153, 766.

33 Kimoto, E; Tanaka, H; Gyotoku, J; Morishige, F; Pauling, L. Ca. Res. 1983, 43, 824.

34 Liotti, F; Guerrieri, P; Menghini, A; Pezzetti, F. J. Ca. Res. Clin. Oncol. 1986, 111, 47.

35 Yagi, K. In: Lipid Peroxides in Biology and Medicine, Academic Press, New York, 1982; p 223.

36 Roffman, E; Wilchek, M. J. Immunol 1986, 137, 40.

37 Dianzani, M. In: Free Radicals, Lipid Peroxidation and Cancer, Academic Press; New York, 1982; p 129.

38 Pottathil, R; Lang, D. Prog. Clin. Biol. Res. 1983, 135, 275.

39 Petkau, A. In: Superoxide Dismutase Vol III, CRC Press; Boca Raton FL, 1985; p 99.

40 Sorenson, J. J. Med. Chem. 1984, 27, 1747.

41 Levinson, W; Oppermann, H; Jackson, J. Biochim. Biophys. Acta 1980, 606, 170.

42 Atkinson, B; Cunningham, T; Dean, R; Somerville, M. Can. J. Biochem. Cell Biol. 1983, 61, 404.

43 Loven, D. Unpublished results.

44 Loven, D; Leeper, D; Oberley, L. Ca. Res. 1985, 45, 3029.

45 Henle, K; Moss A; Nagle, W. Ca. Res. 1986, 46, 175.

46 Ramu, A; Cohen, L; Glaubiger, D. Ca. Res. 1984, 44, 1976.

47 Tobey, R; Enger, M; Griffith, J; Hildebrand, C. Annals New York Acad.Sci. 1982, 397, 256.

48 Endresen, L; Schjerven L; Rugstad H. Acta Pharmacol. et Toxicol. 1984, 55, 183.

DISCUSSION

<u>SORENSON</u>: Has anybody ever studied changes in Cu-Zn SOD activity as a normal cell transforms to a neoplastic cell type?

<u>LOVEN</u>: I believe we will have an answer to that this afternoon from Patricia Egner.

<u>EGNER</u>: With repetitive topical TPA applications to the back of mice, we find approximately 50% reduction in SOD activity. This reduction seems to be specific for Cu-Zn SOD and not Mn SOD.

<u>LOVEN</u>: Oberley studied Morris Hepatoma cells and found that more aggressive growth rate was associated with lower SOD activity. One must not only consider cancer development but also tumor progression.

<u>HARRIS</u>: It's my understanding that rapidly growing cells characteristically have fewer mitochondria. To what extent does a reduced number of mitochondria enter into the observed reduction in Mn SOD?

<u>LOVEN</u>: That's a good point. However, while it's true that in most animal models of tumor cells divide rapidly, most human tumors do not show an increased growth rate. They may show a higher growth fraction but the rate of division is not necessarily higher. In the Morris Hepatoma study they also found a slight decrease in other enzymes in addition to SOD. There seems to be a reduction in both number and SOD activity of mitochondria.

<u>BALL</u>: I don't think I have ever seen a tumor cell that had more than half as many mitochondria as the cell from which it's derived. But that still isn't much of a reduction in energy production capability. Oberley showed that Mn SOD decreased to zero in cancer cells. The Hela cell has no Mn SOD and that's true for a number of other cancer cells which fits the Warberg theory and it's the most attractive possibility.

<u>LOVEN</u>: We have been studying refeeding in cultured CHO cells and what refeeding does to SOD. Refeeding these cells reduces Mn SOD 70%. High concentrations of glucose

induces glycolysis which is consistent with cell division. What's cause and effect in cancer isn't known but they're more glycolytic and definitely geared for growth.

FIELDS: Although there is a reduction in SOD activity, has anybody really measured peroxidative damage? Peroxidative damage produced in vitro with exogenous agents such as Fe or ascorbic acid is suggested as etiologic but isn't it important to find out what peroxidative damage occurs in cancer tissue without stimulation with these agents?

LOVEN: Sato has shown that the thiobarbituric acid reactive material in serum of a cancer patient was elevated so there is some form of lipid peroxidation. The exact site of free radical damage, to the best of my knowledge, has not been investigated. In contrast to that, some of the older literature paradoxically suggests that a tumor cell is more resistant to Fe-induced lipid peroxidation.

CATECHOLAMIDE IRON CHELATORS: ANTIPROLIFERATIVE ACTIVITY ON HUMAN PATHOGENS AND NEOPLASM

Gary T. Elliott
Department of Biopharmaceutical Sciences
University of Arkansas for Medical Sciences
4301 W. Markham Little Rock, Arkansas 72205

INTRODUCTION

A ubiquitous element, iron constitutes five percent of the earth's crust. Iron flavoproteins, hydroperoxidases, cytochrome P_{450}, oxygenases and ribonucleotide reductase are among numerous iron-dependent enzymes.[1] Accessibility of iron by procaryotic and eucaryotic unicellular and multicellular organisms is limited by the poor aqueous solubility of iron. Soluble ferric iron concentrations are only 10^{-18}M in neutral aqueous solutions.[2,3] Bacteria, on the other hand, require iron concentrations of 1×10^{-6} to 5×10^{-8}M for efficient growth.[4] Bacteria and fungi secrete into their environment low molecular weight (500 - 1000 dalton) non-peptide ferric iron specific chelators known as siderophores to accomplish iron assimilation.[5-8] In contrast, most mammalian cells utilize the protein transferrin as their iron transport chelators.[9] One transferrin molecule contains two sites capable of binding one ferric iron atom each. Iron binding is pH sensitive with best activity at pH values above 6.5.[9] SV40-transformed BALB/3T3 cells grown in α-picolinic acid do not have a transferrin requirement.[10] These transformed cells secrete a 1600 dalton peptide known as "Siderophore-Like Growth Factor (SGF)".[11] SGF has been shown to bind ferric iron and deliver it to SV40-transformed BALB/3T3 cells. In non-transformed mammalian cells, DNA synthesis requires the transferrin-dependent delivery of iron.[12,13] Mitotic activity of some cells has been associated with great proliferation of the 90,000 dalton transferrin membrane receptor.

Microbial siderophores on the other hand are non-proteinaceous in nature with the vast majority of the structurally determined chelators falling into either of two chemical classes, namely the hydroxamic acid or catechol variety.[5,6,8,14] Deferoxamine (DF), produced by streptomyces pilosis, exemplifies a hydroxamic acid siderophore. Three hydroxamic acid residues present in the structure allow the formation of a hexacoordinate octahedral high spin complex with ferric iron. With an iron formation constant of 10^{31} at pH 6.0 DF forms tight iron complexes.[15] Enterobactin, a member of the catechol class of siderophores which was isolated from culture supernatant of E.coli, has an iron formation constant of 10^{52} at pH 14. The efficiency of chelate formation is very pH sensitive and falls to 10^{26} at pH 6.0.[16] Chelate formation with DF appears to be less sensitive to pH, owing to pKa's values of 9.03, 9.70 and >11.00.[17] In contrast the catechol hydroxyl pKa's of the catechol iron chelator 2,3 dihydroxybenzoic acid are 11.76 and >13.00.[18]

IRON ASSIMILATION AND MICROBIAL INFECTION

In the last ten years a role for iron assimilation in microbial pathogenicity has been defined.[16-18] A variety of pathogens have been shown to have growth restricting requirements for environmental iron.[19-21] The iron binding proteins lactoferrin and transferrin display bacteriostatic activity on various human pathogens.[22,23] Non-transferrin bound iron concentrations in human serum are extremely low ($\sim 10^{-18}$M); far below levels which will support microbial growth (5×10^{-8} to 1×10^{-6}M).[24] Consequently siderophore production is closely related to within-species differences in virulence of various microbial strains. Administration of iron to septic laboratory animals has been associated with a potentiation of the infectious process and a worsening of the prognosis as indicated by a lowering of the average lethal inoculum size with various bacteria.[28] In the clinical setting, iron overload diseases have been associated with increased incidence of various infections.[29-34] Microbial pathogenicity has been correlated with siderophore production.[35-37] For example, strains of Escherichia coli which have the capacity to produce both the catechol ligand enterobactin and the hydroxamic acid ligand aerobactin show greater pathogenicity in animals and humans than strains which produce only enterobactin.[36] It is a well documented fact that in the face of infection, neoplasm, inflammation, and/or fever the body responds with what has been termed an "iron-withholding defense".[38] Within hours of the initiation of a febrile state, serum total iron binding capacity increases and there is a concurrent increase in hepatic and reticuloendothelial (RE) storage of iron.[39,40] Serum levels of the iron binding protein ferritin increase as do haptoglobin levels.[41,42] Gastrointestinal iron absorption drops by 70% within hours of induction of a febrile state.[43]

IRON ASSIMILATION AND NEOPLASIA

For many years now, it has been accepted that iron-overload is associated with enhanced incidence of various neoplasms.[44-47] Ideopathic hemochromatosis carries with it a 30% incidence of primary hepatic carcinoma.[48] Development of sarcoma at iron-dextran injection sites is well documented.[49]

Beyond the oncogenic effects of iron, it is also well substantiated that neoplasm induces the "sham anemia" also seen with infection and fever as described above.[50] Weinberg in numerous review articles has questioned whether this in vivo iron redistribution represents a host defense mechanism designed to attenuate proliferation of the malignancy.[38,50] The question regarding whether neoplastic cells have a higher iron requirement than their non-malignant precursors has yet to be definitely answered. Lymphosarcoma, safrole-induced neoplasm, plasmacytoma do not appear to concentrate iron.[50-52] On the other hand, lymphoid cells from juvenile ALL patients show substantially higher iron contents than lymphoid cells from normal juveniles.[53] Transformed lymphoid cells have been reported to have 1,000 times the number of transferrin surface receptors found on resting lymphocytes.[13] One might infer from these findings that neoplastic tissues do not concentrate iron in a depot fashion beyond immediate needs, yet, at least in the case of lymphocytic leukemia, baseline metabolic need for iron appears to be higher in neoplastic as opposed to nontransformed cells. In contrast the host appears to attempt rendering iron unavailable to malignant tissue. Large increases in transferrin levels in prostatic fluid have been observed in cases of prostatic carcinoma as compared with normal controls or cases of prostatitis.[54] The existance of a siderotic state seems to be associated with an increase in serum ferritin as though the host was attempting to chelate excess serum iron.[55]

As in the case of bacterial infection, we have generated evidence to suggest that parenteral administration of iron may worsen the prognosis for hosts burdened with a malignancy. DBA-2 mice inoculated intraperitoneally with L1210 cells have been observed to have higher numbers of malignant lymphocytes in peritoneal washings among animals receiving 25mg/kg iron-dextran prior to tumor inoculating as compared with animals receiving dextran alone (two to three fold increases).[56] Although these effects were statistically significant, they were variable with the mouse breeder used. In other experiments it was shown that iron-dextran injections appeared to cause significant increases in the metastatic efficiency of Lewis Lung Carcinoma.[57] The average number of surface visible pulmonary metastatic nodules per mouse increased 300% with iron-dextran treatment (25 mg/kg), although no change in survival was noted. We posed the question of whether iron may play a role in the metastatic process. Much more work in this area needs to be undertaken.

ANTIPROLIFERATIVE ACTIVITY OF IRON CHELATORS

Antiproliferative Antibacterial Activity

In the search for new iron chelators which might be more efficacious at iron-decorporation from siderotic animals and humans, Bergeron et al. developed synthetic methods allowing for the production of a number of natural siderophores and synthetic analogs in large quantities. (Figure 1).[58-61] These compounds were tested by this author and Bergeron for growth stimulating effect in vitro upon a variety of human pathogens including Escherichia coli, Pseudomonas aeruginosa, Staphylococcus aureus and Candida albicans.[62] In no instance was growth stimulation observed. In fact a number of the siderophores induced bacteriostatic activity in the chelator concentrations studied. Initial growth suppression in E.coli and P.aeruginosa was followed by growth resumption at nearly normal rates which the authors speculate may have been associated with the bacterum beginning to produce and utilize endogenous high-affinity siderophore systems. In contrast, S.aureus and C.albicans displayed a continuous growth suppressant effect while in the presence of drug containing media, with maximal bacteriostatic activity once again being produced by Parabactin. Bacteriostatic activity could be destroyed by utilizing the methoxy derivatives of active siderophores e.g. methoxy-parabactin. The methoxy derivatives should have much lower iron-formation constants than the free catechol ligands, supporting the hypothesis that iron chelation was important for antiproliferative activity. Dose response experiments indicated that against Acinetobacten calcoaceticus maximal growth suppression was realized when the moler ratio of chelator to free iron salts in the media was 1:1. As the culture media used in these experiments was free of transferrin, ferritin and other proteins, all iron in the media existed as iron salts or simiple iron-amino acid chelates. The authors hypothesize that such forms of iron would be rapidly chelated by exogenously introduced siderophores, leading to siderophore inactivation. Hence the necessity of a 1:1 molar ratio of ligand to iron for maximal activity.

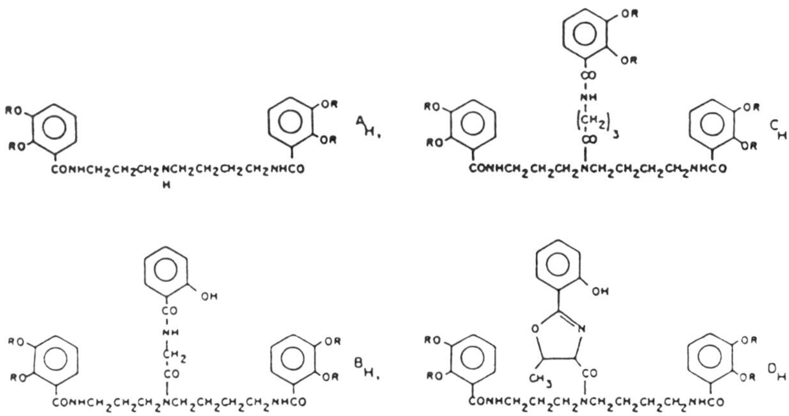

Figure 1. Structures of the iron chelators A_H) Compound II, B_H) Sidglycyl, C_H) Gaba-Sid., D_H) Parabactin.

Antineoplastic and Antiherpetic Activity

There is presently one iron binding compound enjoying clinical use as an antineoplastic agent in the United States, namely hydroxyurea (HU).[63] This drug has been shown to prevent DNA synthesis by interferring with ribonucleotide reductase enzyme.[64] Ribonucleotide reductase (RNR), an iron-containing enzyme, is required for the reduction of ribonucleosides to their deoxy-derivatives.[65] RNR enzyme consists of a homologous pair of dimers.[65] The catalytic site is found in the B_2 subunit and consists of two thiol residues, one tyrosine based hydroxyl radical and a ferric iron atom.

The very short half-life of RNR (6 minutes) is consistent with its modulation effect on the rate limiting step of DNA synthesis.[66] It is not unreasonable to hypothesize that iron chelators capable of entering the cell and distributing to the location of RNR catabolism and/or anabolism could "steal" iron from the enzyme leading to loss of enzymatic reducing potential.

Subsequent of our observation of the antiproliferative activity of various iron chelators on bacteria, we questioned whether they might exert antimitoic activity on other biological systems with a high RNR requirement, such as neoplasm and replicating viruses. In vitro assessment of a number of iron chelators (Figure 1) on L1210 murine leukemia and Herpes simplex Type I indicated varying antireplicative potencies.[67] Forty-eight hour IC_{50} values delineated for L1210 cells in tissue culture of 2.0 uM were observed with both Parabactin and GABA-Sid, while compound II produced a 7 uM IC_{50}. IC_{50} for plaque formation against Herpes simplex I on vero monkey kidney cell monolayers of 0.4 uM, 18 uM and 32 uM for Parabactin, GABA-Sid, and compound II were generated. As expected none of the chelators showed activity against RNA Vesicular stomatitis virus, which is without a RNR replication requirement. Antileukemic and antiherpetic activity could be reversed in vitro by the addition of $FeCl_3$ to the culture media. Good correlations were made between in vitro antineoplastic and antiherpetic activity and the lipid solubility of the various apoferric ligands. Parabactin, the most potent antiproliferative, was also the most lipid soluble compound.

It would appear from this data, that potency may relate with the ability of the ligands to diffuse across cell membranes as the iron free ligand and subsequently exert their iron scavenging activity, causing anti-proliferative effects.

Based upon these observations, it would be reasonable to question whether chelator induced antiproliferative effects on viruses or neoplasms could be elicited in vivo. The average human body contains over 4 gm of iron, not what one would call an iron deficient environment.[17] Yet, we find that virtually all this iron is tightly bound by the iron transport protein transferrin, the iron storage protein ferritin, or heme.[17] Freely soluble iron concentrations in serum are lower than 10^{-18} M.[24] Raymond et al. have looked at the ability of various natural or synthetic siderophores to remove iron from transferrin.[15] Results show that although numerous chelators have much higher iron-formation constants than transferrin's 10^{24}, these ligands only slowly remove transferrin bound iron. For example, Enterobactin with an iron-formation constant of 10^{51} removed only 6% of the total transferrin-bound iron within 30 minutes at a 1:1 molar ratio. These observations have also been noted with the non-catechol siderophores pyochelin or pyoverdin.[68] It is therefore not unreasonable to assume that siderophores administered in vivo could exist in serum for prolonged periods of time without picking up iron and becoming inactivated regarding their antineoplastic or antiviral activity.

Work done by Porter, Bergeron, et al. confirms that Parabactin and Compound (II) exert their antiproliferative activity on L1210 cells, at least in part, via inhibition of RNR.[69] The compounds induced a G_1/S cell cycle block consistent with inhibition of DNA synthesis. The block was found to be reversible by the delivery of $FeCl_3$ to the culture media. Both chelators decreased [^{14}C] thymidine incorporation significantly, while causing only minor changes in [^{3}H] uridine or [^{3}H] leucine incorporation. Pools of dATP and dTTP decreased after exposure to both ligands, as did cytidine-5^{1}-diphosphate reductase activity. No drop in DNA polymerase activity was observed.

Iron chelators have defined a novel class of antiproliferative agents, although their activity in vivo has yet to be examined. Efficacy studies are now underway in a number of laboratories.

REFERENCES

1. Weinberg, E.D. Nutrition and Cancer 1983, 4, 223.

2. Cotton, F.A; Wilkinson, G. In: Advanced Inorganic Chemistry, 4th ed.; John Wiley & Sons, New York, 1980, p 751.
3. Tufano, J.P; Raymond, K.N. J. Am. Chem. Soc. 1981, 103, 6617.
4. Heidinger, S; Braun, V; Pecoraro, V.L; Raymond, K.N. J. Bacteriol. 1983, 153, 109.
5. Neilands, J.B. Ann. Rev. Nutr. 1981, 1, 27.
6. Emery, T. In: Metal Ions in Biological Systems, (Sigel, H., ed.) Vol.7; Marcel Dekker, New York, 1978, p 77-126.
7. Raymond, K.N; Carrano, C.J. Accts. Chem. Res. 1979, 12, 183.
8. Neilands, J.B. Ann. Rev. in Biochem. 1981, 50, 715.
9. Newman, R; Schneider, C; Sutherland, R; Vodenelich, L; Greaves, M. Trends in Biol. Sci. 1982, 1, 397.
10. Fernandez-Pol, J.A. Biochem. Biophys. Res. Commun. 1977, 78, 136.
11. Fernandez-Pol, J.A. Cell. 1978, 14, 489.
12. Larrick, J.W; Cresswell, P. J. Supramol. Struct. Cell Biochem. 1979, 11, 579.
13. Sutherland, R; Delia, D; Schneider, C; Newman, R; Kemshead, J; Greaves, M. Proc. Nat'l Acad. Sci. USA. 1981, 78, 4515.
14. Neilands, J.B. In: Microbial Iron Metabolism (Neilands, J.B., ed.) Academic Press; New York.
15. Audeef, A; Sofen, S.R; Bregante, T.L; Raymond, K.N. J. Amer. Chem. Soc. 1978, 100, 5362.
16. Neilands, J.B. personal communication.
17. Bothwell, T.H; Charlton, R.W; Cook, J.O; Finch, C.A. In: Iron Metabolism in Man, Blackwell Scientific Publications; London, 1979.
18. Worwood, M. In: Iron Excess, Aberrations of Iron and Porphyrin Metabolism, Seminars in Hematology, Vol 14.
19. Bullen, J.J. Rev. infec. Dis. 1981, 3, 1127.
20. Forsberg, C.M; Bullen, J.J. J. Clin. Path. 1972, 25, 65.
21. Simpson, L.M; Oliver, J.D. Infect. Immun. 1983, 41, 644.
22. Chasteen, N.D. Coord. Chem. Rev. 1977, 22, 1.
23. Reiter, B; Brock, J.H; Steel, E.D. Immunology 1975, 28, 83.
24. Weinberg, E.D. Microbiol. Rev. 1978, 42, 45.
25. Flossman, K.D; Muller, G. Acta Biol. Med. Germ. 1980, 39, 327.

26. Owens, W.E; Rolfe, R.D; Finegold, S.M. Abstr. Annu. Meet. Am. Soc. Microbiol. 1982, Vol. 89.
27. Wright, A.C; Simpson, L.M; Oliver, J.D. Infect. Immun. 1981, 34, 503.
28. Holbein, B.E; Jericho, K.W.F; Likes, G.C. Infect. Immun. 1979, 24, 545.
29. Melby, K.L; Slordahl, S; Gutterberg, T.J; Nordbro, S.A.N. Br. Med. J. 1982, 285, 467.
30. Robins-Brown, R.M; Rabson, A.R; Koornhof, H.J. Contrib. Microbiol. Immunol. 1979, 5, 277.
31. Blake, P.A; Merson, M.H; Weaver, R.E; Hollis, D.G; Heublein, P.C. N. Engl. J. Med. 1979, 300, 1.
32. Diamond, L.S; Harlow, D.R; Phillips, B.P; Keister, D.B. Arch. Invest. Med. Supl. 1978, 1, 329.
33. Caroline, L; Kozinn, P.J; Feldman, F; Stiefel, F.H; Lichtman, H. Ann. NY Acad. Sci. 1969, 165, 148.
34. Byles, A.B; D'SA, A. Br. Med. J. 1970, 3, 625.
35. Holzberg, M; Artis, W.M. Infect. Immun. 1983, 40, 1134.
36. Williams, P.H; Carbonetti, N.H. Infect Immun. 1986, 51, 942.
37. Sriyosachati, A; Cox, C.D. Infect. Immun. 1986, 52, 885.
38. Weinberg, E.D. Physiol. Rev. 1984, 64, 65.
39. Greenberg, G.R; Ashenbrucker, H; Louritsen, M.A; Worth, W; Humphreys, S.R; Wintrobe, M.M. J. Clin. Invest. 1947, 26, 121.
40. Cartwright, G.E; Lee, G.R. Br. J. Haematol. 1971, 21, 147.
41. Fairbanks, V.F; Klee, G.G. Prog. Clin. Pathol. 1981, 8, 175.
42. Eaton, J.W; Brandt, P; Mahoney, J.R; Lee, J.T. Science 1982, 215, 691.
43. Cartwright, G.E; Lauritsen, M.A; Jones, P.J; Merrill, I.M; Wintrobe, M.M. J. Clin. Invest. 1946, 25, 65.
44. Milder, M.S; Cook, J.D; Stray, S; Finch, A. Medicine (Baltimore) 1980, 59, 34.
45. Bomfor, A; Williams, R. Q. J. Med. 1976, 45, 611.
46. Berman, C. Adv. Cancer Res. 1958, 5, 55.
47. Robertson, M.A; Harrington, J.S; Bradshaw, E. Br. J. Cancer 1971, 25, 395.
48. Kelly, W. In: Harrison's Principles of Internal Medicine, 10th ed; McGraw Hill; New York NY, 1983; p 530.
49. Weinberg, E.D. Biol. Trace Elem. Res. 1981, 3, 55.

50. Weinberg, E.D. Nutr. and Cancer 1983, 4, 223.
51. Schade, S.G. Cancer Res. 1987, 36, 375.
52. Lipsky, M.M; Hinton, D.E; Goldblatt, P.J; Klaurig, J.E; Trump, B.F. Pathol. Res. Pract. 1979, 178.
53. Carpentieri, U; Myers, J; Thorpe, L; Daeschner, C.W; Haggard, M.E. Cancer Res. 1986, 46, 981.
54. Wilson, E.M; French, F.S; Petrusz, P. Cancer Res. 1982, 42, 243.
55. Batey, R.G; Fong, P.L.C; Shamir, S; Sherlock, S. Dig. Dis. Sci. 1980, 25, 340.
56. Bergeron, R.J; Streiff, R.R; Elliott, G.T. J. Nutr. 1985, 115, 369.
57. Elliott, G.T. In: Potential Therapeutic Applications of Iron Chelators; A Role for Iron in Infection and Neoplasm, Ph.D. Dissertation, University of Florida; Gainsville, FL, 1985, p 58.
58. Bergeron, R.J; Burton, P.S; Kline, S.J; McGovern, K.A. J. Org. Chem. 1981, 46, 3712.
59. Bergeron, R.J; Kline, S.J. J. Am. Chem. Soc. 1982, 104, 4489.
60. Bergeron, R.J; Kline, S.J; Stolowich, N.J; McGovern, K.A; Burton, P.S. J. Org. Chem. 1981, 46, 4524.
61. Bergeron, R.J; McGovern, K.A; Channing, M.A; Burton, P.S. J. Org. Chem. 1980, 45, 1589.
62. Bergeron, R.J; Elliott, G.T; Kline, S.J; Ramphal, R. Antimicrob. Agents and Chemo. 1983, 24, 725.
63. Physician's Desk Reference; Medical Economics Co; Orodell, NJ, 1986; p 1746
64. Moore, E.C; Hurlbert, R.B. Pharmacol. Ther. 1985, 27, 167.
65. Stryer, L. In: Biochemistry, 2nd ed; Freeman & Co; New York, NY, 1981; p 524
66. Eriksson, S; Graslund, A; Skog, S; Thelander, L; Tribukuit, B. J. Biol. Chem. 1984, 259, 11695.
67. Bergeron, R.J; Cavanaugh, P.F; Kline, S.J; Hughes, R.G; Elliott, G.T; Porter, C.W. Biochem. Res. Commun. 1984, 121, 848.
68. Sriyosachati, S; Cox, C. Infection and Immun. 1986, 52, 885.
69. Cavanaugh, P.F; Porter, C.W; Tukalo, D; Frankfurt, O.S; Pavelic, Z.P; Bergeron, R.J. Cancer Res. 1985, 45, 4754.

DISCUSSION

CHIDAMBARAM: Iron complexes of lower members of hydroxyamates have been shown to very rapidly exchange Fe with transferrin so thermodynamic stability doesn't tell us very much. These hydroxyamates are known to inhibit nickel-dependent enzymes such as urease anhydrase. Do you think some of your ligands remove metals from Cu-Zn SOD, ceruloplasmin, or any other metalloproteins?

ELLIOTT (Gary): I would be surprised if they did not inhibit other Fe-dependent enzyme systems. I would be very surprised if they are selective for ribonucleotide reductase. There has been a little work done with cytochrome oxidase because they seem to cause some uncoupling of ATP. We've determined the LD_{50} for our compound to be over a gram per kilo so although they may cause some inhibition of cytochrome oxidase, it's obviously not something that's going to completely uncouple ATP synthesis, which would be much more acutely toxic.

CHIDAMBARAM: Nickel accumulation in kidney has been known for some time. Can hydroxamic acids be used to remove this nickel?

ELLIOTT (Gary): I haven't seen that reported anywhere. I am interested in evaluating our compounds for removal of platinum and plutonium from kidney and other tissues. One problem with hydroxamic acids is that they are so water soluble they have very poor tissue penetration. Our other compounds are much more lipophilic and they may be able to remove toxic metals more effectively.

CHIDAMBARAM: What is the solubility of your Fe complexes?

ELLIOTT (Gary): Parabactin is virtually insoluble in water until it forms an Fe chelate, which is quite soluble in water.

SORENSON: How does this negative charge develop with Fe chelation to make them more water soluble?

ELLIOTT (Gary): All 6 catechols deprotonate with chelation.

SORENSON: Those catechols are nonionic?

ELLIOTT (Gary): With chelation these compounds lose 6 protons.

SORENSON: With 1 ferric ion 6 protons are lost?

ELLIOTT (Gary): Yes.

I've mentioned ribonucleotide reductase inhibiton and there was one other point I wanted to make. Bergeron has nailed down the ribonucleotide reductase inhibition theory. He and Carl Porter have shown that these compounds in fact do not inhibit DNA polymerase. They cause a precipitous drop in the incorporation of thymidine and then, later on, uridine but they have no effect on incorporation of glycine. They also decrease cytosine diphosphate reduction and decrease ATP pools of the neoplasm along with a large decrease in deoxy ATP and something of an obligatory increase in deoxy TTP, all of which is consistent with inhibition of ribonucleotide reductase.

SORENSON: Is it established that ribonucleotide reductase is Fe-dependent?

ELLIOTT (Gary): Yes. There are four subunits. Two allosteric units bound to one another, with Fe(III) at the catalytic site along with a tyrosine residue.

SORENSON: Is there any inhibition of DNA or RNA polymerase, which are supposed to be Zn-dependent enzymes.

ELLIOTT (Gary): Only DNA polymerase was studied. A slight increase in activity was found and suggested to be due to an obligatory increase associated with the shut down of ribonucleotide reductase.

SORENSON: Would these chelators inhibit Cu-dependent enzymes such as Cu-Zn SOD or DBH?

ELLIOTT (Gary): They haven't been studied.

Anticarcinogenic and Radioprotectant Activities of Copper Complexes

EFFECTS OF COPPER COMPLEXES ON MULTISTAGE CARCINOGENESIS

P. A. Egner, B. G. Taffe and T. W. Kensler

Johns Hopkins Medical Institutions

Baltimore, Maryland 21205

The induction of cancer by chemicals involves multiple stages as the target cell progresses from its normal phenotype into a malignant neoplasm. These stages can be operationally defined as initiation, promotion, and progression. Although first described in mouse skin, the applicability of the concept of multistage carcinogenesis to many species and tissues has provided strong presumptive evidence for the likelihood of initiation, promotion, and progression in the etiology of human cancer.[1,2] Cancers caused by environmental agents should be considered, at least in principle, avoidable cancers. The two major strategies for reducing human risk for cancer from environmental exposures are prevention and protection. Prevention entails the elimination of carcinogenic influences and clearly represents the ideal approach to cancer control. However, this solution remains unattainable at present, in large part because the causes of most human cancers simply are not known. Alternatively, protection considers measures designed to interrupt the carcinogenic process without necessarily specific efforts to identify or eliminate carcinogenic influences. Chemoprotection is a means of cancer control by which the development of the disease is subverted through the administration of one or several chemicals. Although a large number of inhibitors of chemical carcinogenesis have been described,[3] copper complexes represent a novel class of cancer chemoprotective agents.

While the underlying mechanisms involved in the neoplastic process remain unresolved, there is ample evidence to suggest that reactive oxygen species participate in all stages of carcinogenesis.[4-7] For example, reactive oxygen can mediate the metabolic activation of carcinogens - a component of initiation. Additionally, some free radical generating compounds are particularly effective progessor agents. However, free radicals are most strongly implicated in tumor promotion. Reactive oxygen detoxifiers and scavengers such as copper complexes, superoxide dismutase [SOD] and phenolic antioxidants inhibit the biochemical and biological actions of tumor promoters *in vitro* and *in vivo.*

DIMINUTION OF SOD ACTIVITY BY TUMOR PROMOTERS

Several years ago Solanki et al.[8] reported that a single application of tumor promoter to the dorsal skin of mice led to a rapid and marked diminution of epidermal SOD activity. Shown in Figure 1, repetitive treatment with the phorbol diester TPA, as occurs during tumor promotion, caused a sustained 40 to 50% decrease in total epidermal SOD activity. This reduction in activity was specific to the copper-zinc form of SOD in that no alteration in the activity of manganese SOD activity was observed. If mice were given only a single dose of TPA, then SOD activity returned to control levels within 1 week. However, after multiple exposures to TPA this recovery was greatly protracted suggesting that a phenotypic alteration had occurred in these cells. As might be anticipated, SOD levels in the resultant papillomas are lower than in the surrounding epidermis. Observations such as these prompted us to ask the question - what would be the effect of restoring the diminished SOD activity in mouse skin on the tumor promotion process?

EFFECTS OF COPPER COMPLEXES ON ODC ACTIVITY

To address this question we chose to utilize the biomimetic SOD, copper (II) $(3,5\text{-diisopropylsalicylate})_2$ [CuDIPS]. CuDIPS is a low molecular weight, lipophilic copper coordination complex that catalyzes the disproportionation of superoxide at rates comparable to native copper-zinc SOD.[9,10] Because of its lipophilic character

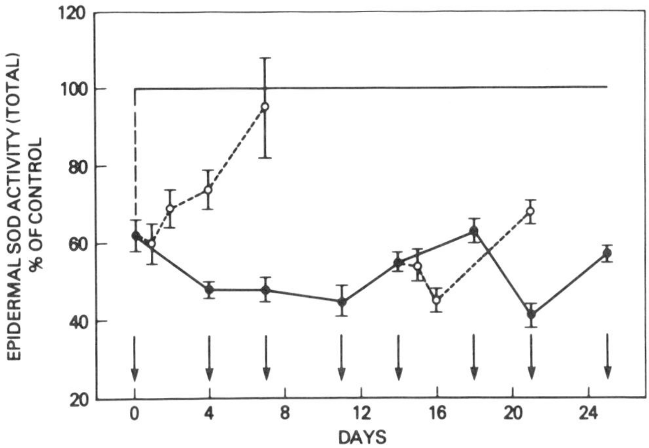

Figure 1. Effect of repetitive TPA treatment on epidermal SOD activity. Epidermis was isolated and assayed for total SOD activity 3 hours after application of 17 nmol TPA. Arrows indicate days of TPA exposure. Closed circles represent mice receiving repeated doses of TPA while open circles represent mice receiving no further treatment with TPA. Values are mean ± S.E. (n = 5).

it presumably penetrates into intracellular sites of reactive oxygen generation and reactivity. One approach for probing the role of reactive oxygen in tumor promotion is to examine the inhibitory effects of CuDIPS - and other antioxidants - on the induction of ornithine decarboxylase [ODC] activity in mouse epidermis by TPA. The induction of ODC activity is a prominent, early and transient event following exposure to epidermal tumor promoters, and measurements of ODC activity offer a useful first approach to the identification and characterization of potential antagonists of tumor promotion.[11] ODC catalyzes the initial and rate-limiting step in polyamine biosynthesis, and elevation of ODC activity and the concommitant increases in polyamine levels are necessary albeit insufficient components of tumor promotion. The topical application of TPA to mouse epidermis results in a rapid and transient induction of ODC activity. This response can be inhibited by the administration of CuDIPS in a dose-dependent fashion.[12] Figure 2, panel B depicts the effects of vary-

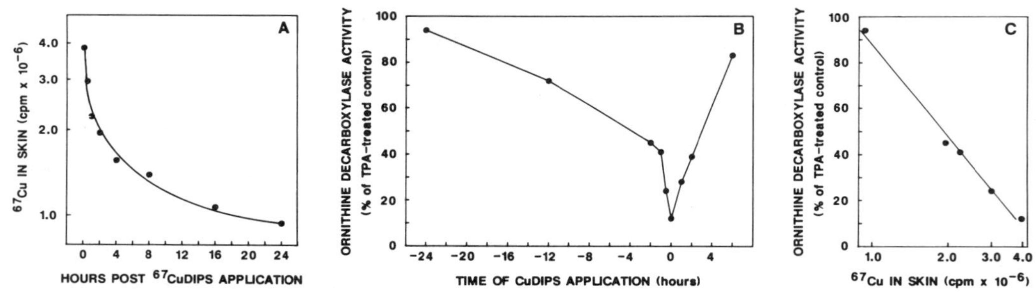

Figure 2. Relationship between elimination of CuDIPS from skin of mice and inhibition of ODC activity. Panel A. Kinetics of elimination of 2 umol ^{67}CuDIPS from skin of mice. Panel B. Effect of time of addition of CuDIPS on TPA-induced ODC activity. Mice received 17 nmol TPA at time 0 and were sacrificed 7 hours later. Mice also received 2 umol CuDIPS at the indicated times before or after TPA treatment. Panel C. Relationship between amount of ^{67}CuDIPS remaining in skin at time of TPA application and inhibition of ODC activity.

ing the time of addition of CuDIPS relative to TPA application on this ODC response. Maximum inhibition occurred when CuDIPS was applied immediately after TPA. Whereas substantial inhibition was maintained if CuDIPS was given within plus or minus 2 hours of TPA, earlier or later applications were only minimally effective. Two conclusions are suggested by these findings. First, the inhibitory effect of CuDIPS is not a manifestation of cell toxicity since in that case inhibition at all time points would be expected. Second, CuDIPS acts as an early antagonist of the phorbol diester-induced response, suggesting a rapid elaboration of superoxide in the skin after TPA treatment.

The dependency of inhibition by CuDIPS on dose scheduling reflects the kinetics of absorption and elimination of the complex in the skin. Panel A depicts the kinetics of elimination of CuDIPS following topical application of ^{67}CuDIPS. There was a rapid phase of elimination with an apparent t½ of 30 minutes and a significant fraction of the dose was removed with this fast half-life. Panel C compares the relationship between amount of CuDIPS present in the skin and percent inhibition of induction of ODC as a function of time of addition as described in panels A & B. Clearly, inhibiton of the ODC response was directly dependent on the tissue concentration of CuDIPS at the time of TPA application. It should be noted that CuDIPS does not directly inhibit ODC activity in the assay reaction. Thus, this action <u>in vivo</u> presumably reflects an inhibition of the signal transduction pathway leading to enzyme induction. Other copper complexes also inhibit the ODC response. Both copper hydrocortisone and copper indomethacin,[9] which possess SOD-like activity, were effective inhibitors of TPA-mediated induction of ODC <u>in vivo.</u> These two complexes stand in contrast to CuDIPS in that the ligands alone, hydrocortisone-21-phosphate and indomethacin, have antipromoter properties unto themselves. Yet in both cases the inhibitory effects against ODC induction were greater in the form of a copper complex. Derivatives of CuDIPS which are lacking in SOD activity were inactive as inhibitors of ODC induction. These include the ligand alone, diisopropylsalicylate [DIPS], the corresponding zinc complex, ZnDIPS, and $CuEDTA^{2-}$. These findings underscore the probable role of the disproportionation of superoxide in the inhibitory actions of these copper complexes.

EFFECTS OF COPPER COMPLEXES ON TUMOR FORMATION

Tumors can be produced in mouse skin through a variety of multistage and complete carcinogenesis protocols. The prototypical initiation - promotion regimen for inducing tumors involves applying a single sub-threshold dose of a polycyclic aromatic hydrocarbon carcinogen such as dimethylbenz[a]anthracene [DMBA] followed 10 days later by twice-weekly applications of TPA for 20 or more weeks. Promotion can also be divided into stages by treating with TPA for 2 weeks for stage I promotion followed by 16 or more weeks of twice-weekly applications of an incomplete promoter such as retinoyl phorbol acetate [RPA] for stage II promotion. Both of these protocols yield a high incidence of tumors, often approaching 100%. Tumors can also be elicited through complete carcinogenesis protocols. In this case, tumors are induced by a single application of a large dose or by repeated applications of an initiating dose of a carcinogen such as DMBA. In these complete carcinogenesis systems the yield of tumors is lower, but these tumors are more frequently malignant. We have utilized all of these protocols in the evaluation of the anticarcinogenic activity of copper complexes.

Figure 3, panel A, presents the effects of pretreatment with CuDIPS on initiation and promotion in the skin of CD-1 mice. Application of CuDIPS 15 minutes before each biweekly treatment of TPA to mice initiated with DMBA reduced tumor development, 60% and 87% in terms of tumor incidence and multiplicity, respectively, after 24 weeks. Analogs lacking SOD-mimetic activity, namely DIPS and ZnDIPS, did not inhibit promotion.[12,13] In contrast to the profound antipromoting activity of CuDIPS, treatment with CuDIPS 15 min prior to the single initiating dose of DMBA produced no significant effect on either tumor inicidence or multiplicity. However, the inhibitory effects of this biomimetic SOD against tumorigenesis induced by complete carcinogenesis protocols with DMBA were pronounced. Shown in panel B, treatment with CuDIPS prior to each weekly application of the initiating dose of DMBA resulted in a 25% reduction in tumor incidence and a 60% decrease in the number of papillomas per mouse after 28 weeks of carcinogen exposure. Panel C depicts the effects on tumorigenesis of CuDIPS applied 15 minutes prior to a single high dose of DMBA. These mice received no further

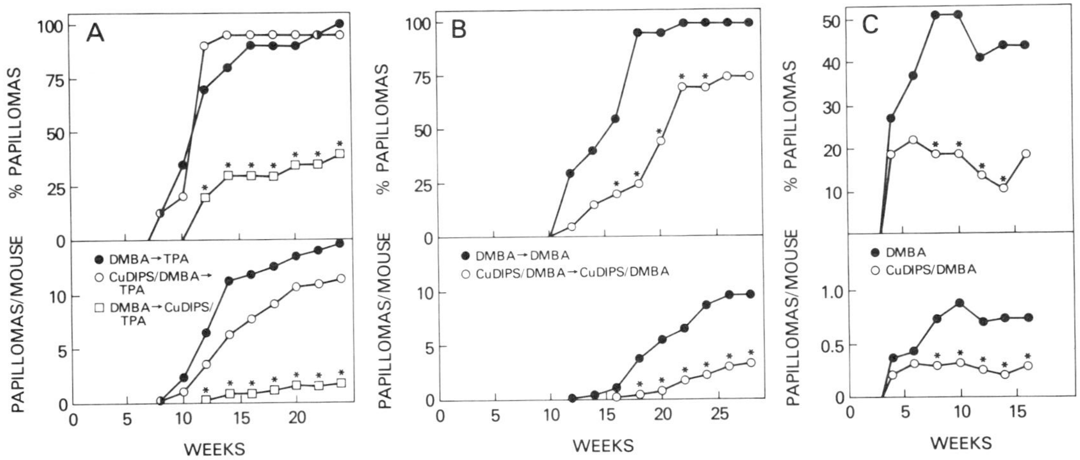

Figure 3. Effects of CuDIPS on tumorigenesis in the skin of CD-1 mice induced by the initiation - promotion regimen or the complete carcinogenesis process. Panel A. Effects of CuDIPS on initiation or promotion. Mice were treated with 2 umol CuDIPS either 15 min before initiation with a single 0.2 umol dose of DMBA or 15 min before twice-weekly promotion with 4 nmol TPA. Panel B. Effect of CuDIPS on complete carcinogenesis by repetitive weekly doses (0.2 umol) of DMBA. Panel C. Effect of CuDIPS on complete carcinogenesis by a single (3.6 umol) dose of DMBA. CuDIPS (0.2 umol) was applied 15 min before DMBA. Reprinted from reference 13.

TABLE I

EFFECT OF CuDIPS ON TWO-STAGE TUMOR PROMOTION IN CD-1 MOUSE SKIN

GROUP	TREATMENT		TUMOR FORMATION					
			After 8 weeks		After 12 weeks		After 16 weeks	
	I	II	%	yield	%	yield	%	yield
1	Ether/TPA[a]	Ether/RPA[b]	55	3.3	70	5.7	75	8.7
2	Ether/TPA	Ether/Acetone	0[d]	0[d]	0[d]	0[d]	0[d]	0[d]
3	Ether/Acetone	Ether/RPA	0[d]	0[d]	40	0.8[d]	60	2.1[d]
4	CuDIPS[c]/TPA	CuDIPS/RPA	5[d]	0.3[d]	20[d]	1.1[d]	25[d]	1.3[d]
5	CuDIPS/TPA	Ether/RPA	30[d]	0.7[d]	50	2.0[d]	65	4.4[d]
6	Ether/TPA	CuDIPS/RPA	25[d]	0.4[d]	25[d]	0.7[d]	35[d]	1.9[d]

[a] 10 nmol twice-weekly: 2 weeks.

[b] 10 nmol twice-weekly: 14 weeks.

[c] 4 μmol 30 min. prior to each application of phorbol diester.

[d] Differs from group 1: $p<0.05$.

treatment. At 16 weeks after dosing a 58% reduction in tumor incidence and a 67% reduction in tumor yield was observed. Inasmuch as CuDIPS inhibits the covalent binding of DMBA to epidermal DNA,[13] these anticarcinogenic effects of CuDIPS may be mediated through actions on xenobiotic metabolizing systems, particularly cytochrome P-450 reductase.[14]

Two-stage promotion can be produced by the sequential applications of TPA and RPA. Table I presents tumor incidence and multiplicity data at 8, 12, and 16 weeks of promotion. Shown in group 1 the complete regimen of initiation and stages I and II promotion provided a high yield of tumors. By contrast, shown in groups 2 and 3, initiation followed by either stage I or stage II promotion alone was not very effective. The relative effects of CuDIPS against stages I and II of promotion are shown in groups 4 through 6. CuDIPS was quite effective against stage II promotion, although stage I was modestly inhibited as well. CuDIPS applied through both stages I and II was very inhibitory. Studies by Schwartz et al.[14] have shown that CuDIPS does not inhibit stage I promotion using these promoters in NMRI mouse; however, these investigators did not examine the possible effects on stage II promotion. Phenolic antioxidants such as BHA and BHT are most active against stage II promotion as well.[7] Thus, reactive oxygen may be involved in regulating cell proliferation and terminal differention in the basal keratinocytes, hallmarks of stage II promotion. The pronounced effects of CuDIPS[12] and antioxidants[16] on the ODC response are consistent with this view.

SUMMARY

The multiple effects of CuDIPS as an inhibitor of tumor initiation in SENCAR mice[17] and tumor promotion and complete carcinogenesis in CD-1 mice[12,13] serve to distinguish this compound from many cancer chemoprotective agents. The biomimetic SODs are useful probes for studying mechanisms of oxygen radical generation, action and damage in mouse epidermis. Although copper complexes have not to our knowledge been examined for anticarcinogenic action in other tissues, the recent findings that systemic administration of CuDIPS to rodents inhibits carbon tetrachloride hepatotoxicity[18] and is radioprotective[19] strongly suggest these agents may have applicability at a number of organ sites. The use of selective scavengers and detoxifiers of reactive oxygen would appear to have promise as chemoprotective agents during the sequential stages of carcinogenesis.

ACKNOWLEDGEMENTS

This work was supported by grants from the American Cancer Society (SIG-3) and NIH (ES 00454 and CA 36380).

REFERENCES

1. Farber, E. Amer. J. Pathol. 1982, 2, 271.
2. Pitot, H.C.; Goldsworthy, T.; Moran, S. J. Supramol. Struc. Cell. Biochem. 1981, 17, 133.
3. Wattenberg, L.W. Cancer Res. 1985, 45, 1.
4. Kensler, T.W.; Trush, M.A. Envir. Mut. 1984, 6, 593.
5. Kensler, T.W.; Trush, M.A. In: Superoxide Dismutase, Vol III; CRC Press, Boca Raton, FL, 1985, 191.
6. Cerutti, P.A. Science 1985, 227, 375.
7. Slaga, T.; Solanki, V.; Logani, M. In: Radioprotectors and Anticarcinogens, Academic Press, NY, 1983, 471.
8. Solanki, V.; Rana, R.S.; Slaga, T.J. Carcinogenesis 1981, 2, 1141.
9. Sorenson, J.R.J. J. Med. Chem. 1976, 19, 135.
10. Lengfelder, E.; Weser, U. Bull. Eur. Physiopath. Resp. 1981, 17, 73.
11. O,Brien, T.J. Cancer Res. 1976, 36, 2644.
12. Kensler, T.W.; Bush, D.M.; Kozumbo, W.J. Science 1983, 221, 75.
13. Egner, P.A.; Kensler, T.W. Carcinogenesis 1985, 6, 1167.
14. Werringloer, J.; Kawano, S.; Chacos, N.; Estabrook, R.W. J. Biol. Chem. 1979, 254, 11839.
15. Schwartz, M.; Peres, G.; Kunz, W., Furstenberger, G.; Kittstein, W.; Marks, F. Carcinogenesis 1984, 5, 1663.
16. Kozumbo, W.J.; Seed, J.L.: Kensler, T.W. Cancer Res. 1983, 43, 2555.
17. Solanki, V.; Yotti, L.; Logani, M.K.; Slaga, T.J. Carcinogenesis 1984, 5, 129.
18. Joyave, J.L.; Steinhauser, L.S.; Dillehay, D.L.; Born, C.K.; Hamrick, M.E. Biochem. Pharmacol. 1985, 34, 3915.
19. Sorenson, J.R.J. J. Med. Chem. 1984, 27, 1747.

DISCUSSION

HERNANDEZ: You indicated that free radicals are coming from PMNs. Have you measured PMN infiltration in your mouse model?

EGNER: Yes. We've applied a phorbol diester(TPA) and determined that at 16 hours after the last application, we have maximal PMN influx. This inflammatory response seems to be a very important part of the promotion sequence. We've also found PMN recruitment following phorbol diester application. If we apply a second dose of TPA, which causes a tremendous outpouring of O_2^- and then apply carcinogen immediately after TPA application, we can double or triple the 7,8-diol binding to the epidermis as compared to vehicle-treated control levels.

HERNANDEZ: Are you planning studies with neutrophil anti-serum, to deplete animals of neutrophils, to determine if PMNs are the only source of these oxygen radicals?

EGNER: I personally wasn't aware of that technique, but that offers alot of exciting possibilities.

REED: Determining the amount of ^{67}Cu in skin versus inhibition of TPA-induced ODC activity is a very important future study. I was a little surprised that it did not appear to be a linear relationship.

EGNER: We were also surprised by this apparent biphasic result and we repeated it several times. We plan to study this when we can get an additional supply of ^{67}Cu.

REED: Two to five umol of Cu(II)(3,5-dips)$_2$, that's 1 to 2.5 mg, was applied to a small area of skin. How much of that stays on skin and how much is absorbed and systemically distributed? Does rapid absorption and distribution from the site of application require the use of these large amounts of compound?

EGNER: We don't actually know. There seems to be a fine line between getting Cu(II)(3,5-dips)$_2$ to function as an antipromoter and also as an anticarcinogen, and not having it effectively scavenge radicals. We find we need a 2 umol dose as opposed to smaller doses. Also, after topical application of Cu(II)(3,5-dips)$_2$, mice lick the site of application. We have housed them individually. This presented a tremendous problem when we used ^{67}Cu. I don't really know how much of it they lick off and is systemically distributed following gastrointestinal absorption or how much is absorbed from the site of application. We thought about anesthetizing these animals, or restraining them for a period of time to determine how much of the dose is absorbed at the site of application. When we did those experiments with ^{67}Cu(II)(3,5-dips)$_2$ we systematically measured the amount of radioactivity in every tissue and organ we took. We had very high expectations of determining exactly where the isotope was going but they persisted in trying to lick their backs.

LOVEN: Does DMBA require liver activation to become a carcinogen or is it activated in skin cells?

EGNER: I don't know. I assume that it's activated by P_{450}.

REINERS: Keratinocytes, epidermal cells, are capable of activating DMBA and other aromatic hydrocarbons.

LOVEN: Does DMBA undergo redox cycling?

REINERS: DMBA does not but the diol can be oxidized to a quinone, which is reversible.

SORENSON: Did you just add the ^{67}Cu(II), that you obtained from Los Alamos, to 3,5-dips to make your ^{67}Cu(II)(3,5-dips)$_2$?

EGNER: It was a nightmare. We always use your procedure to make ^{67}Cu(II)(3,5-dips)$_2$. We assumed that all we had to do was just put everything together. It was a terrible problem with the change in solubility of their isotope compound. Los Alamos sent us ^{67}Cu in a solvent that I think contains something like propylene glycol or something. I am still not sure what it was. It was something that was completely incompatible with our method of making Cu(II)(3,5-dips)$_2$.

MODULATION OF BENZOPYRENE DEPENDENT MUTAGENESIS AND XANTHINE OXIDASE ACTIVITY BY COPPER COMPLEXES

JOHN J. REINERS, JR.

THE UNIVERSITY OF TEXAS SYSTEM CANCER CENTER

SCIENCE PARK - RESEARCH DIVISION

SMITHVILLE, TX 78957

INTRODUCTION

Chemical induced murine skin carcinogenesis can be operationally separated into sequential stages. Murine skin tumors can be experimentally produced by a single topical application of a subthreshold dose of carcinogen (initiation) if followed by repetitive treatments with nontumorigenic tumor promoting agents (promotion). Initiation is thought to encompass those events that damage or modify DNA; whereas promotion includes those processes facilitating the expression of the latent effects of initiation. The superoxide dismutase (SOD) biomimetic Cu(II) $(3,5\text{-diisopropylsalicylate})_2$ (CuDIPS) has been reported to inhibit both the initiation[1,2] and promotion stages[2,3] of murine skin multistage carcinogenesis. The mechanism(s) by which CuDIPS inhibits initiation and promotion are speculative, and are the subject of this article.

MATERIALS AND METHODS

Liver homogenates and microsome preparations were prepared by previously published procedures[4]. Epithelial homogenates were prepared from murine dorsal skin. Freshly excised skin pelts were placed dermis side down on a glass plate and the epidermis was repeatedly scraped with a razor blade. The scrapings were suspended in 0.5 ml 50 mM $NaKPO_4$, pH 7.0 and homogenized with a Brinkmann polytron.

The homogenate was centrifuged for 15 min at 13,500xg and the supernatant fluid was recentrifuged for 5 min at 13,500xg. The resulting supernatant fluid was used immediately for xanthine dehydrogenase/oxidase assays.

The Ames Salmonella typhimurium revertant assay, and the cytochrome P-450 reductase assay, as they pertain to the reported study, have been previously described in detail[4].

Xanthine dehydrogenase and xanthine oxidase activities were collectively assayed by a procedure described by Ghezzi et al[5]. Reaction mixtures (final volume 40 μl) contained enzyme preparation, 40 nmoles NAD, 2 μmoles Tris-HCl (pH 9.0), and 10 μl of [8-^{14}C]hypoxanthine (9μCi/ml, 50-53 mCi/mmole). Total XO activity was assayed in the absence of NAD. Reactions were initiated by the addition of enzyme preparation and incubated for 5-10 min at 37°C. Reactions were terminated by the addition of 20 μl of 1M perchloric acid and centrifuged for 2 min at 13,500 xg. Five μl of the supernatant fluid were spotted on cellulose TLC plates, prespotted with 2.5μl of 10^{-2} M solutions of unlabelled hypoxanthine, xanthine, and uric acid. TLC plates were developed with a $CH_3(CH_2)_2CH_2OH:CH_3OH:H_2O:NH_4OH$(25%)(60:20:20:1) mixture, and reference metabolites were located with a U.V. light. The area of the TLC plates corresponding to xanthine and uric acid was excised and analyzed by liquid scintillation counting. Specific activity is expressed as nmoles of hypoxanthine oxidized to xanthine plus uric acid/min/mg protein. Protein content was determined with the Bio-Rad protein assay reagent using bovine serum albumin as a standard.

RESULTS

Inhibition of Benzo[a]pyrene Mutagenesis and P-450 Reductase

Topical application of CuDIPS prior to initiation of murine skin with the procarcinogen 7,12-dimethylbenz[a]anthracene (DMBA) has been reported by two groups to inhibit tumorigenesis[1,2]. DMBA must be metabolized in order to be tumorigenic and both groups provided indirect evidence implicating a CuDIPS-dependent effect on DMBA metabolism. The procarcinogen benzo[a]pyrene (BP), like DMBA, must be metabolized in order to be tumorigenic or mutagenic in the Ames Salmonella typhimurium assay[6]. We have used the Ames

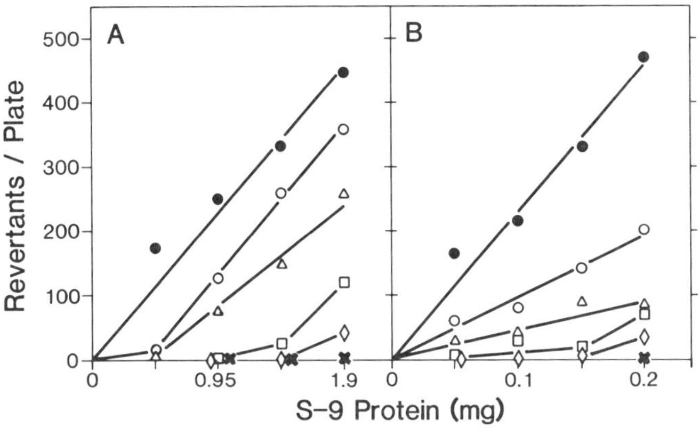

Fig. 1. Inhibition of BP mutagenesis by CuDIPS. S-9 homogenates prepared from normal murine liver (A) and mice pretreated with Aroclor 1254 (B) were used in Ames assays supplemented with solvent (●) or 15 μM (○), 30 μM (△), 45 μM (□) and 60 μM (◇) CuDIPS. No mutagenesis above the spontaneous background was detected in the absence of NADP and glucose 6-phosphate (X). Data are from reference 4.

assay as a tool for indirectly surveying the effects of CuDIPS and its analogs on the metabolic activation of BP.

Supplementation of liver homogenate S-9 preparations with CuDIPS strongly inhibited BP-dependent mutagenesis in the Ames assay (Fig. 1). Inhibition was dose dependent and independent of whether liver homogenates were prepared from normal or Aroclor 1254 pretreated mice. The requirement for a NADPH generating system for mutagenesis suggests that the cytochrome P-450 dependent mixed-function monooxygenase system (MFO) is responsible for the metabolic activation of BP. Neither diisopropylsalicylate (DIPS) or Zn(II)(diisopropylsalicylate)$_2$ (ZnDIPS), analogs of CuDIPS lacking SOD activity,[2,3] inhibited BP mutagenesis (Fig. 2A). $CuSO_4$, on a molar basis, was almost as effective as CuDIPS at inhibiting BP mutagenesis.

Mixed-function monooxygenase-dependent metabolism of xenobiotics requires reduction of the P-450 cytochromes by cytochrome P-450 reductase. Analogous to the results which

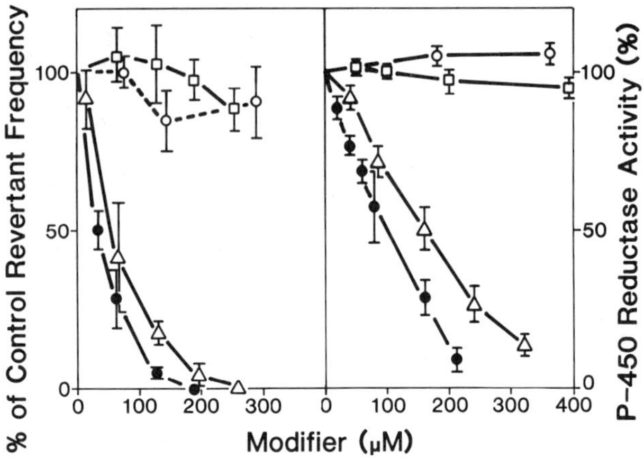

Fig. 2. Inhibition of BP mutagenesis (A) and P-450 reductase (B) by CuDIPS. S-9 liver homogenates (A) or microsomes (B) prepared from Aroclor 1254-treated mice were used in Ames or P-450 reductase assays supplemented with CuDIPS (●), $CuSO_4$ (△), ZnDIPS (□), and DIPS (○). Data are from reference 4.

were obtained in the Ames assay, CuDIPS inhibited P-450 reductase activity in microsomal preparations prepared from Aroclor 1254 treated mice, whereas DIPS and ZnDIPS had no effect at comparable concentrations (Fig. 2B). $CuSO_4$, like CuDIPS, also inhibited P-450 reductase activity in a dose responsive manner.

Dissociation of Antimutagenic and SOD Activities of CuDIPS

Both CuDIPS[3] and $CuSO_4$[7] are capable of disproportionating the reactive oxygen species superoxide. In order to determine whether the SOD activity of CuDIPS was responsible for its inhibitory effects, bovine liver Cu,Zn-SOD was surveyed for antimutagenic-antireductase activities (Fig. 3). Bovine liver Cu,Zn-SOD, at a SOD titer representing a 100 fold excess of CuDIPS-SOD mimetic activity, had no effect on either BP mutagenesis or P-450 reductase activity. Under the assay conditions employed, the CuDIPS mutation and P-450 reductase inhibition curves were displaced by a factor of approximately 3.

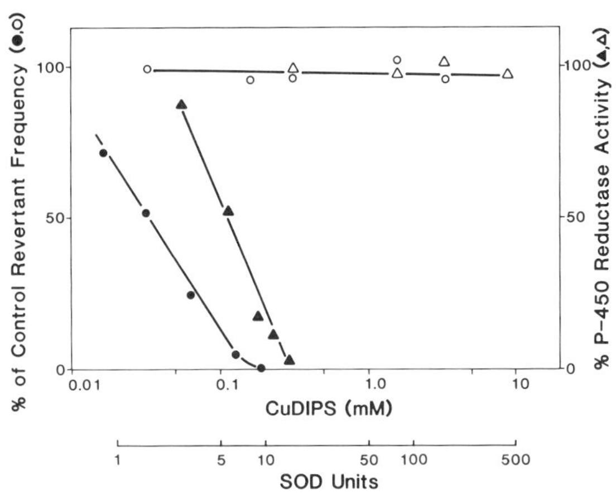

Fig. 3. Inhibition of BP mutagenesis and P-450 reductase by CuDIPS but not bovine SOD. Liver S-9 homogenates prepared from Aroclor 1254-treated mice were either used in an Ames assay (circles) or P-450 reductase assays (triangles) supplemented with either CuDIPS (closed symbols) or bovine SOD (open symbols). Data are from reference 4.

Tumor Promoter Dependent Elevation of Epidermal Xanthine Oxidase Activity.

Although the critical molecular and biochemical events intrinsic to the processes of promotion are unknown, a substantial data base has accumulated implicating a role for the reactive oxygen species superoxide in skin tumor promotion[3,4,8,9]. To date, the source(s) of the reactive oxygen species relevant to tumor promotion in murine skin are speculative.

Xanthine dehydrogenase (XD) and xanthine oxidase (XO) are purine catabolism pathway enzymes responsible for the conversion of hypoxanthine to xanthine, and xanthine to uric acid. As a dehydrogenase XD uses NAD as an oxidant, whereas XO uses O_2 as an oxidant, resulting in the production of superoxide. Della Corte and Stirpe[10] originally demonstrated that rat liver XD could be converted to XO by a variety of *in vitro* treatments. Recently, it was demonstrated that the XD to XO conversion is not solely an *in vitro* phenonomen. Elevated XO activity, mediated by the

Treatment	nmole/min/mg XO + XD		XO / (XO + XD) (%)
Acetone	1.74	4.08	45.5
Acetone + CHX	1.53	3.01	52.5
TPA	5.80	8.34	70.0
TPA + CHX	1.84	3.29	57.2

Table 1. TPA-dependent elevation of xanthine dehydrogenase and oxidase activities in murine epidermis. Epidermal homogenates were prepared from the dorsal skins of mice 48 h after treatment with TPA or acetone. Cycloheximide (CHX) was applied topically just prior to TPA.

conversion of XD to XO, is associated with intestinal ischemia[11] and ethanol associated myocardial and hepatic toxicity[12]. The findings that the XD-XO inhibitor allopurinol can eliminate or lessen the severity of tissue injury in the aforementioned examples[12] suggest that XO is the causative factor of the cytotoxicity.

Topical application of a promoting dose of TPA (2μg) to the dorsal skin of SENCAR mice resulted within 48 h in a 2-fold increase in total epidermal XD plus XO activity, and a 3-fold increase in XO activity (Table 1). Particularly noteworthy is the increase in the XO/XD+XO ratio, which is indicative of a conversion of XD to XO. Topical application of the protein synthesis inhibitor cycloheximide just prior to TPA treatment inhibited the TPA dependent increases in XO activity and the XO/XD+XO ratio. Collectively these data suggest that TPA induces the synthesis of epidermal XD and facilitates the conversion of XD to XO.

The relative <u>in vivo</u> promoting activities of TPA and its analogs could be correlated with their abilities to elevate XO activity. A dose response increase in XO specific activity was measured for topical applications of 0.5-2 μg TPA (Table 2). In SENCAR mice, 0.5 μg of TPA is a marginal promoter, whereas 2 μg of TPA is the standard and optimal dose. The weak tumor promoter phorbol-12,13-dibenzoate resulted in a modest elevation of XO at a dose of 20μg per application. The nontumor promoting agent phorbol had no effect on XO activity.

Treatment	Dose(μg)	% of control
Acetone	-	100
TPA	0.5	167
	1.0	188
	2.0	291
Phorbol	2.0	98
	20.0	94
4-O-methyl-TPA	2.0	120
	20.0	101
Phorbol-12,13-dibenzoate	2.0	149
	10.0	179

Table 2. Elevation of xanthine oxidase in murine epidermis by TPA and its analogs. Epidermal homogenates were prepared from the dorsal skins of mice 48 h after topical treatment.

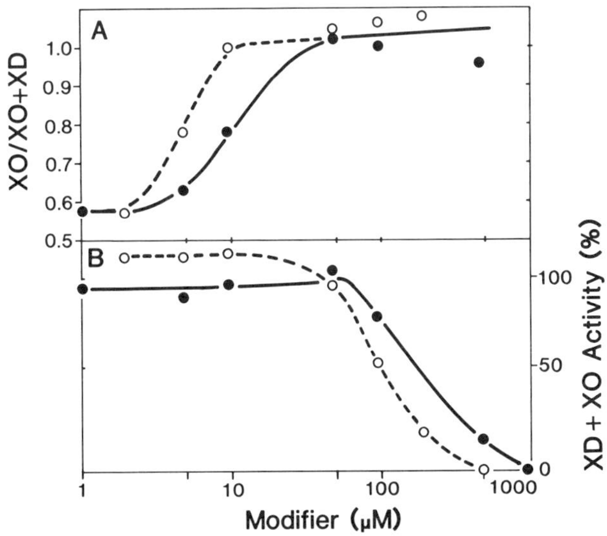

Fig. 4. CuDIPS and $CuSO_4$ dependent conversion of xanthine dehydrogenase to xanthine oxidase (A) and inhibition of oxidase (B). Epidermal homogenates were incubated at room temperature for 10 min with various amounts of CuDIPS (●) or $CuSO_4$ (○) prior to assay. At concentrations greater than 50 μM of either agent XO is the predominate activity.

CuDIPS-Dependent Conversion of XD to XO and Inhibition of XO Activity

Depending upon the concentration of CuDIPS, supplementation of an epidermal homogenate with CuDIPS had variable effects on XD and XO activities (Fig. 4). In the reported experiment, approximately 42% of the total XD plus XO activity was XD. At CuDIPS concentrations $< 5\mu M$, no effects on the XO/XO+XD ratio were detected. However, at greater concentrations, the ratio approached unity in a dose responsive manner. This suggests that CuDIPS can cause the conversion of XD to XO. At concentrations of CuDIPS greater than $50\mu M$, a dose dependent inhibition of XO activity was noted. Because XD is presumably converted to XO by 50 μM CuDIPS, XO is actually the enzymatic species being inhibited at the higher concentrations. Like CuDIPS, $CuSO_4$ also facilitated the conversion of XD to XO and could inhibit XO activity.

DISCUSSION

CuDIPS is a potent inhibitor of BP mutagenesis as scored in an Ames assay. The _in vitro_ antimutagenic properties of CuDIPS were independent of its salicylate structure since neither DIPS nor ZnDIPS inhibited BP mutagenesis. Rather, the antimutagenic properties of CuDIPS were due to the presence of copper, but independent of the copper-dependent SOD-mimetic activity.

The requirement of a NADPH generating system for mutagenesis in the Ames assay suggests that the MFO system is responsible for the metabolic activation of BP. An important component of the MFO system is cytochrome P-450 reductase. Inhibition of reductase activity would prevent reduction of the cytochrome P-450 components of the MFO system, resulting in an inhibition of xenobiotic metabolism. The antimutagenic activities of the CuDIPS analogs correlated with their abilities to inhibit P-450 reductase. Furthermore, the dose dependencies for inhibition of mutagenesis and reductase activities were very similar. Collectively these data suggest that the _in vitro_ antimutagenic activity of CuDIPS is probably the consequence of an inhibition of P-450 reductase activity. Although it is speculative to extrapolate from _in vitro_ data, it is conceivable that the reported anti-initiating properties of CuDIPS[1,2] may be related to its ability to modulate procarcinogen metabolism by inhibiting P-450 reductase activity.

Topical application of the tumor promoter TPA to murine epidermis resulted in the synthesis of XD, and the conversion of XD to XO. Functioning as an oxidase, XO uses molecular oxygen as an oxidant with the resulting production of superoxide. Although there is a strong precedent for the role of superoxide in tumor promotion[8,9], and the *in vivo* promoting activities of TPA analogs correlated with their abilities to elevate XO activity, any discussion of a role for elevated XO activity in skin tumor promotion is speculative at this time. Similarly, although CuDIPS could inhibit XO activity in epidermal homogenates at concentrations greater than 50 μM, it is unknown whether this inhibitory activity is responsible for the reported anti-promoting properties of CuDIPS.

ACKNOWLEDGEMENT

The reported studies were supported by PHS grant CA 40823, awarded by the National Cancer Institute, Department of Health and Human Services.

REFERENCES

1. Solanki, V; Yotti, L; Logani, M; Slaga, T. Carcinogenesis 1983, 5, 129.
2. Egner, P; Kensler, T. Carcinogenesis 1985, 6, 1167.
3. Kensler, T; Bush, D; Kozumbo, W. Science (Wash) 1983, 221, 75.
4. Reiners, Jr., J; Brott, E; Sorenson, J. Carcinogenesis. 1986, 7, In Press.
5. Ghezzi, P; Bianchi, M; Mantovani, A; Spreafico, F; Salmono, M. Biochem. Biophys. Res. Commun. 1984, 119, 114.
6. Reiners, Jr., J; Crowe, P; McKeown, C; Nerland, D; Sonnenfeld, G. Carcinogenesis 1984, 5, 125.
7. Rabani, J; Klug-Roth, D; Lilie, J. J. Phys. Chem. 1973, 77, 1169.
8. Kozumbo, W.; Trush, M.; Kenser, T. Chem. Biol. Interaction 1985, 54, 199.

9. Kensler, T; Trush, M. Environ. Mutagen. 1984, 6, 593.
10. Della Corte, E; Stirpe, F. Biochem. J. 1972, 126, 739.
11. Granger, D; Rutili, G; McCord, J. Gastroenterology 1981, 81, 22.
12. Oei, H; Zoganas, H; McCord, J; Schaffer, S. Res. Comm. Chem. Path. Pharm. 1986, 51, 195.

DISCUSSION

ELLIOTT (Graham): Is it correct that a chemical which stimulates cell activity could be a promoter and chemical that inhibits cell activation would inhibit tumor formation?

REINERS: No. There are a lot of proliferative agents, specifically hyperplastic agents, that cause skin growth etc. They're not promoters. All promoters show a hyperplastic activity but not all hyperplastic agents are promoters. We still don't know what is absolutely necessary for promotion.

ELLIOTT (Graham): So pure activation of a cell with, for example, increased O_2^- formation doesn't necessarily result in tumor formation. Is there any relationship at all between an increased metabolic rate, as expressed for example by increased oxygen radical formation, and the ability to promote tumors?

REINERS: I'm not really sure I can answer that. It comes down to, based upon published and unpublished work of Prof. Kenseler, being clear that the oxygen radical story is not sufficient to totally explain promotion. It can be associated with the second stage of promotion, not the first stage.

ELLIOTT (Graham): Was the time course for the increase in xanthine oxidase activity in the skin related to any change in cell infiltration?

REINERS: Yes. Kinetics of elevation and recovery perfectly paralleled hyperplasia that one gets following topical application of TPA.

ELLIOTT (Graham): So you found a permanent change in a

certain cell population which exists as long as cells live?

REINERS: No, I don't think it's permanent. What is occurring during that time period is a massive wave of epidermal proliferation, and it's impossible to make any comments about what happens in individual cells.

HERNANDEZ: Craven, Pfanstiel, and DeRubertis (J. Clin. Invest. 1986, 77, 850.) studying carcinoma of the colon found that Cu(II)(3,5-dips)$_2$ was very effective in protecting against cell proliferation. Carcinomas of the colon are associated with chronic inflammation. The theory is that PMNs are in colonic tissue for so long, because of the chronic inflammation, that they are responsible for cancer development. There is still a lot of research to be done in that area but it seems as though there is going to be a close correlation between chronic inflammation and cancer. I have just passed from the theory that xanthine oxidase is the most important, to the theory that PMN's are the most important. I used Cu(II)(3,5-dips)$_2$ in an intestinal preparation and I don't know whether the dose that I used was inhibiting xanthine oxidase or if it was scavenging O_2^-. We used 25 mg/kg with 3 kilogram cats, and I don't think we were achieving very high concentrations. When you go from your in vitro system to an in vivo system, would you like to use Cu(II)(3,5-dips)$_2$ concentrations that are xanthine oxidase inhibitors or that are SOD-like in activity? If you have a circumstance that is inducing the conversion from xanthine dehydrogenase to xanthine oxidase, it would be deleterious and cause pathology.

REINERS: I think it's going to be very difficult to figure out an appropriate in vivo dosage. I haven't done any in vivo experiments and measured effects of Cu(II)(3,5-dips)$_2$ on xanthine oxidase. I have done some in vitro experiments that, as of yet, are unpublished. In thinking about them I think I now understand what is happening. Experiments I am eluding to are experiments that were done with cultured keratinocytes. I'm practicing heresy by saying this, but I have very good data that suggest that at very very low concentrations of Cu(II)(3,5-dips)$_2$, similar to doses used by Prof. Kensler, I can induce ornithine decarboxylase (ODC) in cultured keratinocytes. If I use higher concentrations of Cu(II)(3,5-dips)$_2$, I don't see this induction of ODC. Based upon the literature, the current thought is that Cu(II)(3,5-dips)$_2$ should be an outstanding inhibitor of that enzyme. That's not what I find but I'm using very very low concentrations. Based upon the work that Sue Fisher has done and the fact that xanthine oxidase will essentially result in induction of ODC, what's happening, at least in my in vitro culture system, is that low doses of Cu(II)(3,5-dips)$_2$ convert the dehydrogenase into the oxidase and then the oxidase is inducing ODC. If high concentrations of Cu(II)(3,5-dips)$_2$ are used along with TPA as a stimulator of ODC, then you can inhibit. Here is a paradox where a low concentration of Cu(II)(3,5-dips)$_2$ does one thing and a very high concentration has the opposite effect. I think dosing in animal experiments is going to be very difficult. I'm not exactly sure which parameter to use to select a dose.

REED: In your presentation you concentrated on the role of P_{450} in metabolizing xenobiotics. However, this is also a key system in the metabolism of endogenous compounds including synthesis and transformations of steriod hormones and aracadonic acid products. If the reductase involved in these transformations is affected by Cu(II)(3,5-dips)$_2$, might this play a role in the use of Cu(II)(3,5-dips)$_2$ as an antipromoting agent?

REINERS: It could possibly. As long as the reductase is being inhibited and steroids etc. are metabolized by the P_{450} system, Cu(II)(3,5-dips)$_2$ may affect these metabolic processes. It's going to inhibit metabolism. I don't think it's going to make any difference in terms of what xenobiotic is presented. Its metabolism is going to be altered. Because its metabolism is altered, of course, there are going to be physiological effects.

(Editor's note: Cu(II)(3,5-dips)$_2$ may have been shown to modulate P_{450} reductase or Fe(III) oxycytochrome requiring process in vivo using pharmacological doses. Rats pretreated with BAP so as to induce P_{450} and then treated with paracetamol to induce liver toxicity have been protected against paracetamol-induced toxicity with Cu(II)(3,5-dips)$_2$. (Wendel, A; Heidinger, S. Res. Comm. Chem. Path. Pharmacol. 1980, 28, 473.). It is not known as to whether or not pharmacologic doses of Cu(II)(3,5-dips)$_2$ will affect P_{450} activity in non-BAP-induced situations.)

RADIOPROTECTANT EFFECTS OF $CU(II)(3,5\text{-}DIPS)_2$

Hamid Salari,[1,2] Max L. Baker,[2] John B. Barnett,[3] Lee S. F. Soderberg,[3] William M. Willingham[4] and John R. J. Sorenson[1]

[1]Department of Biopharmaceutical Sciences, College of Pharmacy, [2]Department of Radiology and [3]Department of Microbiology and Immunology, College of Medicine, University of Arkansas for Medical Sciences, Little Rock, Arkansas 72205
[4]Department of Natural Sciences, University of Arkansas at Pine Bluff, Pine Bluff, Arkansas 71601

INTRODUCTION

Ionizing radiation of oxygenated aqueous solution is known to produce superoxide as a secondary product of radiation.[1-3] Part of the cellular injuries caused by radiation can be attributed to the acute increase of superoxide. Addition of Cu-Zn superoxide dismutase (Cu-Zn SOD) to the extracellular medium can significantly protect cultured cells against radiation damage.[4] Adding Cu-Zn SOD to cultures of myoblasts protected them from a radiation-induced decrease in viability and altered morphology[5] as well as chromosomal damage.[6]

Radioprotective effects of Cu-Zn SOD have also been demonstrated *in vivo*. Mice given Cu-Zn SOD parenterally show increased radioresistance. Petkau et al.[7] reported that a dose of 35mg/kg of Cu-Zn SOD when given intravenously to Swiss mice one hour before irradiation increased the radiation $LD_{50/30}$ from 6.27Gy to 7.0Gy. Intrapertoneal injection of 35 mg/kg Cu-Zn SOD to Sliciddy mice was less effective.[8] Because copper(II)(3,5-diisopropylsalicylate)$_2$ [$Cu(II)(3,5\text{-}DIPS)_2$] has SOD-mimetic activity[9] and it is a lipid soluble copper complex, capable of crossing lipid membrane barriers, it was tested as a radioprotectant. $Cu(II)(3,5\text{-}DIPS)_2$ was found to be effective *in vivo* when 60% of the pretreated mice survived a lethal dose of radiation.[10]

In preparing for a more detailed evaluation of the radioprotectant activity of $Cu(II)(3,5\text{-}DIPS)_2$, its acute toxicity was determined both *in vitro* and *in vivo*. A concentration of 10ug of $Cu(II)(3,5\text{-}DIPS)_2$ per ml of culture medium produced no death in cultured mouse L-929 cells. Probit analysis of acute toxicity data obtained with 8 to 10 week old C57BL/6 female mice revealed that 80mg of $Cu(II)(3,5\text{-}DIPS)_2$ per kilogram of body weight (mg/kg) was a nontoxic dose suitable for subsequent *in vivo* radioprotectant studies.

Five ug of $Cu(II)(3,5\text{-}DIPS)_2$ per ml of culture medium protected L-929 cells against doses of ionizing radiation ranging from 1 to 6 Gy (1Gy=100rad) and 80mg/kg of the compound protected mice against doses of radiation ranging from 7 to 10 Gy. The dose modifying effect in both cases was 1.2.

MATERIALS AND METHODS

3,5-DIPS was obtained from Aldrich Chemical Company. $Cu(II)(3,5\text{-}DIPS)_2$ was synthesized according to techniques described by Sorenson.[11] A uniform and stable suspension of this copper complex was prepared by first wetting the complex with propylene glycol (final concentration in suspension 4%) and then suspending the wetted powder in a 1.4% solution of polyvinyl alcohol in sterile saline (PG, PVA-saline). We routinely obtained samples before and after treating each group of mice. These samples are then analyzed for copper to determine the actual dose used in treatment. Using this procedure it has been determined that the actual dose given was 80 ± 5 mg/kg for all treatments to date. For cell culture studies 3,5-DIPS and $Cu(II)(3,5\text{-}DIPS)_2$ were dissolved in 100% ethanol to make the appropriate addition to culture medium.

L-929 mouse fibroblasts were originally maintained in Eagle's minimal essential medium (MEM, GIBCO, Grand Island, NY) supplemented with 10% calf serum and antibiotics (100 units penicillin/ml and 100 ug streptomycin/ml, GIBCO).

Cells were allowed to grow at 37°C in humidified air containing 5% Co_2 using standard tissue culture techniques. The cell line and methods for the survival experiments have been described previously.[12-14]

Toxicity studies of these compounds were accomplished by adding ethanol (1% final concentration in medium) solutions of 3,5-DIPS or Cu(II)(3,5-DIPS)$_2$ to L-cell cultures in pH adjusted Hank's balanced salt solution (HBSS) to give a final concentration ranging from 0.5 to 30 ug/ml or 10 to 50 ug/ml, respectively.

Irradiation of cultures was performed in a J. B. Shepherd Associates ^{137}C irradiator providing a dose-rate of 1.7 Gy/min at the center of the irradiation chamber. Irradiation was delivered 30 min. after or immediately before exposing cells to medium containing 5 ug/ml Cu(II)(3,5-DIPS)$_2$. Controls were exposed to medium containing 1% ethanol. Non-irradiated cells were similarily treated to determine plating efficiency, which ranged from 45 to 60% for these experiments, and compound toxicity.

Compounds were allowed to remain in contact with L-929 cells for 2 hours with each treatment and then the HBSS was replaced with complete growth medium containing 16% calf serum. Cell cultures were then returned to the incubator at 37°C for 2 weeks prior to colony counting. After the 2 week growth interval colonies were fixed, stained, and counted.[12,14]

Adult female C57BL/6 mice (8 to 10 weeks old) weighing 18-21 grams were obtained from the National Center for Toxicological Research, Redfield, Arkansas. Mice were housed at a density of five per cage in plastic cages and given food and water ad libitum, and maintained on a 12 hour light/dark cycle to entrain any circadian rhythms which could influence the radiation response. Prior to irradiation, mice were randomly divided into two groups (15 mice/group), vehicle-treated or Cu(II)(3,5-DIPS)$_2$-treated.

Determinations of radiation $LD_{50/30}$ and dose modifying effect of Cu(II)(3,5-dips)$_2$ were done with a ^{137}Cs irradiator (J. L. Shephard and Associates) at room temperature, with a dose-rate of 1.7 Gy/min. A single study was conducted at room temperature with ^{60}Co unit

(Theratron 780, Atomic Energy of Canada Ltd) with a field size of 14 X 20 Cm^2 and a dose-rate of 0.4 Gy/min. Dosimetry was performed using NBS traceable Baldwin-Farmer ionization chambers.

$Cu(II)(3,5\text{-}DIPS)_2$ was administered in 0.5 ml of vehicle by subcutaneous (s.c.) injection in the neck region. In these experiments control mice received a similar volume of vehicle (PG, PVA-saline). Probit analysis[15] was used to calculate the lethal dose for 50% of the mice. All deaths occurred within 5 days postinjection.

Probit analysis was also used to calculate radiation lethality value and the effect of the drug on this value. Mice were injected s.c. with vehicle or 80 mg/kg Cu(II)-$(3,5\text{-}DIPS)_2$ 3 hours before irradiation and the rate of death followed for 30 days to determine $LD_{50/30}$ values.

RESULTS AND DISCUSSION

$Cu(II)(3,5\text{-}DIPS)_2$ is virtually insoluble in water, so that the most acceptable and convenient form for administration was an aqueous suspension with PG, PVA-saline. Figure 1 presents drug toxicity results. These indicate that the LD_{50} value for $Cu(II)(3,5\text{-}DIPS)_2$ was 260 mg/kg and that all mice destined to die did so within 5 days. For subsequent experiments, 80 mg/kg Cu(II)(3,5-$DIPS)_2$ was administered subcutaneously 3 hours before irradiation. This dose has never been found to produce acute toxicity. The effect of $Cu(II)(3,5\text{-}DIPS)_2$ on radiation-induced lethality is shown in Figure 2. The $LD_{50/30}$ value as determined by probit analysis for vehicle-treated mice was 7.92 Gy with 95% confidence limit of 7.57-8.25 Gy, while for $Cu(II)(3,5\text{-}DIPS)_2$ treated mice the value was 9.04 (8.62-9.55)Gy. Dose modification factor (DMF), calculated as the ratio of $LD_{50/30}$ values, was found to be 1.15 for $Cu(II)(3,5\text{-}DIPS)_2$.

Figure 3 shows percent survival 40 days after irradiation. Radioprotection with $Cu(II)(3,5\text{-}DIPS)_2$ was noted at all doses of radiation (7.0 to 10.0 Gy) and showed an increase in survival ranging from 21 to 46% compared with vehicle treated mice. These results were obtained at a dose rate of 1.7 Gy/min.

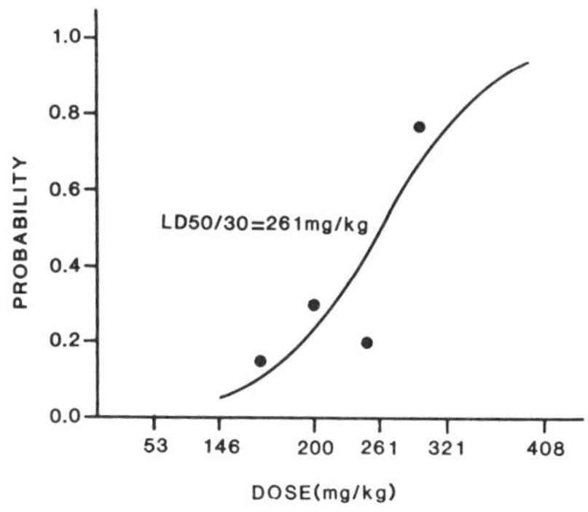

Figure 1
The LD50/30 obtained for C57BL/6 mice given Cu(II)-(3,5-DIPS)$_2$ subcutaneously.

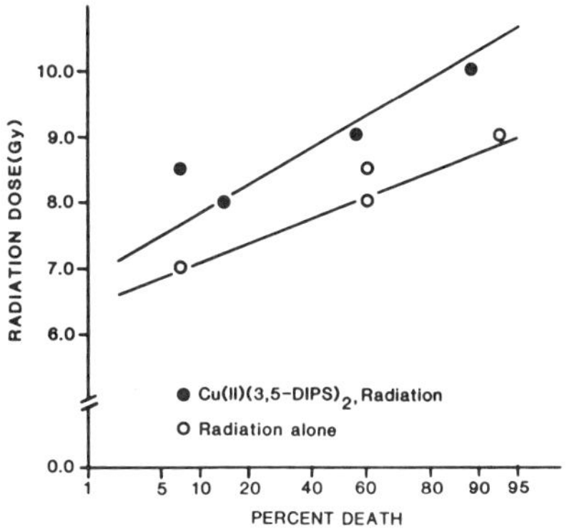

Figure 2
Percent lethality (probit scale) versus radiation dose for C57BL/6 mice given 80 mg/kg of Cu(II)(3,5-DIPS)$_2$ or vehicle 3 hours before irradiation.

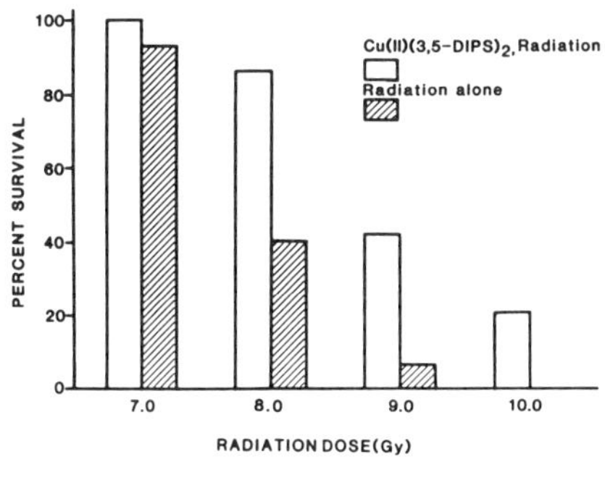

Figure 3
Percent survival vs radiation dose for C57BL/6 mice given 80 mg/kg of Cu(II)(3,5-DIPS)$_2$ or vehicle 3 hours before irradiation.

When the dose-rate was lowered to 0.4 Gy/min using the ^{60}Co unit, the 80 mg/kg Cu(II)(3,5-DIPS)$_2$ treatment produced 73% survival in comparison to 13% survival for control mice as shown in Figure 4.

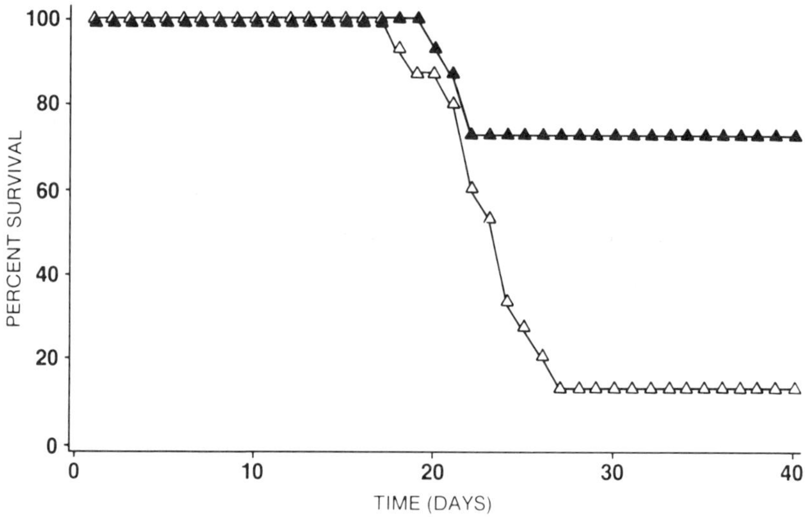

Figure 4. Survival of C57BL/6 mice exposed to 10 Gy ^{60}Co irradiation and treated with Cu(II)(3,5-DIPS)$_2$ (▲) or vehicle (△), given subcutaneously 3 hours before irradiation.

Figure 5 shows results of toxicity studies with Cu(II)(3,5-DIPS)$_2$ and 3,5-DIPS acid on L-cells using the single cell colony forming assay. A concentration of 10 ug of Cu(II)(3,5-DIPS)$_2$ per ml of culture medium caused no cell death while 20 ug of the complex per ml of medium show a marked loss in colony formation (e.g., 50% reduction). However, this complex was less toxic than 3,5-DIPS acid which produced 40% and 99% reduction in colony formation when cells were exposed to a concentration of 10 ug or 30 ug of the acid per ml of medium, respectively.

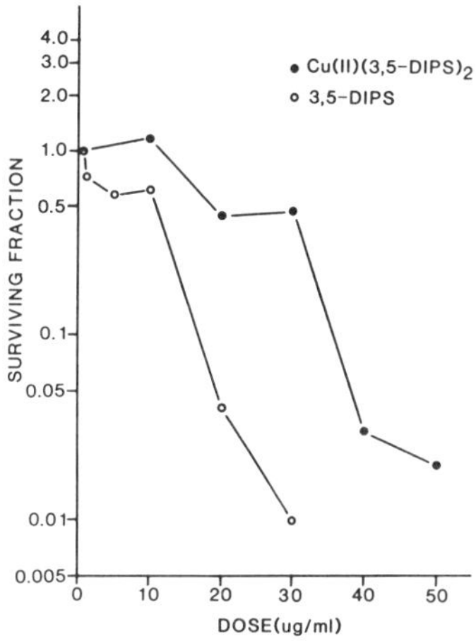

Figure 5
Effects of various concentrations of 3,5-DIPS and $Cu(II)(3,5\text{-}DIPS)_2$ on survival of L-929 cells.

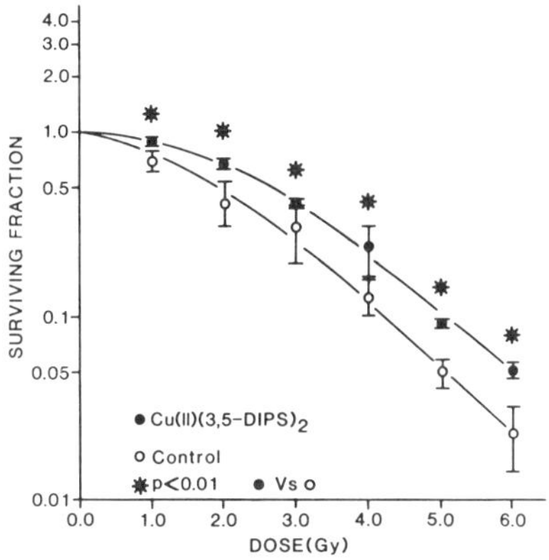

Figure 6
Survival curves showing the radioprotectant effect of $Cu(II)(3,5\text{-}DIPS)_2$ on L-cells treated before irradiation.

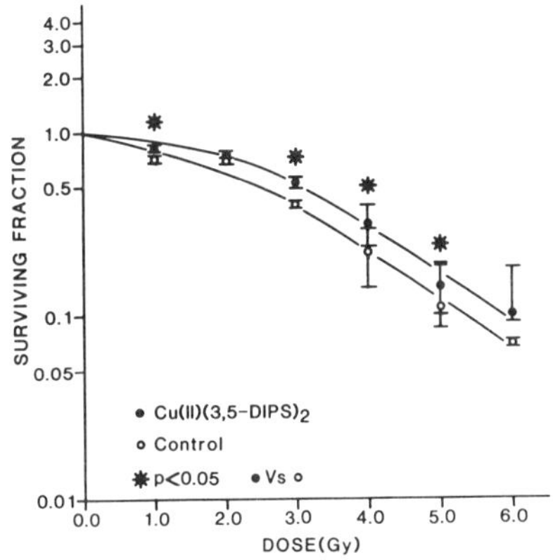

Figure 7
Survival curves showing the radioprotectant effect of $Cu(II)(3,5\text{-}DIPS)_2$ on L-cells treated after irradiation.

Figures 6 and 7 show that treatment of L-cells with medium containing 5 ug/ml $Cu(II)(3,5\text{-}DIPS)_2$ either 30 min. before (Fig. 6) or immediately after irradiation (Fig. 7), increased survival at all radiation dose levels. Statistic analyses (Student t-test) revealed that survival for post-irradiation treatment was significantly increased at radiation doses of 1, 3, 4, and 5 Gy and the survival for pre-irradiation treatment was significant at all doses of irradiation.

These remarkable radioprotectant activities of Cu(II)(3,5-DIPS)$_2$ may not be solely due to its disproportion of superoxide but they may also be related to its ability to react with hydroxyl radical, hydroperoxyl radical, hydrated electrons, and hydrogen atoms. Simple copper complexes are known to disproportionate superoxide and hydroperoxyl radicals, reduce hydroxyl radical and they reduced by hydrated electrons and hydrogen atoms at diffusion limited rates of reaction.[16-19]

CONCLUSION

Cu(II)(3,5-DIPS)$_2$ protects mice against lethal doses of ionizing radiation. As described elsewhere[20] and in these proceedings, increased survival of irradiated mice following treatment with Cu(II)(3,5-DIPS)$_2$ could be associated with rapid recovery of bone marrow cell and spleen cell immunocompetency as well as significant early increases in numbers of IL-3 responsive spleen cell precursors.[21] Cu(II)(3,5-DIPS)$_2$ also protected L-929 mouse fibroblast against lethal doses of ionizing radiation. The magnitude of survival for post-irradiation treated cells was smaller than pre-irradiation treatment, however, any protection post-irradiation merits further study. These observations of radioprotection offers a new approach to the development of more effective radiation protecting agents.

ACKNOWLEDGEMENT

We are indebted to the Arthur Armbrust Cancer Research Fund, International Copper Research Association, Max and Victoria Dreyfus Foundation, Elisa U. Pardee Foundation, and Denver Roller Corporation for financial support and the National Center for Toxicological Research for the mice used in these studies.

REFERENCES

1. Behar, D; et al. J. Phys. Chem. 1970, 74, 3209.
2. Czapski, G. Ann. Rev. Phys. Chem. 1981, 22, 171.
3. Okada, S. In: "Radiation Biochemistry", Okada, S. Ed.; Academic Press; New York, 1970; vol. 2, p 1.
4. Nordenson, I; Bekman, G; Beckman, L. Hereditas 1976, 82, 125.
5. Michelson, A.M; Buckingham, M.E. Biochem. Biophys. Res. Commun. 1975, 65, 886.
6. Nordenson, I. Hereditas 1978, 89, 163.
7. Petkau, A; et al. Biochem. Biophys. Res. Commun. 1975, 65, 886.
8. Mitsuyuki, Abe; et al. Int. J. Radiation Oncology Biol. Phys. 1981, 7, 205.
9. Fridovich, I. Science 1978, 875.
10. Sorenson, J.R.J. J. Med. Chem. 1984, 27, 1747.
11. Sorenson, J.R.J. J. Med. Chem. 1976, 19, 135.
12. Dalrymple, G.V; Sanders, J.L; Baker, M.L. Nature 1967, 216, 708.
13. Dalrymple, G.V; Sanders, J.L; Baker, M.L. J. Theor. Biol. 1968, 21, 368.
14. Dalrymple, G.V; Sanders, J.L; Baker, M.L; Wilkinson, K.P. Radiat. Res. 1969, 37, 90.
15. Hardison, C.D; Quads, D; Langston, R. D. In: "Nine Functions for Probability Distribution," SUGI Supplemental Library User's Guide; Sas Institute Inc.; Cary, NC, 1983.
16. Weser, U; Lengfelder, E; Sellinger, K.H; Schobtz, L. In: "Inflammatory Disease and Copper"; J.R.J. Sorenson, Ed.; Humana Press, Clifton, NJ, 1982; p 173.
17. Rabani, J; Klugroth, D; Lilie, J. J. Phys. Chem. 1973, 77, 1169.
18. Farhataziz; Ross, A.B. In: "Selected Specific Rates of Reactions of Transients from Water in Aqueous Solution"; U.S. Government Printing Office; Washington DC, 1977.
19. Anbar, M; Neta, P. Int. J. Appl. Rad. Isotop. 1967, 18, 493.
20. Soderberg, L.S.F; Barnett, J.B; Baker, M.L; Salari, H; Sorenson, J.R.J. Submitted for publication.
21. Soderberg, L.S.F; Barnett, J.B; Baker, M.L; Salari, H; and Sorenson, J.R.J. In: Biology of Copper Complexes, J.R.J. Sorenson, Ed; Humana Press, NY, (this proceedings).

DISCUSSION

LOVEN: You mentioned you switched between ^{60}Co and ^{137}Cs in your dose rate studies. The problem is that ^{60}Co is a more energetic gamma emitter and it's going to have a different depth dose distribution versus ^{137}Cs. A better way to do the dosing would be to move ^{60}Co source to achieve different dose rates.

SALARI: We were attempting to obtain data that could be compared with data obtained earlier with a ^{60}Co source that was used to deliver this dose-rate. We did adjust the distance to obtain this dose-rate.

LOVEN: But you worked with Co and Cs didn't you? What I am saying is that Co is a different energy than Cs. You were studying a $LD_{50/30}$ dose which means you were looking at effects on bone marrow. If you use different sources you are going to get different doses of radiation penetrating to the bone marrow with those 2 types of radiation sources.

BAKER: Both radiations were done under conditions of maximum back scatter which should eliminate the depth dependency. Obviously the buildup is different for the two different types of radiation. Our dosimetery should have taken care of that.

SORENSON: The new data obtained with ^{60}Co was gotten so that we could compare these results with results we had gotten earlier with a ^{60}Co source that delivered the same dose-rate. The dose of $Cu(II)(3,5\text{-dips})_2$ was the variable in these two studies.

LOVEN: Your cell survival data were significantly different for treated and non-treated cells at different radiation doses. Were the DQs significantly different?

SALARI: We haven't calculated DQ values yet. That's an oversight on our part.

LOVEN: Did you use a feeder cell layer or did you just do straight plating in your cell survival studies? There may be a cell survival effect there as well. People at Armed Forces Radiobiology Research Institute have shown the number of dead cells plated can influence survival of living cells.

SALARI: No. We did not use a feeder layer.

LOVEN: The varying number of dead cells may still have drug attached or associated with them. They may be slowly releasing it over time, or if dead cells never take up any drug then survivors are exposed to more drug.

HARRIS: In your mouse studies, you found 100% survival for 10 days after irradiation and then survival decreased to 0 with time. Were you able to determine the cause of death?

SORENSON: With these doses of radiation, the cause of death is due to bone marrow suppression and loss of immuno-compentency. They ultimately die of overwhelming infection.

IWAKAWA: You showed percent survival for mice treated 3 hours before irradiation and 24 hours before irradiation. Some mice in the 3 hour pretreatment group died within the 10 day period following irradiation. Do you have any idea as to why they died?

SORENSON: The answer is we don't know. But, when we gave the dose that Commander Brian Gray intended to give, 250 mg/kg, we killed all of our mice. We did that several times and then finally came to the conclusion that Brian had not actually given the dose that he had intended to give. That's why we did the LD_{50} study. When we looked at his survival data it was clear that the 3 hour pretreatment group had much more mortality in the 10 day post-irradiation period than the non-treated or the 24 hour pretreatment groups. What we suspect happened was that Brian or whoever worked with him made a suspension that had partially settled and the 24 hour pretreatment group was treated first with less compound than intended. It is likely that the 3 hour pretreatment was treated with the same suspension, but now the suspension was more concentrated. However, there was still some settling of the suspension and while the intended dose was not given, more compound was given to this group than the 24 hour pretreatment group. We suggest that the dose we now use, 80 mg/kg, is probably close to the dose given to the original 24 hour pretreatment group. We no longer find acute toxicity in that 10 day posttreatment period. We also analyze doses we give by atomic absorption spectrophotometry to be certain that the intended dose is given. Does that answer your question?

IWAKAWA: Yes. Is the radiation-induced death due to the gastrointestinal syndrome?

SORENSON: With these radiation doses death is likely to be due to the hematopoietic syndrome although we have not ruled out the gastrointestinal syndrome as a cause of some deaths.

GREENAWAY: How was the structure of Cu(II)(3,5-dips)$_2$ ascertained? Was there an x-ray crystal structure determination? I would not expect that structure by comparison with Cu(II)(salicylate)$_2$ which is essentially the same. There are 2 possible crystal structures. One is a dimer and the other is a monomer where only the caboxyl group bonds along with a couple of water molecules. The hydroxyl group doesn't bond so it's not a chelate compound. I guess we need to do an x-ray crystal structure. The point being that if it's not a chelate then it probably does not hold together quite as well as it would if it were a chelate, and it may be more susceptible to exchange with other competing ligands.

SORENSON: That material is a brown powder and we have not been able to grow a crystal of it. We would like to get an x-ray crystal structure and we would be happy to supply compound to anyone with equipment capable of analyzing a powder. It may very well be a mononuclear carboxylate complex with water bonded in the axial positions. The isolated brown powder never contains 2 molecules of water. Following gentle drying at 50°C and 15 mm Hg overnight for 24 hours the powder analyzes for 0.5 or 1 water. This is our standard operating procedure. This partially hydrated compound is more easily suspended in comparison with the anhydrous powder. Anhydrous complexes are very difficult to resuspend for pharmacologic evaluations. We show the square planar structure as a matter of reasonable convenience but we would like to know the structure of this material with certainty. Everybody who's worked with Cu(II)(3,5-dips)$_2$ knows that it's actually a binuclear complex when it is dissolved in ether, alcohol, propylene glycol, or something like that. It goes immediately to a green compound and we know the structure of this green complex with certainty based upon Prof. Greenaway's ESR analysis and Prof. Cordes' x-ray crystal structure determination.

IWAKAWA: Can I think of Cu(II)(3,5-dips)$_2$ as a radiation dose modifier?

SORENSON: That seems reasonable at this point. I think it was Prof. Dean Loven that made a valid point in that we should have seen something remarkable when we changed our radiation source. We did. When we used the softer ^{60}Co source for the delivery of 40 rads/minute, 13% of the non-treated mice given a total dose of 1000 rads survived whereas 73% of the treated animals exposed to the same radiation dose and dose-rate survived. The percent survival increased for non-treated and treated mice exposed to the less energenic ^{60}Co source.

GENERAL DISCUSSION

SORENSON: Jim (Crook), when we began our radiation protection experiments, every mouse injected with Cu(II)(3,5-dips)$_2$ died within the 10 day post-irradiation period. We had used the dose of 250 mg/kg, the dose we thought Brian Gray had used. That dose plus radiation caused death before the irradiated non-treated animals died. We've come a long way in figuring out what happened. We hope we can do that in the cancer work as well. As to what is the cause of radiation-induced death, mice irradiated with doses of up to 1000 rads of ionizing radiation die as a result of bone marrow and spleen cell death which results in immuno-incompetency and an inability of the host to combat infection. Isn't that right Jim?

CROOK: Yes. That's right. If you go above 1000 rads death is due to the gastrointestinal syndrome.

SORENSON: If doses of 1000 to 5000 rads are used, mice die quicker because those doses of ionizing radiation cause denudation and bleeding or ulcers of the gastrointestinal tract that allow bacteria to more rapidly invade a host that is also immunologically incompetent. Cu(II)(3,5-dips)$_2$ has antiulcer activity and gastric wound healing activity. Dr. Iwakawa's suggestion that we examine the gastrointestinal tract of these animals is a good suggestion, especially for animals exposed to 1000 rads. Prof. Chang who is the pathologist collaborating with us is really anxious to do that as well.

CROOK: The most sensitive parameter of whether someone is going to recover from acute radiation syndrome is the rapidity and the extent of lymphopenia development. Absolute lymphocyte counts done over 48 hours is the best predictor you have because you're really acting as your own radiation biological monitor in that aspect. I think the last paper is going to be particularly pertinent. I anxiously await those results we've heard about over the past four days as to whether there are or are not changes in either up regulation or down regulation of lymphocytes. I would be intrigued to see if those lymphocytes, especially helper cells and things like that, are going to be affected by Cu(II)(3,5-dips)$_2$.

SORENSON: Speaking for Prof. Soderberg and the rest of our group, we do hope that we will be funded to do these studies. (Editor's note: We were funded by NCI!)

HARRIS: We heard 2 or 3 presentations in our symposium dealing with the use of radioactive isotopes of Cu, ^{67}Cu and ^{64}Cu. ^{67}Cu is particularily useful for metabolic studies since it has a half-life of about 80 hours. It would help us to know just exactly what arrangements need to be made to get this isotope for research purposes and experiences of those who have gone through that procedure in using it.

FIELDS: ^{67}Cu can be purchased with very high specific activity and it is inexpensive. Payment for shipping is the only expense. However, one must make detailed plans for its use, particularily with regard to definite delivery date and time, how quickly it can be gotten into the laboratory, establishing, with certainty, its purity, and how it can be quickly used in animal experimentation. When I made all of these arrangements with the Los Alamos Laboratory the radioactive material was really good. The specific activity was fantastic because there is hardly any cold ^{63}Cu. There are no problems in purchasing ^{64}Cu from New England Nuclear. The service is very nice. They provide it all year long and if they cannot make it due to holidays they will phone in advance. The problem is that its half-life is only 12 hours and the specific activity is really low.

CROOK: For those of you who are interested in ^{67}Cu or other isotopes, the Oak Ridge Isotope Sales Office should be your first stop in trying to find out a number of things. They are responsible for selling isotopes from all the national laboratories and they can tell you what's available, what the cost will be, and when it's available. They will send you the catalog and a sales list. Once you have decided what you want, they'll tell you which lab it will come from and you can communicate directly with that lab if you want. I think there's a \$130.00 or \$150.00 shipping-container charge. The other thing is timing. I think by law the national laboratories can only provide these isotopes during certain months of the year. If it's available through a commercial provider, the national laboratories are prohibited by law from providing that because they can't compete with a private supplier.

BARNEA: I've used ^{64}Cu and I agree that it should be checked for purity. In checking a sample we found we had a lot of radioactivity at 1.1 MEV when we scanned the spectrum and the physicist at Los Alamos suggested we had a zinc contaminated sample. Since then we've had very good experience with the ^{64}Cu. I am now informed that the ^{67}Cu has very high specific activity but it may contain a lot of ^{64}Cu which is much higher in energy and you must protect yourself much better. Now they are making it about once a month until December and you can get the material within 48 hours from the moment they produce it. The minimum charge is $200 per order plus shipping. The people you deal with are very cooperative and very helpful.

LOVEN: Everyone is concentrating on radioactive isotopes. What about using stable isotopes and trying neutron activation or x-ray emission? The nice thing about stable isotopes is you don't have to use it the next day.

UNKNOWN: Off the top of my head, there are at least 3 stable isotopes and 2 of them are not worth anything and the other is mediocre. Isotope abundance is very low and measurement requires some sophisticated equipment. It certainly is not a technique that can be used by the person who makes measurements on a sporadic basis. I agree that it offers the advantage of not having to worry about radiation dangers. That's why it's being considered for human nutrition balance studies. Detection, sensitivity, and specificity are still limiting factors.

PICKART: We collaborated with the Berkley USDA lab a number of years ago in using x-ray photo-electron spectrophotometry to determine Cu in a number of biological systems including cells. It's a complicated technique but it is capable of determining very low levels. The only problem is that this equipment is generally not readily available.

UNKNOWN: What about using trace amounts of other similar elements like silver. Is there a good isotope of silver that could be used?

KEEN: I don't see how that would be useful. If you give

silver it's going to bind to Zn-thioneines as well as Cu-thioneines.

<u>SORENSON</u>: Pat (Egner), what was the form of the ^{67}Cu that you got? Did you say it had some propylene glycol in it?

<u>EGNER</u>: We have had difficulties using ^{67}Cu. I honestly do not remember what the vehicle was. The only thing I do know is that we tried to obtain it 4 times. We were successful twice. The first time we tried to use it and we couldn't get it to complex with 3,5-dips acid and Los Alamos called to say it was junk and not to use it. Another time it was considerably contaminated with ^{64}Cu, so we could't use it.

<u>SORENSON</u>: Did you say it was pasty?

<u>EGNER</u>: It was like an oily solution.

<u>SORENSON</u>: I thought it was suppose to be the nitrate?

<u>UNKNOWN</u>: I think it comes as the chloride in hydrochloric acid.

<u>EGNER</u>: You're right, it is hydrochloric acid. We had to dilute it. That's one reason we had trouble making the complex with 3,5-dips acid.

<u>SORENSON</u>: Is it in water?

<u>EGNER</u>: No, I'm sure it's in hydrochloric acid. It was thick.

<u>HARRIS</u>: When I was in Australia, we had access to some very good ^{64}Cu, it was approximately 100 times hotter than normally available ^{64}Cu. I wonder if our manufacturers could make it of that quality with regard to specific activity?

<u>LOVEN</u>: When you are using something like 3,5-dips acid or other drugs and you have reason to believe the complex is rather stable, it may be possible to label your drug with a stable isotope which might give some useful initial information.

ACCELERATED RECOVERY OF IMMUNOCOMPETENCE WITH RADIOPROTECTIVE CU-DIPS

L. S. F. Soderberg[1], J. B. Barnett[1], M. L. Baker[2]

H. Salari,[2] and J. R. J. Sorenson[3]

Departments of Microbiology and Immunology[1], and Radiology[2], College of Medicine, and the Department of Biopharmaceutical Sciences,[3] College of Pharmacy, University of Arkansas for Medical Sciences, Little Rock, Arkansas

INTRODUCTION

Copper(II)(3,5-diisopropylsalicylate)$_2$ (Cu-DIPS) provided substantial radioprotectant activity when administered 3 hours prior to exposure to gamma irradiation[1]. The present study investigated immune parameters of Cu-DIPS-mediated radioprotectant activity. It was found that Cu-DIPS did not prevent the destruction of mature lymphocytes, but accelerated the recovery of both the numbers of cells in the spleen and the responsiveness of those cells.

METHODS

Experimental Design

C57BL/6 mice were injected subcutaneously with a single dose of 80 mg/kg Cu-DIPS or vehicle (4.0% propylene glycol used as a wetting agent, and alcohol in saline). After 3 hours about one half of the mice in both groups were exposed to 8.0 Gy (1.7 Gy/min) using a ^{137}Cs

irradiator. Assays of immune reactivity were performed on animals in each of the four treatment groups on days 7, 14, 24 and 42. The vehicle-treated and irradiated group had a 40% survival rate at day 42 while the Cu-Dips treated and irradiated group had a 85% survival rate.

Mitogen Reactivity

Animals were sacrificed and spleen cell suspensions were prepared and washed 3 times. Microcultures of 2 X 10^5 cells/culture in RPMI 1640 with 5% fetal calf serum and 0.05 mM 2-mercaptoethanol were exposed to concanavalin A (Con A, 2 ug/ml), phytohemagglutinin (PHA 25 ug/ml), or lipopolysaccharide (LPS, 25 ug/ml). Triplicate cultures were labeled with ^{3}H-thymidine (1 uCi/ml) over the final 4 hr of 72 hr incubation.

Plaque-Forming Cell Assay

Mice in each treatment group were immunized intraperitoneally with 2 X 10^8 sheep red blood cells (SRBC) on days 2, 9, 19, or 37 after treatment. Five days later the immunized mice were sacrificed and their spleen cells evaluated for direct (IgM) plaque-forming cell by a standard slide assay[3].

RESULTS

The effect of Cu-DIPS on spleen weight following irradiation was examined. As shown in Figure 1, spleen weights dropped to 27-28% of normal 7 days after irradiation and there was no difference between Cu-DIPS and vehicle groups. At 14 days after irradiation, vehicle-treated spleens increased marginally to 35% of controls, while the Cu-DIPS group was at 83% of, and not statistically different from, the unirradiated controls. At 24 days there was a great deal of variability in both irradiation groups. In general the very sick-looking mice had very large spleens. By 42 days, when deaths due to irradiation had already occurred, there were no extraordinarily large spleens. The very large spleens at day 4 had large amounts of connective tissue and were not matched by large numbers

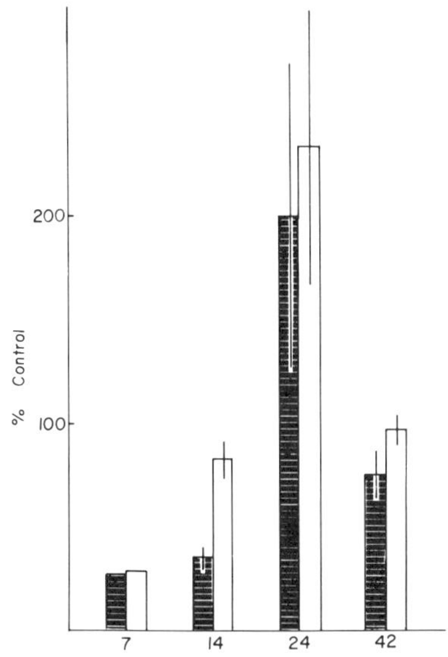

Figure 1. Spleen weights at various times after irradiation. Dark, striped bars: vehicle-treated; open bars: Cu-DIPS-treated, control values are shown in Table 1.

of recoverable cells per spleen (Figure 2). Seven days after irradiation less than 2% of control numbers of spleen cells could be recovered in either Cu-DIPS or vehicle group. Consequently more mice were needed, particularly at 7 days, to provide sufficient viable cells for immune assays. The numbers of recoverable spleen cells increased at significantly more rapid rate in Cu-DIPS-treated mice, which still only achieved 60% of control values at 42 days when spleen weights were normal. For *in vitro* immune assays, cultures were prepared using equal numbers of cells to highlight Cu-DIPS-mediated changes in immune activity independent of changes in cell numbers. Figure 3A shows spleen cell reactivity to Con A per culture and Figure 3B shows total reactivity per spleen. Spleen cells from both groups of irradiated mice were unresponsive to the T cell mitogen, Con A, at times earlier than 24 days. At 24 days spleen cells from Cu-DIPS-treated mice had recovered 65%, but since deaths due to radiation had already occurred, only the most radiation-resistant mice were left for assay. Thus, the 42 day vehicle group was higher due to attrition of weak animals. The total Con A responsiveness per spleen (Figure 3B) was zero until day 24 when Cu-DIPS treated mice had a 16-fold greater capacity to respond to Con A.

Spleen cell responses to the T cell mitogen, PHA, were also evaluated (Fig. 4). There was less of an increase due to Cu-DIPS in PHA reactivity than with Con A. In individual cultures, PHA responsiveness recovered to 125-150% of control values by day 42. This suggested that PHA responsive cells were populating the spleen to a greater proportion than other cell types. On a per spleen basis, PHA responsiveness recovered to 75% (Cu-DIPS) and 50% (vehicle) of control capacity. The greater recovery of PHA than Con A supports contentions that these mitogens effect different T cell subpopulations or maturational states.[4]

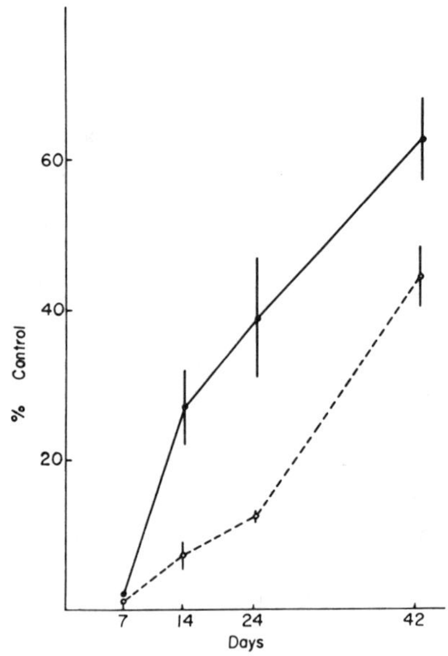

Figure 2. Spleen cellularity at various times after irradiation. Solid line: Cu-DIPS-treated; dashed line: vehicle-treated. Control values are shown in Table 1.

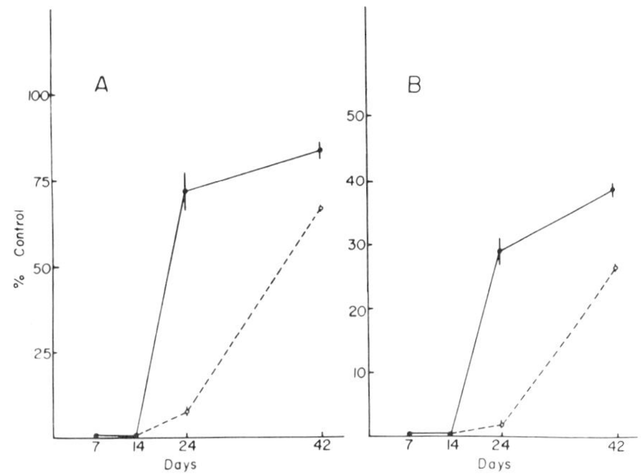

Figure 3. Con A responsiveness at various times after irradiation. A: Activity per culture; B: Activity per spleen; Solid lines: Cu-DIPS; dashed lines: Vehicle. Control values are shown in Table 1.

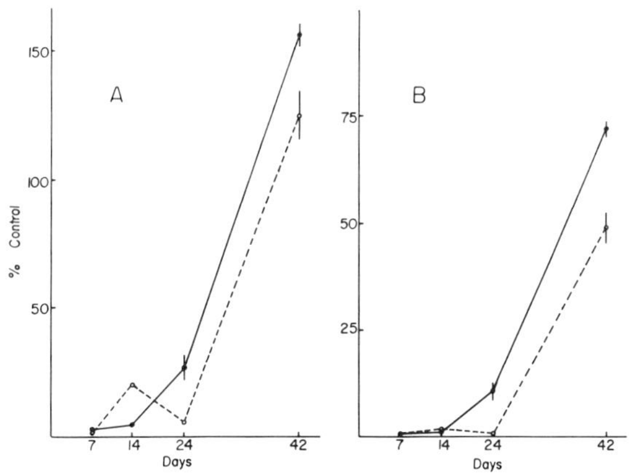

Figure 4. PHA responsiveness at various times after irradiation. A: Activity/culture; B: Activity/spleen; Solid lines: Cu-DIPS; dashed lines: vehicle. Control values are shown in Table 1.

B cell responsiveness of the spleen cells was evaluated using the mitogen, LPS (Fig. 5). As with the T cell mitogens, all cultures were unresponsive to LPS 7 days after irradiation. Unlike the T cell responses, cultures recovered substantial LPS reactivity by day 14, although the Cu-DIPS group was not different from the vehicle group. By 24 days, the Cu-Dips group was twice as responsive as the vehicle group and Cu-DIPS enhanced the total spleen capacity to respond to LPS by 8-fold.

We tested the antibody responsiveness of irradiated mice using a plaque-forming cell (PFC) assay (Fig. 6). The T-dependent antigen, SRBC, stimulated 5-fold more PFC at 14 days and levels, which were not significantly different from normal at 24 and 42 days in cultures from Cu-DIPS-treated mice (Fig. 6A). Full antibody responsiveness was recovered in Cu-DIPS-treated mice 42 days after irradiation (Fig. 6B), while vehicle-treated mice recovered only to 34% of control capacity.

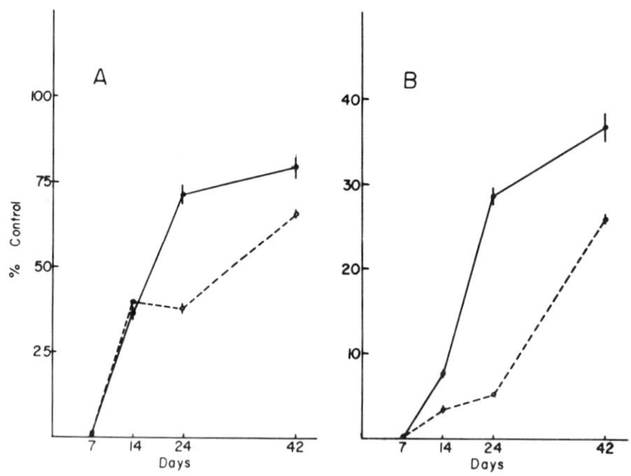

Figure 5. LPS responsiveness at various times after irradiation. A: activity/cultures; B: activity/spleen; Solid lines: Cu-DIPS; Dashed lines: vehicles. Control values are shown in Table 1.

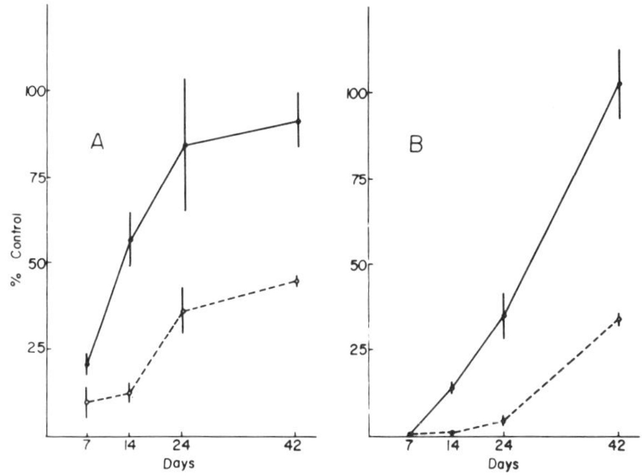

Figure 6. PFC responsiveness at various times after irradiation A: activity/culture; B: activity/spleen Solid lines: Cu-DIPS; dashed lines: vehicle. Control values are shown in Table 1.

Table 1

Unirradiated Control Values

	Exp. 1	Exp. 2	Exp. 3
Spleen weight (mg)	77.5±0.5	80.5±5.5	94.0±5.1
Cells/Spleen (x10^7)	7.25±1.55	11.2±0.4	8.62±1.22
Con A (CPM)	89,739±4953	59,271±1575	61,796±827
PHA (CPM)	47,615±1010	31,613±1456	40,055±1150
LPS (CPM)	40,631±5019	35,548±575	35,578±3764
PFC/10^5 spleen cells	396±51	105±10	243±21

DISCUSSION

Cu-DIPS treatment did not protect mature lymphocytes from the destructive effects of radiation. Spleen cell numbers and responsiveness in all assays were near or at zero, 7 days after exposure to 8.0 Gy. The radiosensitity of mature lymphocytes[5] has been described at length as has the loss of spleen mass.[6,7] Pretreatment with Cu-DIPS accelerated both the recovery of spleen cellularity and the recovery of lymphocyte responsiveness following irradiation. Spleen weights and the numbers of cells per spleen recovered at a significantly greater rate in Cu-DIPS-treated mice following irradiation. As described elsewhere,[8] Cu-DIPS-treatment caused dramatic early increases in the numbers of IL-3-responsive precursors in the spleens of irradiated mice. Such precursors are multipotent progenitor cells (CFU-S) and are probably responsible for increases in spleen cellularity.

Independent of the effects of Cu-DIPS on increasing cell numbers, Cu-DIPS signficantly accelerated the recovery of both T and B cell mitogen responsiveness and T-dependent antibody responses. While Cu-DIPS-induced increases in spleen cellularity were observed at 14 days after irradiation, Cu-DIPS-treatment did not effect Con A or LPS responsiveness until 24 days. The delay in enhancing mitogen responses was very likely due to the period required for the progenitor cells to differentiate into lymphocytes. However, antibody responsiveness recovered from irradiation earlier than mitogen responsiveness, suggesting that the PFC assay may be more sensitive to the enhancing effects of Cu-DIPS.

ACKNOWLEDGEMENTS

We are indepted to the Arthur Ambrust Cancer Research Fund, International Copper Research Association, Max and Victoria Dreyfus Foundation, Elisa U. Pardee Foundation, and Denver Roller Corporation for financial support.

REFERENCES

1. Salari, H., Baker, M. L., Barnett, J. B., Soderberg, L. S. F., Willingham, W. M., and Sorenson, J. R. J. In: Biology of Copper Complexes; Ed. J. R. J. Sorenson; Humana Press, NJ, 1987 (this proceedings).
2. Barnett, J. B., Soderberg, L. S. F., and Menna J. H. Toxicol. Let 1985, 25, 173.
3. Yeh, N.-H. and Soderberg, L. S. F. J. Biol. Response Modif. 1982, 1, 267.
4. Stobo, J. D., and Paul, W. E. J. Immunol. 1973, 110, 362.
5. Markoe, A. M. Prog. exp. Tumor Res. 1980, 25, 219.
6. Vasin, M. V., L'rova, T. S., Antipor, V. V. and Davydor, B. I. Radiobiol. 1977, 17, 113.
7. Galliechio, V. S., Chen, M. G., Watts, T. D., and Gamba-Vitalo, C. Exp. Hematol. 1983, 11, 553.
8. Soderberg, L. S. F., Barnett, J. B., Baker, M. L., Salari, H., and Sorenson, J. R. J. In: Biology of Copper Complexes Ed. J. R. J. Sorenson; Humana Press, NJ, 1987 (this proceedings).
9. Schrader, J. W. Ann. Rev. Immunol. 1986, 4, 205.

DISCUSSION

LOVEN: How long after radiation did you wait to measure CFUS in the recipient animal? You said you did a spleen cell assay for the interleukin-3 (IL-3) responsive cells.

SODERBERG: Those animals were examined on days 7, 14, and 24 but it's not the same as CFUS assay. This is an in vitro colony forming assay. It's suggested that IL-3 stimulates the same CFUS precurser cells.

LOVEN: Does Cu(II)(3,5-dips)$_2$ alone have any effect on cell kinetics in the absence of radiation?

SODERBERG: The only effect that we saw due to Cu(II)(3,5-dips)$_2$ treatment alone, was an increase in the cellularity of the spleens, which was not reflected by an increase in T-cell or B-cell reactivity.

LOVEN: What about in terms of being able to recover these IL-3 responsive cells?

SODERBERG: There were no significant differences in the IL-3 responsiveness.

LOVEN: Have you done any type of radiation survival curve for these IL-3 responsive cells?

SODERBERG: No. Just mouse survival due to treatment.

LOVEN: You haven't determined survival of these cells in vitro following radiation?

SODERBERG: No.

IWAKAWA: When you determined significance compared with control groups, what was the level of significance and what statistical method was used?

SODERBERG: I'm talking about significant at $p<0.05$ or lower by the Student t-test.

IWAKAWA: I don't think you can combine those data and compare them as a group, because each time a different number of mice died, right?

SODERBERG: That's a good point. I was thinking about presenting all these data as histograms but this seemed easier.

IWAKAWA: Have you done spleen pathology?

SODERBERG: We are very anxious to do spleen pathology. We have not done that yet.

Other Biological Effects of Complexes and Complexing Agents

pH-DEPENDENT SUPEROXIDE DISMUTASE-MIMETIC REACTIVITY OF COPPER(II)ETHYLENEDIAMINETETRAACETATE

William M. Willingham[1] and John R. J. Sorenson[2]

[1]Department of Natural Sciences, University of Arkansas at Pine Bluff, Pine Bluff, Arkansas 71601; [2]Department of Biopharmaceutical Sciences, College of Pharmacy, University of Arkansas for Medical Sciences, Little Rock, Arkansas 72205

SUMMARY

Superoxide dismutase (SOD) mimetic activity of Copper(II)ethylenediaminetetraacetate [Cu(II)EDTA] was studied in the pH range of 6.0 to 8.0 using the xanthine/ xanthine oxidase superoxide generating system with nitroblue tetrazolium. Cu(II)EDTA disproportionated superoxide within the pH range studied without affecting the synthesis of uric acid. This disproportionation by Cu(II)EDTA without inhibition of xanthine oxidase is offered as evidence that the addition of EDTA to biological preparations for the purpose of complexing copper and preventing copper mediated disproportionation is not a valid practice.

INTRODUCTION

Interest in copper containing cupreins as protective enzymes arose after the report of McCord and Fridovich[1] that a cuprein had superoxide dismutase reactivity. Mann and Keilin[2], Markowitz et al.[3], Kimmel et al.[4], Porter and Folch[5] had isolated cupreins but McCord and Fridovich were the first to show that erythrocuprein disproportionated superoxide according to the following equation:

$$O_2^{\dot{-}} + O_2^{\dot{-}} + 2H^+ \longrightarrow O_2 + H_2O_2 \qquad (1)$$

It has also been shown that small molecular weight copper complexes[6,7,8,9] as well as inorganic copper(II) compounds[10,11] disproportionate superoxide in a manner similar to superoxide dismutase. EDTA is commonly added to biological systems to scavenge inorganic copper or loosely bonded copper in order to prevent its disproportionation of superoxide. The suggestion that Cu(II)EDTA did not disproportionate superoxide was originally offered by McCord and subsequently supported by Fridovich[1] and Fridovich[14].

The notion that Cu(II)EDTA does not disproportionate superoxide appeared somewhat unusual to us in light of the fact that many copper complexes show this reactivity. It is known that superoxide disproportionating complexes must have available sites for superoxide bonding as shown in the following reaction sequence.

$$L_4Cu(II) + O_2^{\dot{-}} \longrightarrow L_4Cu(II)O_2^{\dot{-}} \longrightarrow$$

$$L_4Cu(I)O_2 \longrightarrow L_4Cu(I) + O_2 \qquad (2)$$

$$L_4Cu(I) + O_2^{\dot{-}} \longrightarrow L_4Cu(I)O_2^{\dot{-}} \longrightarrow$$

$$L_4Cu(II)O_2^{2-} \xrightarrow{2H^+} L_4Cu(II) + H_2O_2 \qquad (3)$$

We thought that such sites would become available on Cu(II)EDTA under acidic conditions as depicted in Figure 1; therefore, we investigated this complex for SOD-mimetic activity in the 8 to 6 pH range.

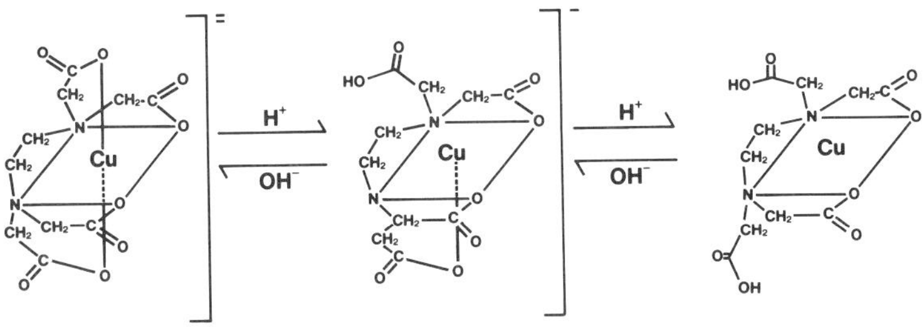

Figure 1. pH-Dependent Structure of Cu(II)EDTA

To examine the possibility that the observed pH related change in SOD-mimetic reactivity of Cu(II)EDTA might be due to inhibition of the conversion of xanthine to uric acid by xanthine oxidase, we evaluated uric acid synthesis under the same conditions used to determine SOD-mimetic reactivity.

EXPERIMENTAL PROCEDURE

Spectrophotometric assays were done with a Shimadzu Spectronic 200 uv-visible spectrophotometer. Beef liver catalase was obtained from ICN Pharmaceuticals Inc., nitroblue tetrazolium chloride monohydrate was supplied by Aldrich Chemical Co; xanthine oxidase (0.52 U/10 g protein) and xanthine (Grade V) were purchased from the Sigma Chemical Co. $CuSO_4 \cdot 5H_2O$ was obtained from the Baker Chemical Co., and Na_2H_2EDTA was purchased from Alpha.

Superoxide Dismutase Mimetic Reactivity

Inhibition of nitroblue tetrazolium (NBT) reduction by superoxide produced with the xanthine oxidase superoxide generating system at various pH values was determined by measuring absorbance at 540 nm 5 minutes after initiating the reaction with the addition of xanthine. The reaction mixture consisted of 52 ml 0.05 M sodium phosphate buffer (pH 7.8), 20 ml of 1% gelatin solution, 8 ml of phosphate buffer solution containing 32 mg of NBT, 8 ml of phosphate buffer solution containing 4 units of xanthine oxidase and

1 ul of catalase solution containing 316,000 units of enzyme per ml. A 1,000 uM stock solution of Cu(II)EDTA was prepared by dissolving 3.716 mg in 5 ml of distilled deionized water at pH 7.0 and diluting to 10 ml. pH adjustments were made using either 1 percent HCl or 1 percent NaOH. Appropriate aliquots of stock solution were used to obtain the desired concentrations.

Individual cuvettes contained 1.3 ml of buffer, 0.5 ml of the gelatin solution, 0.2 ml of NBT solution, 0.2 ml of deionized distilled H_2O, or 0.2 ml of deionized distilled H_2O containing Cu(II)EDTA, 0.2 ml of the xanthine oxidase solution and 0.5 ml of xanthine solution containing 5 mg of xanthine per 50 ml of solution. The xanthine addition resulted in a final liquid volume in each cuvette of 2.9 ml. This addition was also used to start the reaction. The absorbance observed with no Cu(II)EDTA added was taken as 100 percent reduction of NBT or no SOD-mimetic reactivity.

Assay for Uric Acid Synthesis by Xanthine Oxidase

The reaction mixture and the method of initiating uric acid synthesis with xanthine were the same as that for SOD-mimetic activity except the NBT was omitted. Uric acid synthesis was monitored by observing the change in absorbance at 292 nm, the λmax for uric acid.

Synthesis of Cu(II)EDTA

The Cu(II)EDTA was synthesized according to the method of Kirschner[12] with minor modifications. The Na_2H_2EDTA (16.8 g) was added slowly to a $CuSO_4.5H_2O$ solution (12.6 g in 100 ml of H_2O) with constant stirring and heating at 85°C. The hot solution was filtered and allowed to cool over a four hour period. The blue solid that separated on cooling was collected by filtration and partially dried at 15 mm Hg in a heated, 50°C, desiccator overnight. It decomposed on heating from 245 to 265°C, and was identified as $[Cu(C_{10}H_{14}N_2O_8)].H_2O$. Analysis calculated: C, 32.30; H, 4.34; N, 7.54. Found: C, 32.69; H, 4.28; N, 7.46.

Results and Discussion

As shown in Figure 2, Cu(II)EDTA had an IC_{50} value greater than 70 uM, at pH 8. However, the IC_{50} value decreased in a pH related manner, reaching a value of 8.7 uM at the lowest pH studied. It is well to note that even at neutral pH, Cu(II)EDTA has sufficient SOD-mimetic activity to be classified as active according to the criteria of Roberts and Robinson[13].

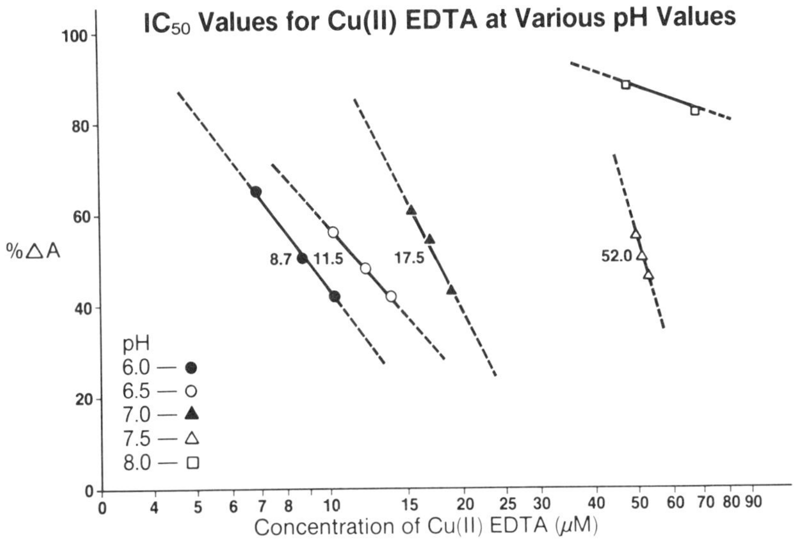

Figure 2. The SOD-mimetic activity of Cu(II)EDTA

Table I reveals that not only was there no inhibition of xanthine oxidase, but there was an apparent increase in uric acid synthesis. Additionally, from Table I, it can be observed that the rate of change in absorbance at 292 nm is constant within the Cu(II)EDTA concentration range studied at each pH.

Data presented here are consistent with the mechanism of superoxide disproportionation suggested by Fridovich[14] for Cu-ZnSOD. Cu(II)(Salicylate)$_2$, a small molecular weight complex, has also been shown to disproportionate superoxide by this same mechanism[15]. Moreover, these data support our original contention that sites become available on Cu(II)EDTA under acidic conditions. The fact that the synthesis of uric acid was not inhibited in this study enriches the notion that the reduction in absorbance of NBT at 540 nm when Cu(II)EDTA is added to the solution is due to SOD-mimetic reactivity of Cu(II)EDTA and not the inhibition of xanthine oxidase.

TABLE I

Effect of Cu(II)EDTA on uric acid synthesis by xanthine oxidase in the pH range of 8 to 6.

Concentration of Complex in Cuvette (uM)	Average A	%ΔA	Average ΔA/Min
		pH 8.0	
0.0	0.243	0	0.24
55.2	0.266	9	0.23
62.2	0.383	58	0.25
69.0	0.395	62	0.25
		pH 7.5	
0.0	0.466	0	0.23
50.0	0.461	-1	0.21
51.7	0.472	1	0.21
53.5	0.498	7	0.21
		pH 7.0	
0.0	0.240	0	0.15
13.8	0.243	1	0.16
15.5	0.268	12	0.16
17.2	0.292	22	0.16
19.0	0.336	40	0.16
		pH 6.5	
0.0	0.442	0	0.14
10.4	0.453	3	0.14
12.1	0.482	9	0.14
13.8	0.486	10	0.14
		pH 6.0	
0.0	0.333	0	0.12
6.8	0.376	13	0.12
8.6	0.457	37	0.10
10.4	0.462	39	0.11
12.1	0.480	44	0.10

In light of these results, we are suggesting that investigators exercise care in assessing superoxide obligate reactions when the reactions take place in the presence of Cu(II)EDTA. We also call into question the mediated use of EDTA to inhibit copper superoxide disproportionation and those published reports wherein EDTA was added to rule out the possibility that observed disproportionation was copper-dependent.

ACKNOWLEDGEMENTS

We are indebted to the MARC Faculty Fellowship program, Arthur Armbrust Cancer Research Fund, International Copper Research Association, Max and Victoria Dreyfus Foundation, Elsa U. Pardee Foundation, and Denver Roller Corporation for financial support of this work.

REFERENCES

1. McCord, J; Fridovich, I. J. Biol. Chem. 1969, 244, 6049
2. Mann, T; Keilin, D. Proc. Roy. Soc. Ser. B Biol. Sci. 1939, 126, 303
3. Markowitz, H; Cartwright, G; Wintrobe, M. J. Biol. Chem. 1959, 234, 40
4. Kimmel, J; Markowitz, H; Brown, D. J. Biol. Chem. 1959, 234, 46
5. Porter, H; Folch, J. J. Neurochem. 1957, 1, 260
6. Deuschle, U; Weser, U. Inorganica Chimica. Acta 1985, 107, 275
7. Cashin, C; Lewis, J; Burden T. British J. of Rheumatology 1985, 24, 137
8. Brigelius, R; Spottl, R; Bors, W; Lengfelder, E; Saran, M; Weser, U. Febs. Letters 1974, 47, 72
9. Younes, M; Weser, U. Febs. Letters 1976, 61, 209
10. Rabani, J; Klug-Roth, D; Lilie, J. J. Phys. Chem., 1973, 77, 1169
11. Halliwell, B. Febs. Letters 1975, 56, 34
12. Kirschner, S. JACS 1956, 78, 2372
13. Roberts, N; Robinson, P. British J. of Rheumatology 1985, 24, 128
14. Fridovich, I. Adv. Enzymol. 1974, 41, 35
15. deAlvare, L; Goda, K; Kimura, T. Biochem. Biophys. Res. Comm., 1976, 69, 687

DISCUSSION

LOVEN: Larry Oberley always advocated the use of DTPA instead of EDTA. Is the Cu(II) complex of DTPA similarily affected by pH as far as SOD-like activity is concerned?

WILLINGHAM: I am aware of its use from literature reports along with EDTA but we have not studied its Cu(II) complex in this regard.

DAMERON: Does it have SOD-like activity in the presence of BSA or an amino acid mixture or does it only have SOD-like activity in a clean system?

WILLINGHAM: We haven't done that. This work was done to examine the validity of the use of EDTA to complex with Cu in biological systems and prevent SOD-like activity suggested to be associated with Cu in those systems. The influence of peptides and amino acids in our system certainly merits our attention.

SORENSON: Two days ago, Prof. Reed showed that $Cu(II)(3,5\text{-}dips)_2$ disproportionated hydrogen peroxide. Do you think that there might be any disproportionation of hydrogen peroxide by Cu(II)EDTA?

WILLINGHAM: There might be. That's the next experiment we are going to do. Data we obtained in examining the effect of this complex on xanthine oxidase show an increase in the production of uric acid. Catalase-like activity of Cu(II)EDTA could account for that observation. This complex may not only have SOD-like activity but catalase-like activity as well. We plan to determine the catalase-like activity of the Cu(II)EDTA complex.

GREENAWAY: Do you know whether O_2^- actually has to bond to Cu or whether the mechanism could be an outer sphere electron transfer?

WILLINGHAM: For the complex we are working with or for Cu complexes in general?

GREENAWAY: For the complex you are working with specifically. Your data suggests that it does need to bond to Cu, since you're saying you have to have a free coordination site for the reaction to proceed at an appreciable rate. Is it known whether this reaction requires inner sphere coordination of O_2^- directly to Cu, or whether electron transfer can occur through an outer sphere mechanism?

WILLINGHAM: Inner sphere coordination is required for disproportionation by Cu-Zn SOD and small molecular weight Cu complexes.

GREENAWAY: Is is known how the reduction potential of O_2^- changes with pH? The range between pH 8 and pH 6 may influence the amount of HO_2 and O_2^- present and I wonder whether some of the difference in oxidation and reduction of O_2^- might be due to pH?

SORENSON: The pK_a for dissociation of HO_2 is 4.7. Fred (Greenaway), do you imagine a reduction or oxidation being catalized by the ligand that might be attached to Cu as an outer sphere reduction or oxidation process?

GREENAWAY: I don't know. It occurred to me that it might be a possibility. Young and Lipard (J. Am. Chem. Soc. 1980, 102, 4920.) studied the reaction of O_2^- with Cu(II)-(salicylate)$_2$ at different pH values and suggested that the basicity of O_2^- had some role in the reaction process.

SORENSON: I think it was Bielski (Handbook of Methods for Oxygen Radical Research, R.A. Greenwald, ed., 1985, 77.) who showed that Cu(II) complexes can disproportionate HO_2. So, apparently, the Cu(II) site can compete for bonding with H^+ in HO_2 and may even use the proton in catalyzing the disproportionation of HO_2.

RISK: I really like the good solid perspective of an inorganic chemist explaining how things work in biological systems. However, as a biologically based person, I'm wondering about the Cu amino acid complexes which are quite significant in intercellular and intracellular transport. If you look at the orientation of Cu in those complexes, might they function in a similar fashion? Do they have the proper structure to react like your Cu(II)EDTA model?

WILLINGHAM: I would think so. A number of amino acid complexes of Cu have been shown to disproportionate O_2^-. It's a question as to what complexes are going to be in solution. When investigators add EDTA they assume that EDTA will remove ionic Cu or amino acid bonded Cu and prevent its disproportionation of O_2^-. That assumption may be correct as far as the complexing of Cu but it may be false with regard to the elimination of SOD-like activity.

RISK: Prof. Barnea's model has impressed me in terms of the central nervous system. As I understand it, Cu is binding to the nitrogen of histidine. From your prospective of inorganic chemistry, can you see how the Cu would look complexed there as opposed to a carboxylic group that you've been studying in EDTA and if it might have a similar mechanism of action?

WILLINGHAM: Yes. It is known to disproportionate O_2^- and HO_2. Of those endogenous ligands in plasma, histidine is a component of the most abundant Cu complexes. It's difficult to remove Cu from histidine. Cu-Zn SOD uses 4 histidine residues to bond and hold both Cu and Zn in its protein matrix. The imidazole rings of those histidine residues are very important as far as the bioactivity of Cu-Zn SOD is concerned.

FIELDS: Do I understand correctly that yours was an in vitro system where only Cu and EDTA were present in solution? There were no other metals. In a physiological environment like plasma, we have other metals that are present and presumably EDTA will bind those other metals, not only Cu. Is EDTA only specific for Cu or will it bind any metal which is present in plasma?

WILLINGHAM: Yes, it will form chelates with other metals and yes, this is an in vitro system.

FIELDS: So actually if you add EDTA to plasma, you could not tell which metal was bound to EDTA.

WILLINGHAM: That's true I guess, but it's known that Cu will disproportionate O_2^- and what those who use it want to do is to tie up the Cu. We have shown that if you add EDTA to a Cu containing system and Cu(II)EDTA is formed, it will disproportionate O_2^- depending upon the pH. If the pH is 8.0 or 10.0 then you will not find SOD-like activity. If the pH is between 6.0 and 7.0 there will be SOD-like activity.

ELLIOTT (Gary): To expand on Dr. Fields' question, among the biological metalloelements, Fe has the highest formation constant with EDTA. It's about 10^{31}. I would imagine the stability of the Cu complex is going to be substantially less than that, about 10^{25}. One would expect on a thermodynamic basis that EDTA would probably complex with Fe preferentially which of course would be devoid of activity.

SORENSON: Did you say the Fe EDTA complex would have no disproportionating reactivity?

ELLIOTT (Gary): Is that not correct?

SORENSON: Fe and Fe complexes also disproportionate O_2^-. So if indeed Fe EDTA was formed, and that was really causing the problem that people dump in millimolar quantities of EDTA to take care of, it is also likely to be pH dependent in it's disproportionating reactivity. Also, the stability constant for Fe(III)EDTA is several orders of magnitude less than the stability constant for Cu(II)EDTA.

CHIDAMBARAM: For the catalytic activity of Cu(II)EDTA, you indicated it has to be reduced to Cu(I). Do you have any evidence that Cu goes through an oxidation-reduction process during the disproportionation reaction?

WILLINGHAM: De Alvare, Goda, and Kimura (Biochem. Biophys. Res. Comm. 1976, 69, 687.) have provided strong evidence in support of a redox mechanism using ESR. They demonstrated that Cu(II) complexes are reduced to Cu(I) which, in turn, is oxidized to Cu(II) in disproportionating O_2^-.

CHIDAMBARAM: As a follow-up to Prof. Loven's comment, we tried to distinguish between complexing properties of EDTA and DTPA for Fe. Their stability constants are roughly the same but they have different kinetic properties. Fe(III)-EDTA can be oxidized faster where as the Fe(III)DTPA complex undergoes oxidation very slowly. So kinetics may be an important parameter in considering complexes of EDTA and DTPA.

GREENAWAY: Remember that we are talking about an oxidation reduction reaction and quite apart from just the geometry of the compound we have to think of the effect of these ligands on the oxidation-reduction potential. Fe in solution has a reduction potential of +0.77 volts and this drops all the way down to -0.12 volts on complexation with EDTA. You go a long way in the wrong direction once you add EDTA. It may be just a delicate play on how far the reduction potential drops upon complexation of these metalloelements, just as much as the geometry of the compound. I don't know the figures for Cu(II)EDTA, but it would be interesting to look at the reduction potential of Cu(II)EDTA as a function of pH and you might consider doing that.

NUTRITIONAL IRON BIOAVAILABILITY AND CHARACTERIZATION OF THE IRON BINDING COMPONENTS OF PINTO BEANS

M. V. Chidambaram, M. Nuccio
and G. W. Bates

Departments of Biochemistry and Biophysics
and Animal Science, Texas A&M University,
College Station, TX 77843

INTRODUCTION

Beans constitute a significant portion of the diet of people of developing nations. Many seeds of vegetable origin such as lentils and beans are rich in carbohydrates and also contain a relatively high concentration of trace elements such as iron, zinc and copper[1]. The bioavailability of iron in a meal is governed by complex factors. These include: (1) the chemical properties of iron-binding food components (2) the presence of heme and non-heme iron, (3) randomization of food iron during cooking and digestion and (4) food interactions that influence iron complexation and oxidation-reduction processes. Iron bioavailability is an area of public health concern and has attracted the attention of scientists from many disciplines.

In vitro methods are useful in predicting the iron bioavailability from meals. It is generally accepted that iron solubility and reduction to the Fe^{2+} state will enhance absorption. In vitro methods can readily measure these parameters during simulated digestion. Methods to separate soluble and insoluble iron include centrifugation and dialysis. Miller et al.[2] have employed the latter. The amount of dialyzable iron is used as a measure of bioavailability. Their results have been supported by comparison with the literature on human iron uptake[3]. Rao and Prabhavathi[4] using the centrifugation approach compared the results of in vitro iron solubilization with human iron uptake using a variety of meals. A reasonable agreement between "ionizable" iron and iron uptake was observed. Kojima et al.[5] studied iron solubilization from pinto beans and the effects of chelators, reducing agents and various food combinations. Foods and beverages known to enhance or inhibit human iron absorption had comparable effects on iron solubilization and reduction in vitro.

In this paper we report on the advantages of ferrozine (Fz) over bathophenanthroline sulfonate (BPS) in the quantification of Fe^{2+} and the fractionation and partial characterization of iron binding components of digested pinto beans.

MATERIALS AND METHODS

Ferrozine (Fz) was obtained from Sigma Chemical Corp. A 10 mM stock solution of Fz at pH 6.0 was prepared. Radioactive ^{59}Fe as $FeCl_3$ was supplied by New England Nuclear. All other chemicals were of the highest reagent grade and not further purified.

Dialysis membrane (nominal molecular weight cut off <14,500) was supplied by Spectrapor Corporation and pretreated as follows. Tubing was boiled in distilled water containing 5% $NaHCO_3$ and 2% EDTA for an hour. After thorough washing with distilled water, they were stored in 40% ethanol under refrigeration. The pinto bean supernatant (5 or 10 ml) was dialyzed at 4 ^{o}C against a large excess (2 liters) of 100 mM acetate buffer at pH of 6.0. Two or three changes of buffer were made.

A Sephadex G-50 column (2.5 x 90 cm) was prepared and equilibrated with a 100 mM acetate buffer at pH 6.0 at 4 ^{o}C. The column was calibrated with blue dextran which eluted at 195 ml (Vo) and bromophenol blue which eluted at 800 ml (Vt). The flow rate was adjusted to 30 ml/hr. Concentrated pinto bean supernatant was prepared without enzyme additions for the chromatographic studies.

Studies of iron exchange between iron binding components of pinto bean supernatant and apotransferrin were carried out using electron paramagnetic resonance (EPR) spectroscopy (IBM Bruker Model ER-200D). To 0.5 ml of pinto bean supernatant was added $NaHCO_3$, apotransferrin (70 µM) and Hepes buffer at pH 7.5. The formation of iron-transferrin was monitored by the absorbance at 465 nm and the EPR signal intensity at g' =4.3.

Protein estimation was carried out by the method of Lowry et al.[6] Radioactivity of the samples containing ^{59}Fe were measured using a Gamma-Trak model #1191 (TM Analytic) system. SDS gel (10%) electrophoresis was carried out by the method of Laemmli[7]. Molecular weight standards were used to estimate relative size of the pinto bean proteins.

Pinto beans were purchased from a local supermarket. Iron contents of the beans and digestate were determined using a Perkin Elmer Model 4000 atomic absorption spectrophotometer. The iron content of dry beans was found to be 67 ppm. One hundred grams of cooked pinto beans (20 gm dry weight) was found to contain 1.34 mg iron.

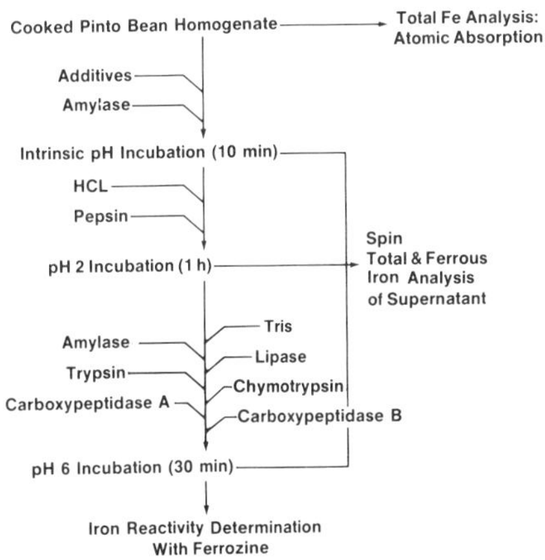

Fig. 1 Flow chart of the in vitro protocol developed to study iron bioavailability from pinto beans. The concentrations of enzymes (W/V) used in the pinto bean slurry are: amylase: 0.04-0.1%; pepsin: 0.5%; lipase: 0.02%; trypsin: 0.1%; carboxypeptidase A (2.5mg%) & B (0.25mg%).

The bean slurry (a 50 ml aliquot prepared from a mixture of 50 g cooked beans and 150 ml water) was subjected to an in vitro digestion as shown in the flow chart above. The incubations with enzymes at pH 6, 2 and 6 roughly simulated in vitro the digestive process of the mouth, stomach and duodenum, respectively. A 1.5 ml aliquot of bean slurry was obtained at various times during incubation and centrifuged at 1000 rpm for 1 min. To 0.5 ml of a clear pinto bean supernatant, 0.25 ml of 10 mM Fz was added. Then 0.5 ml of 0.1 M Hepes (pH 7.9) was added to bring the pH of the reaction mixture to 7.5. This measured the Fe^{2+} in pinto bean supernatant. To measure the total iron, 0.1 ml of 0.1

M dithionite was added to the pinto bean supernatant, prior to the Fz addition. The molar absorptivity of the $Fe(Fz)_3$ complex is 27,000 $M^{-1} cm^{-1}$ at its absorbance maximum 563 nm. This was used to calculate the Fe^{2+} concentrations.

RESULTS

COMPARISON OF THE CHROMOGENS BATHOPHENANTHROLINE SULFONATE AND FERROZINE

A Beer's law type plot for the detection of Fe^{2+} in a mixture of Fe^{2+} and Fe^{3+} by the chromogens BPS and Fz is shown in Fig. 2.

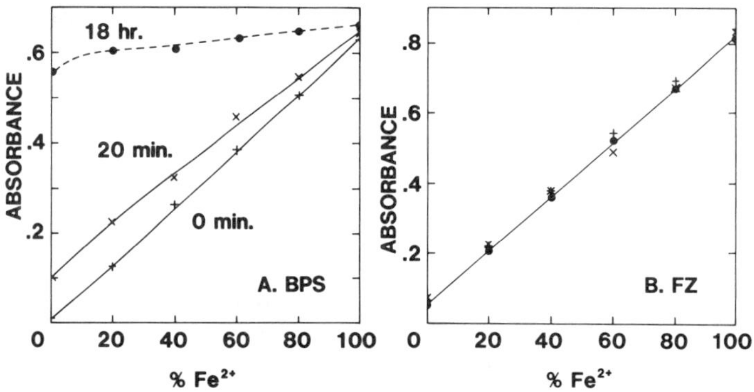

Fig. 2 The detection of Fe^{2+} presented in a mixture of $FeSO_4$ and $FeCl_3$ (kept at pH 2.0), using (A) 714 µM BPS and (B) 714 µM Fz at pH 7.5. The concentrations of Fe^{2+} and Fe^{3+} are systematically varied to achieve a constant total concentration of iron. For Fe-BPS complex the absorbance is measured at 534 nm and for the Fe-Fz complex, at 563 nm.

With BPS, absorbance measured at 534 nm shows a linear dependence on the concentration of Fe^{2+} and a molar absorptivity of 22,140 M^{-1} cm^{-1}. However, a 20 minute delay introduced in measurements causes a significant increase in the absorbance readings. This is caused by a fraction of Fe^{3+} being reduced to Fe^{2+}. When the samples are measured after 18 hrs in the presence of BPS, virtually all the Fe^{3+} is reduced and detected by BPS. Additional experiments have shown that this apparent reduction by BPS did not require buffer or light. Similar experiments were carried out using Fz as the chromogen at pH 7.5. Contrary to what is observed with BPS, Fz did not reduce Fe^{3+} over an 18 hr period. A Beer's law plot is found to be linear with a molar absorptivity of 27,000 M^{-1} cm^{-1}, in agreement with published values[8].

FRACTIONATION AND REACTIVITY OF IRON BINDING COMPONENTS IN PINTO BEAN SUPERNATANT

Cooked pinto beans were subjected to an in vitro digestion using the protocol described above and the pinto bean supernatant analyzed for iron. At the end of pH 2 and pH 6 incubations about 20 µM and 15 µM iron, respectively, were detected by the ferrozine assay. At pH 2 approximately 50% of the iron is found in the Fe^{2+} state suggesting the presence of endogenous reductants. The ultraviolet spectrum of pinto bean supernatant obtained at the end of an in vitro digestion shows absorption peaks at 265 nm and 340 nm. The intense absorption at 265 nm may be due to protein with a low tyrosine content, or the presence of phenylalanine or adenine nucleotides.

The changes brought about by dialysis in absorbance at 265 nm, and the iron and protein contents are shown in Table 1. In the dialyzed preparation about 72% decrease in absorbance at 265 nm and a 58% decrease in iron content are noted.

TABLE 1

Key properties of pinto bean supernatant:

BEAN SUPERNATANT	A265 (b)	Protein (c) µg/ml	Total Iron (d) µM
Undialysed	42.6 (n=1)	1256 (n=3)	79.7 (n=7)
Dialysed (a)	12.5 (n=7)	592.1 (n=3)	34.1 (n=5)
% decrease on dialysis:	71	53	57

(a) Concentrations are corrected for volume changes that occur during dialysis. (b) Expressed as absorbance unit. (c) Expressed as µg/ml of pinto bean supernatant by the method of Lowry et al.[6] (d) Calculated as µM by the ferrozine assay.

Approximately 70% of the protein, presumably of low molecular weight, is also found to be dialyzable. These changes confirm the presence of both dialyzable and nondialyzable iron binding components in pinto bean supernatant.

The presence of iron binding and Fe^{2+} oxidizing activities in pinto bean supernatant can be detected via competition between Fz and those agents for added Fe^{2+}. To the dialyzed and undialyzed preparations of pinto bean supernatant at pH 6.0 10 μM Fe^{2+} is added. The percentage of Fe^{2+} detected at varying concentrations of Fz is shown in Fig. 3. The concentration of Fz needed to detect 100% of the Fe^{2+} added is found to be 100 μM and 250 μM for the dialyzed and undialyzed preparations, respectively. Fe^{2+} oxidation or binding agents are at least partially dialyzable.

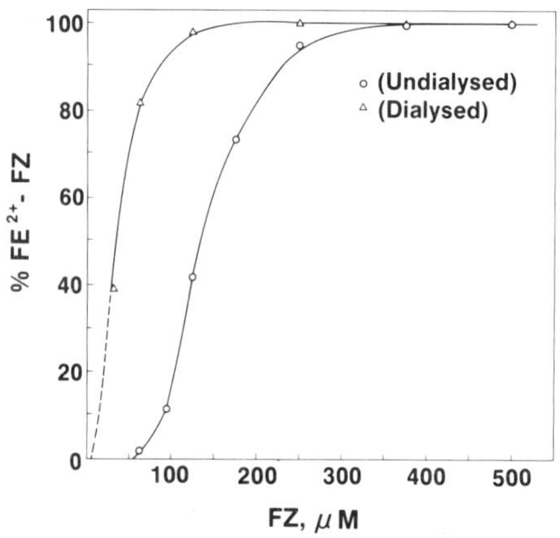

Fig. 3 The percent of Fe^{2+}-Fz complex observed, upon the addition of 10 μM Fe^{2+} to samples of pinto bean supernatant (undialyzed and dialyzed), as a function of Fz concentration. The absorbance change used in the calculation is that obtained within one minute after the addition of 10 μM Fe^{2+}.

THE CHROMATOGRAPHIC SEPARATION WITH SEPHADEX G-50

A Sephadex G-50 column was used to fractionate the various iron binding components in pinto bean supernatant. A typical chromatogram is shown in Fig. 4. Three major protein fractions are resolved. One, presumably of high molecular weight, elutes just after the void volume (elution volume 240 ml). The two lower molecular weight species appear at 475 ml and 525 ml. Chromatography of the dialyzed preparation shows only one significant peak with an elution volume of 240 ml corresponding to the high molecular weight species. To locate the iron binding components, pinto bean supernatant was incubated with ^{59}Fe and the resulting mixture was chromatographed. The radioactivity is found to be associated specifically with two components in pinto bean supernatant (see Fig. 4C).

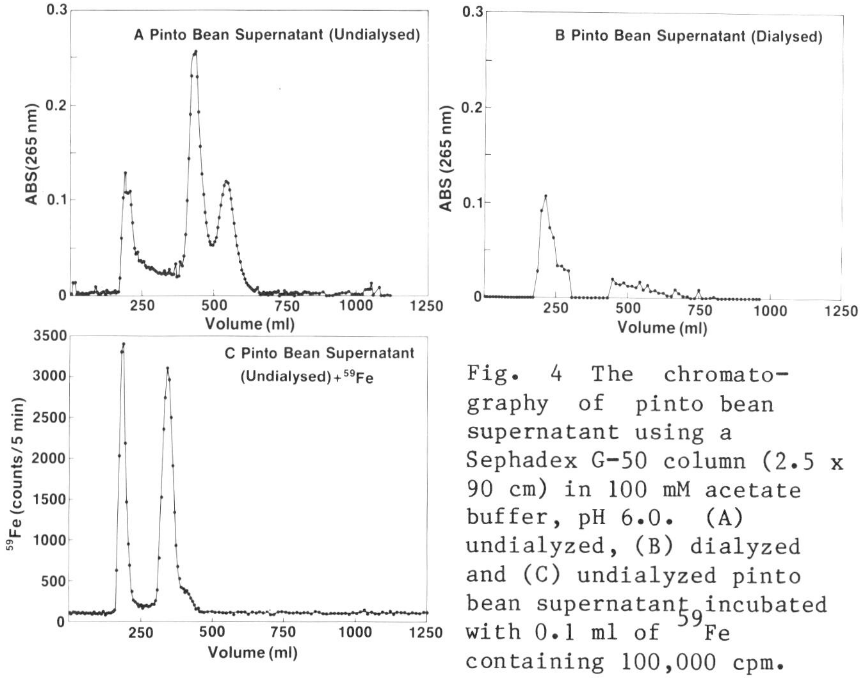

Fig. 4 The chromatography of pinto bean supernatant using a Sephadex G-50 column (2.5 x 90 cm) in 100 mM acetate buffer, pH 6.0. (A) undialyzed, (B) dialyzed and (C) undialyzed pinto bean supernatant incubated with 0.1 ml of ^{59}Fe containing 100,000 cpm.

SDS gel electrophoresis and densitometry of both pinto bean supernatant samples identifies a protein with a molecular weight of 49,000. The low molecular weight, dialyzable iron binding component in pinto bean supernatant appears to have a molecular weight near or less than 14,500.

IRON DONATION FROM PINTO BEAN SUPERNATANT TO APOTRANSFERRIN

Transferrin, the serum iron binding protein, is able to sequester two Fe^{3+} at the specific iron binding sites. For structural and functional information on transferrin, the reader is referred to a recent review[9]. The kinetics of iron exchange between monomeric iron chelates and apotransferrin are well characterized[10]. To probe the chemical nature of the iron binding components, we have reacted pinto bean supernatant with apotransferrin. EPR and visible spectroscopy were used to monitor the iron transfer reaction. The EPR spectrum of iron saturated transferrin obtained after this iron exchange reaction is compared with that of the bean supernatant (Fig. 5).

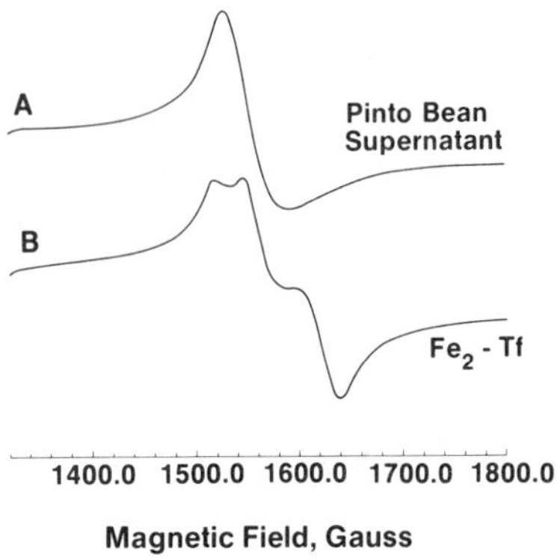

Fig. 5 The EPR spectra of (A) pinto bean supernatant and (B) iron-transferrin resulting from the exchange reaction as detailed in the Materials and Methods section. The concentration of reagents in the final reaction mixture are: pinto bean soluble iron, 53 µM; apotransferrin, 50 µM; HCO_3, 20 mM; and Hepes, 10 mM; pH 7.5.

Preliminary experiments indicate that 60% of the iron in pinto bean supernatant exchanges rapidly (half life = 2 min.) with apotransferrin while the iron bound to the high molecular weight species (a dialyzed preparation) exchanges the iron only slowly (half life = 4 hrs.). Iron complexes in pinto bean supernatant exhibit different exchange reactivities. Much of the Fe^{3+} in the digested pinto bean supernatant is chelated to low molecular weight ligands and is readily available to react with apotransferrin. We presume that this would be of relatively high bioavailability.

DISCUSSION

Our results indicate that Fz is the chromogen of choice in the quantification of Fe^{2+}. Advantages include a high molar absorptivity of $Fe(Fz)_3$ complex, its non-interaction with Fe^{3+} and lower cost. The in vitro methodology reported here is designed to mimic the digestive process. Broader application of this method to study iron mobilization from pinto beans is described elsewhere[11]. Soluble iron is assumed to be more bioavailable than sediment-associated iron. Our results indicate about 58% of the iron binding components are dialyzable. The predominant non-dialyzable fraction, which retains 42% of iron, is found to have a molecular weight of 49,000. Phytoferritin, a storage protein found in dry peas and lentils, has been isolated and characterized[12]. It is thought to consist of 24 subunits with molecular weights in the range of 20,000. The isolation of native ferritin from pinto beans would be useful in determining whether cooking degrades ferritin to dialyzable iron binding components. Iron bioavailabliity is influenced by redox reactions, chelation, adsorption and polymerization. Fe^{2+} is generally believed to be of higher bioavailability, however, Fe^{3+} bound to low molecular weight components may be available for absorption. Transferrin was chosen as a probe to study iron exchange reactions with pinto bean supernatant. Iron-transferrin yields a unique three component EPR spectrum near g' = 4.3. That can be utilized for quantification. An important advantage in using the EPR technique is that samples can be cloudy, opaque or even solid materials in complex food mixtures.

Under our experimental conditions approximately 60% of iron exchanged rapidly with transferrin. This agrees closely with the amount of dialyzable iron of 54% we found in pinto bean supernatant. Low molecular weight monomeric complexes of iron are expected to be most reactive. In conclusion, use of the improved in vitro method reported here can provide an approach to predicting iron bioavailability from foods. Further characterization of iron binding components and their iron release mechanisms will improve our ability to manipulate diets to improve iron bioavailablity.

ACKNOWLEDGEMENTS

This study was supported by a research grant (AM-32163) from the National Institutes of Health. Support was also obtained from a grant in aid from Gerber Products Company of Fremont, Michigan, USA. Matt Jordan carried out the EPR measurements and Mrs. Joyce Fuller provided excellent clerical support.

REFERENCES

1. USDA, Agricultural Handbook, Revised, 1984, # 8-11.
2. Miller, D. D.: Schricker, B. R.: Rasmussen, R. R.: Van Campen, D. Am. J. Clin. Nutr. 1981, 34, 224
3. Schricker, B. R.: Miller, D. D.: Rasmussen, R. R.: Van Campen, D. Am. J. Clin. Nutr. 1981, 34, 2257
4. Rao, N.: Prabhavathi, T. Am. J. Clin. Nutr. 1978, 31, 169
5. Kojima, N.: Wallace, D.: Bates, G. W. Am. J. Clin. Nutr. 1981, 34, 1392
6. Lowry, O.: Rosebrough, N.: Farr, A.: Randall, R. J. Biol. Chem. 1951, 193, 265
7. Laemmli, U. K. Nature 1970, 227, 680
8. Zak, B.: Baginski, E. S.: Epstein, E.: Weiner, L. M. Ann. of Clin. Lab. Sci. 1971, 1, 14
9. Chasteen, N. D. in: Advances in Inorg. Biochem. Elsevier Press; New York, 1983; vol. 5; p 202.
10. Cowart, R. E.: Swope, S.: Loh, T. T.: Chasteen, N. D.: Bates, G. W. J. Biol. Chem. 1986, 261, 4607
11. Reddy, M. B.: Chidambaram, M. V.: Fonseca, J.: Bates, G. W. Clin. Phys. Biochem. 1986, 4, 78
12. Crichton, R. R.: Ponce-Ortiz, Y.: Koch, M. H. J.: Parfait, R.: Stuhrmann, H. B. Biochem. J. 1978, 171,349

DISCUSSION

SORENSON: It looks as though the change in pH of chyme as it moves into the intestine is advantageous for absorption of Fe(II). I assume that there are ligands in the digest that form complexes with Fe(II). Is it known what they are and how they might facilitate absorption with regard to solubility?

CHIDAMBARAM: No one has characterized these ligands. However, the greater the amount of Fe(II), the greater the absorption since the more absorbable oxidation state is Fe(II).

SORENSON: Are Fe(II) complexes generally more covalently bonded than Fe(III)?

CHIDAMBARAM: Yes, Fe(II) bonds better with nitrogen and the pi ligands so that intuition is correct.

<u>SORENSON</u>: Is anything this good being done with regard to components of the U.S. diet and bioavailability.

<u>CHIDAMBARAM</u>: The U.S. recommended daily allowance is something like 18 mg and the amount of ascorbate was about 60 mg. The idea came about in Mexico City where they have this bean problem. Rather than changing the dietary structure we tried to enhance availability using enhancers to insure that the people will like the food.

<u>FIELDS</u>: What is the reason for enhancing bioavailability of Fe when meat is present in the diet?

<u>CHIDAMBARAM</u>: Meat Fe is in heme and heme Fe is actively absorbed while non-heme Fe is slowly absorbed in a passive process, which is a severe restriction of Fe bioavailability.

<u>FIELDS</u>: Do you know if things like simple sugars affect Fe bioavailability?

<u>CHIDAMBARAM</u>: Yes. Phytate and carboxylic acid ligands also affect bioavailability. I think I have seen a couple of reports where availibility is limited by fiber in the diet.

<u>PICKART</u>: Is a mineral supplement or ascorbic acid recommended in areas where diets are associated with very low mineral uptake?

<u>CHIDAMBARAM</u>: Ascorbic acid increases Fe absorption. I think supplementation or fortification is the way to go. Boxed cornflakes are always fortified with Fe but the availability question has to be answered. If we fortify can we get it out of the food?

<u>PICKART</u>: In terms of the world population, what's the number of people that are having problems like this?

<u>CHIDAMBARAM</u>: About a billion people suffer from Fe deficiency and 300 million suffer from Fe overload.

<u>RISK</u>: Have you looked at binding of other metals to these ligands?

<u>CHIDAMBARAM</u>: Yes, but not with any concerted effort.

RISK: Would these two compounds that you have studied be involved in metal transport to root nodule bacteria which are characteristics of legumes?

CHIDAMBARAM: I haven't worked in this area nor have I followed that literature.

FIELDS: Just don't get over excited about taking ascorbic acid because although it has been shown to increase absorption of Fe it is well established that it decreases absorption of Cu so we have a problem.

SORENSON: Dr. Fields, is it generally taught, from the nutrition point of view, that perhaps vitamin C should be taken at a time other than the time essential metallo-elements are taken or supplemented?

FIELDS: Yes, this is true. It has been shown that if ascorbic acid is not present at the same time as the diet containing Cu is consumed then there is no problem with absorption. However, dietary Fe absorption improves when ascorbate is given at the same time.

BALL: If Mg was given 75 minutes before eating its absorption was less than when it was given 75 minutes after eating.

ELLIOTT (Gary): We always used hydroxylamine in the bathophenanthroline assay for Fe(II) to avoid the oxidation of Fe(II) to Fe(III). Did you use hydroxylamine in that assay?

CHIDAMBARAM: No, I used dithionite which is a little stronger reducing agent than hydroxylamine hydrochloride but both would do the same job. They are equally effective in reducing Fe(III) to Fe(II).

ELLIOTT (Gary): If we use the bathophenanthroline sulfonate (BPS) assay, how long should we wait before making the spectrophotometric measurement?

CHIDAMBARAM: With ferrozine you can wait for some time because it doesn't interact with food components. With BPS, it's my experience that you have to make the measurement as quickly as possible.

Poster Presentations

A MODEL STUDY FOR TRANSPORT OF METALLO-DRUGS WITHIN LIPID BILAYERS

William E. Antholine,[1,3] Witold K. Subczynski,[1,2] James S. Hyde,[1] and David H. Petering[3]

National Biomedical ESR Center, Medical College of Wisconsin, Milwaukee, WI.[1]

Department of Biophysics, Institute of Molecular Biology, Jagiellonian University, Krakow, Poland.[2]

Department of Chemistry, University of Wisconsin-Milwaukee, Milwaukee, WI 53201.[3]

Introduction.

While many studies are concerned about the transport of metal ions through membranes, our work is the first, to our knowledge, that uses a lipid soluble copper complex to probe membrane structure and motion (1). Previous work has shown that 3-ethoxy-2 oxobutyraldehyde bis (N^4-dimethylthiosemicarbazonato) Cu(II), $CuKTSM_2$, is relatively inert with respect to reduction by thiols and with respect to competition for either the metal or the ligand from cellular constituents (2,3). The toxicity of $CuKTSM_2$ against Ehrlich ascites tumor cells is greatly reduced from the toxicity of CuKTS for which protons replace the methyl groups on the amine (Fig. 1).

Substitution of methyl groups for hydrogens on the amine results in an increase in the partition coefficient between octanol and water and a decrease of about 100 mV for the reduction potential. The rate for reduction of $CuKTSM_2$ by thiols decreases by about two orders of magnitude from the rate for reduction of CuKTS. Thus cells

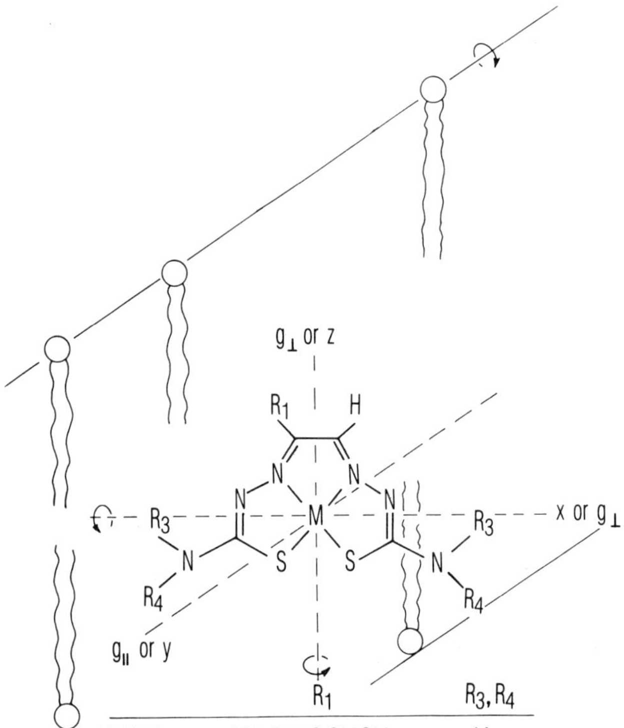

	R1	R3, R4
H_2KTS	$CH_3CH_2OCHCH_3$—	H
H_2KTSM2	$CH_3CH_2OCHCH_3$—	CH_3
H_2PTSM2	CH_3—	CH_3

Fig. 1. Structures of bis(thiosemicarbazonato) copper (II) complexes in a lipid bilayer.

retain a red hue upon addition of $CuKTSM_2$ because $CuKTSM_2$ is relatively inert.

Recently we have shown that $CuKTSM_2$ can be used as a probe of dimyristoylphosphatidylcholine (DMPC) vesicles (1). The partition coefficient for $CuKTSM_2$ is greater than 50:1 and probably about 200:1 in favor of the hydrophobic lipid bilayer over the polar water phase. ESR spectra in oriented membranes indicate that $CuKTSM_2$ is oriented with the plane of the complex normal to the bilayer surface, that is, intercalated with its square plane parallel to the acyl chains in the lipid bilayer. The fluid/gel phase transition temperature of the bilayer can be determined by parameterizing the resolution of the nitrogen hyperfine structure in the ESR spectra for $CuKTSM_2$ as a function of temperature (1). Also, the effect of bimolecular collisions between $CuKTSM_2$ and stearic acid spin labels suggests that $CuKTSM_2$ is distributed throughout the bilayer and diffuses faster than the lipids. This work focusses on whether $CuKTSM_2$ is a special probe of membrane structure or whether many copper complexes could be expected to probe the structure and motion within the lipid bilayer.

Results and Discussion.

Nuclei were isolated from Ehrlich ascites cells by standard procedures. $CuKTSM_2$ (0.5mM) was added to the nuclei and the ESR spectrum was recorded at room temperature in an ESR flat cell (Fig. 2). The ESR spectrum consists of $CuKTSM_2$ in at least two regions, a region where $CuKTSM_2$ tumbles rapidly and a slow motion region. The ESR spectrum for fast tumbling $CuKTSM_2$ is modeled by $CuKTSM_2$ in paraffin oil (PFO, light paraffin oil from MCB Manufacturing Co.) at 120°C and the ESR spectrum for $CuKTSM_2$ in the slow motion region is modeled by $CuKTSM_2$ in PFO at 0°C (Fig. 2). Paraffin oil is particularly suitable because the solvent consists of irregular lengths of carbon chains which surround the copper complex much like acyl chains in the lipid bilayer. In the region on the high field side indicated by a vertical line in Fig. 2, there is maximum spectral intensity for $CuKTSM_2$ with fast motion and minimal intensity for $CuKTSM_2$ with slow motion. The substantial intensity in this region for $CuKTSM_2$ added to nuclei confirms the presence of a region in which $CuKTSM_2$ is in fast motion. The peak-to-peak, signal heights for the other lines ($M_I = -3/2$ and $\pm 1/2$) for $CuKTSM_2$ in fast motion decrease in height. Therefore the spectrum for $CuKTSM_2$ in fast motion cannot account for the height of the most intense line in the spectrum for $CuKTSM_2$ added to nuclei. Futhermore, the spectrum for $CuKTSM_2$ added to nuclei does not match any of a continuum of spectra for $CuKTSM_2$ in PFO between 120°C and -60°C (unpublished spectra).

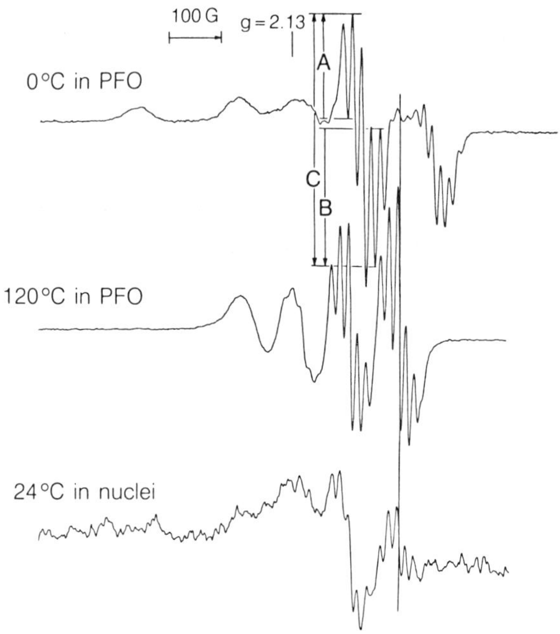

Fig. 2. ESR spectra for $CuKTSM_2$ (1 mM) in light paraffin oil, PFO, at 0°C (top spectrum) and 120°C (middle spectrum). ESR spectrum (bottom) for $CuKTSM_2$ (0.10 mM) after addition to whole nuclei obtained from Ehrlich ascites tumor cells. Note the amplitudes A, B, and C which are used to calculate the Cu-motion parameter are indicated in the top spectrum.

For these reasons, there are at least two regions with different motions for $CuKTSM_2$ added to nuclei. A similar result was found for $CuKTSM_2$ added to Ehrlich ascites cells where $CuKTSM_2$ is presumed to be within the outer membrane (2).

How unique is $CuKTSM_2$? Can many copper complexes be used to probe lipid bilayers? It appears to us that small changes in the structure of the ligand affect the probe. For example, if the $CH_3CH_2OCH(CH_3)$- group is replaced by CH_3-, the resulting complex $CuPTSM_2$ also partitions favorably into the lipid phase of DMPC bilayers, but rotational motion of the complex is faster as measured by the Cu-motion-parameter (Figure 3).

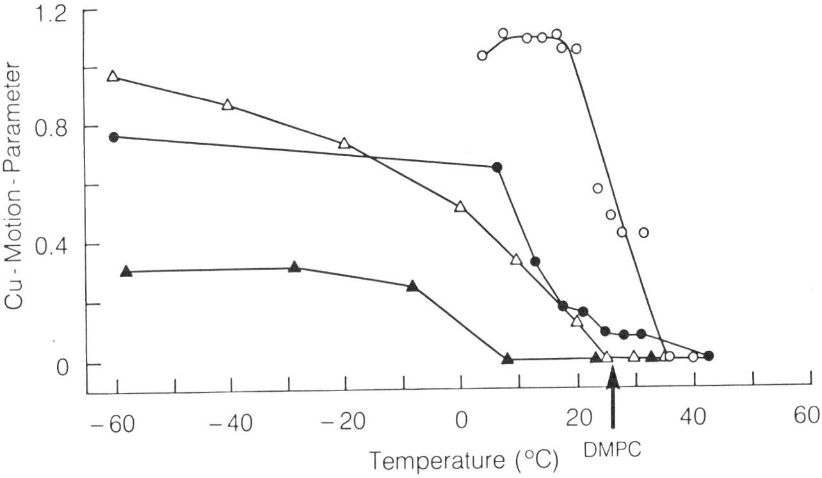

Fig. 3. Plot of Cu-motion parameter, (A + B)/C, versus temperature for $CuKTSM_2$ in lipid bilayers L-α-dimyristoyl phosphatidylcholine, DMPC,(○); CuKTSM, in paraffin oil (PFO), (△); $CuPTSM_2$ in DMPC bilayers (●); and $CuPTSM_2$ in PFO.

The amplitudes A, B, and C used to calculate the Cu-motion parameter are defined in Fig. 1. As the rotational motion increases, the resolution of the nitrogen hyperfine structure decreases and the Cu-motion parameter decreases (Fig. 3). The rotational motion for $CuPTSM_2$ is faster than the motion for $CuKTSM_2$ in both DMPC lipid bilayers and paraffin oil. In addition the largest changes in the Cu-motion-parameter for $CuPTSM_2$ do not occur at the phase transition temperature as observed for $CuKTSM_2$.

The results of this study indicate that the immobilization of $CuKTSM_2$ which is observed when the complex is taken up into the membrane fraction of nuclei or Ehrlich cells may have a simple explanation. The lipid structure of the membrane may prevent rapid tumbling of the drug, and thereby give rise to the anisotropy in the ESR spectrum characteristic of slowly tumbling or frozen paramagnetic species.

Several questions are raised by this finding. The

first is how to rationalize the orientation of $CuKTSM_2$ in the membrane. Previous evidence demonstrates that the planar structure "intercalates" between nonpolar phospholipid acyl chains within the membrane (1). The extended pi bonding system of the coordinated ligand, the ethoxyethyl side chain, and the methyl groups give the structure the nonpolar character to interact favorably with the oriented lipid groups when it, too, is oriented perpendicular to the plane of the membrane.

Secondly one wants to understand the molecular basis for the kinetic barrier in the bilayer in the absence of rapid rotation about the x axis defined in Fig. 1. Once $CuKTSM_2$ is inserted between molecules of phospholipid there is only one obvious hindrance to rotation about the y axis, perpendicular to the lipid side chains, comprising the interior of the membrane. Possibly the large ethoxyethyl group in the R_1 position (Fig. 1) provides a steric drag to rotation. Apart from this, rotation in the xz plane does not require a rearrangement of phospholipid molecules. Substitution of methyl in $CuPTSM_2$ for the ethoxyethyl group in $CuKTSM_2$ increases motion possibly by reducing the drag. Similarly, rotation of the planar complex about the z axis requires no net rearrangement of neighboring phospholipid molecules. Figure 4 illustrates.

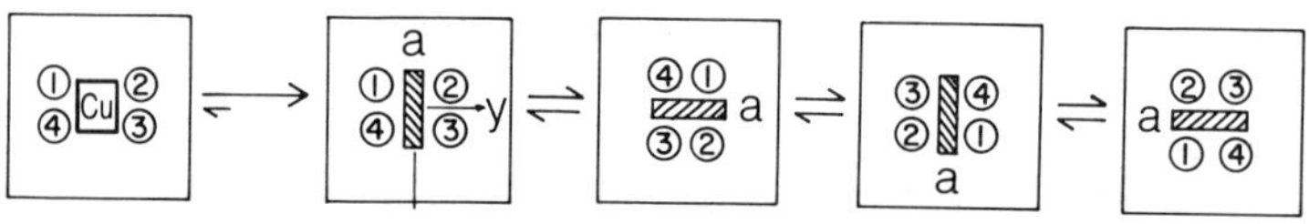

Fig. 4. No change in energy and no major reorganization of "solvent" structure required to change from panels 2 to 5 but rearrangement of lipid is required for the square plane perpendicular to the long axis of the acyl chains (panel 1).

that a symmetric operation on $CuKTSM_2$ and its direct neighbor does not alter the spacial relation of the lipid components. Therefore, there is little difference in energy in any of these component states of the rotational process. Given the freedom of two dimensional diffusion for the molecular constituents of the bilayer, there should be little hindrance to rotation of $CuKTSM_2$ about the z axis.

The analysis illustrated in Fig. 4 immediately reveals that rotation of the copper complex 90^0 about the x axis until it is parallel to the plane of the membrane must disrupt the hydrophobic interactions between drug and lipid and move the lipid molecules out of their initial positions in the two-dimensional field. Both of these events raise the energy of the system and therefore hinder this rotational process.

The parallelism in structure between intercalation in membranes and intercalation of planar complexes in DNA suggests a mode of transport for $CuKTSM_2$ to DNA which avoids contact with the cytoplasm. It is also intriguing to speculate on a mechanism whereby $CuKTSM_2$ is slowly released from the bilayer. More studies are clearly necessary.

Next, one seeks an explanation for the effect of $CuKTSM_2$ upon the transition temperature of the membrane. First, it seems likely that the large ethoxyethyl group as well as the central coordinated copper ion provide the major disturbances to the normal liquid structure of the liposomes. This R_1 moiety sweeps out a large space and is the principal contribution to nonpolarity of the structure. The axial position of the copper atom contains

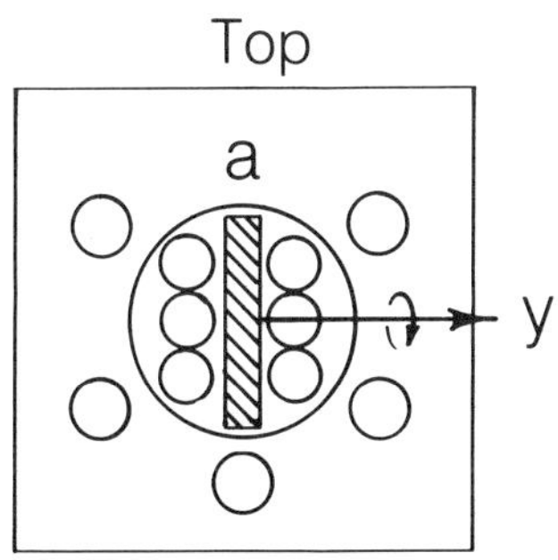

Fig. 5. Lipids (○) which must be moved out to change the static structure to a dynamic structures are within the circle containing the copper complex (▨).

electron density and may partially counteract favorable hydrophobic interactions between ligand and lipid. To the extent that both of these forms disrupt the normal liquid structure of the bilayers, these necessitate an incremental lowering of the transition temperature.

How does one explain that this effect occurs at a very low concentration of $CuKTSM_2$ relative to lipid, i.e., 1 mole %? At any time one estimates that with rapid vertical motion of the complex along the z axis, it is interacting with three lipid molecules per face of the chelate in each bilayer or 12 phosphoglycerides (Fig. 5). To exert a generalized effect on the membrane such as substantial broadening and lower of Tm, one would expect that the perturbation must see a much larger concentration of lipid. However, it is found that D, the diffusion coefficient for $CuKTSM_2$, is about 10 times faster than for the fatty acid spin labels and thus probably for the constituent lipids (1). Thus one drug molecule effectively influences $\geq$120 molecules of the bilayer. At 1 mole% $CuKTSM_2$, the entire membrane feels the presence of its addition.

Acknowledgements. Support of this work by NIH Grants GM35472, CA22184 and RR01008 is gratefully acknowledged.

References.

1. W. K. Subczynski, W. E. Antholine, J. S. Hyde, and D. H. Petering, J. Am. Chem. Soc. (in press).

2. W. E. Antholine, R. Basosi, J. S. Hyde, S. Lyman, and D. H. Petering, Inorg. Chem. 23, 3543 (1984).

3. Minkel, D. T., and Petering, D. H., Cancer Res. 38, 117 (1978).

COPPER, CERULOPLASMIN, AND SUPEROXIDE DISMUTASE IN TUMOR BEARING MICE

A.G. Fuchs, M.A. Avila, and E.S. de Lustig

Instituto de Oncología "A.H. Roffo"
Avenida San Martín 5481
Buenos Aires, Argentina

INTRODUCTION

One of the secondary effects of changes in copper metabolism is tumor disease. Copper is one of the elements which is presumably involved in the development of tumor in its early growth stages. McAuslan established, in vitro, that copper salts and a copper protein complex triggered migration of endothelial cells.[1] Ziche observed failure of tumor angiogenesis in rabbits fed a hypocupric diet and determined that the accumulation of copper in tissues precedes the process of angiogenesis.[2] Authors who studied the action of metals in tumor tissue found that they contain a greater concentration of copper than did normal tissue.[3] Apelgot observed a decrease in ^{64}Cu concentrations in livers of tumor bearing animals in comparison with normal animals. Moreover there was high activity in the nucleus of the tumor cells.[4] It has also been described that circulating copper and ceruloplasmin are higher in experimental tumor bearing animals[5] and humans with tumor.[6] Good correspondence with tumor stage has also been observed.[7] On the other hand, Boveris et al. studied levels of liver cytosolic superoxide dismutase (Cu-Zn SOD) and other antioxidizing enzymes using measurements of chemiluminiscence. It was found that livers of tumor bearing animals possed decreased antioxidizing enzyme activity and increased chemiluminiscence with respect to normal livers.[8] Studies by Oberley et al. showed that SOD is decreased in tumor cells in comparison to normal cells.[9]

Purpose

Our studies were done to evaluate:

1) Tissue copper expression in two experimental murine mammary adenocarcinomas (M3 and MM3) with different angiogenic and metastatic capacity during tumor growth.

2) Serum copper, ceruloplasmin oxidase activity, and total liver cell SOD activity in mice bearing tumor(MM3) with the highest metastatic and angiogenic capacities during tumor growth.

3) With the purpose of going further into the metabolism of copper in tumor (MM3) bearing mice, activities of two metalloenzymes with capacity to neutralize superoxide anion, the oxidase activity of the ceruloplasmin and the total liver SOD activity, were correlated since liver regulates the metabolism of copper.

MATERIAL AND METHODS

Histochemical studies of altered tissue copper were performed in two mammary adenocarcinomas (M3 and MM3)[10] with different metastatic and angiogenic capacities. Studies on serum copper, ceruloplasmin oxidase activity, and total liver SOD activity were only done with the highest metastatic and angiogenic model (MM3). Animals were killed by orbital exaguination after being anesthezised with ether. Three stages of tumor growth were determined: 1) 0-1-mm (1st stage); 2) 10.1-20 mm (2nd stage); 3) 20.1-30mm (3rd stage). Measurement of tumor was made by vernier calipers and the size estimated according to $\sqrt{d1 \times d2}$.

Tissue Copper

Five tumors (M3 and MM3) taken at random at each growth stage were fixed with 10% formaldehyde, pH 6.9-7.2. Half of the samples (n = 100) were stained with rubeanic acid and the rest with diphenylcarbazone. Hematoxilin stained sections and sections incubated with 1% EDTA were taken as control samples. Location of copper was done by staining thin sections with potassium ferrocyanide. Normal mammary gland tissue was taken as control tissue.[11]

Serum Copper

Serum was obtained from 51 BALB/c mice with MM3 tumors and from 55 normal mice and stored in plastic tubes. Determination of serum copper was done by the Bathocuproeine method using the Merckotest 3319 kit.

Determination of the Ceruloplasmin Oxidase Activity

Serum was taken from 53 BALB/c mice MM3 and from 19 normal mice. The liophilized samples were reconstructed with distilled water and the oxidase activity was measured by the Schosinsky et al. method using o-dianisidine as substrate at 450 nm.[12]

Determination of SOD Activity

Five random livers representing each tumor stage were employed as well as five random normal livers taken as controls. Extraparenchymal tissue was removed and livers cut and washed. Small pieces were exhaustively washed with 0.9% NaCl until the retained blood was eliminated. These samples were homogenized in cold phosphate buffer, 0.1M, and centrifuged at 3000 xg at 4^oC. Total SOD activity was determined in the supernatant by the method of Misra et al. taking into account the enzymatic unit as 50% of inhibition of the oxidation of epinephrine to adrenocrome measured at 480 nm.[13]

STATISTIC ANALYSIS

The x^2 test was used in studies of tissue copper histochemistry. The Student t test was used to compare results of serum copper, ceruloplasmin, and SOD obtained for control values and those of various tumor stages.

RESULTS

Copper in Tissue

Copper was principally located in the nuclear fraction

of M3 and MM3 tumors. MM3 tumors had a greater percentage of positive copper cells than did M3 in the analyzed stages ($p<0.001$). Copper content also differed among stages ($p<0.05$) in the M3 tumors.

The normal mammary gland contained iron and both iron and copper were detected in the cytoplasm.

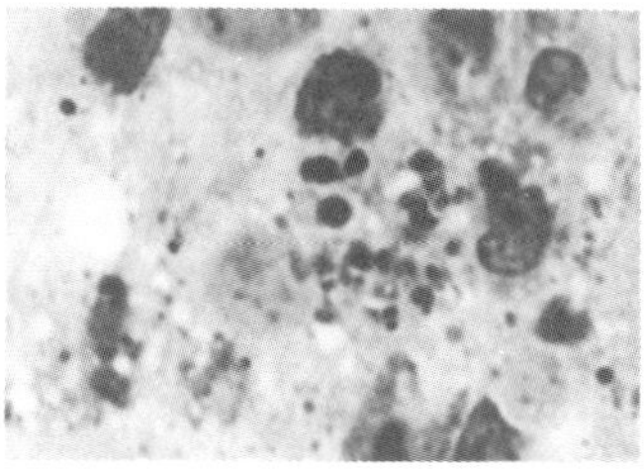

Ultra thin section of normal mammary gland.

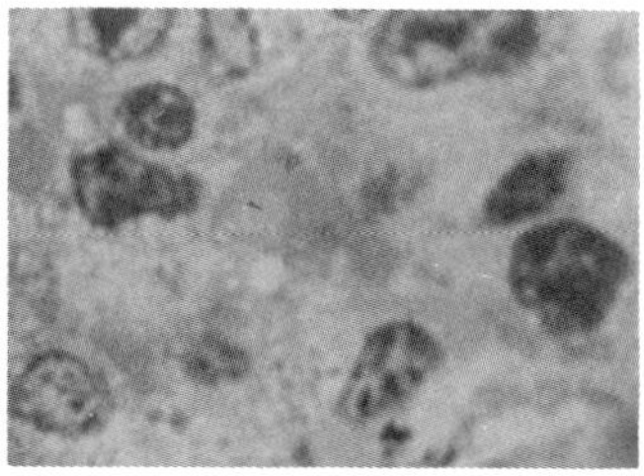

Ultra thin section of MM3 tumor.

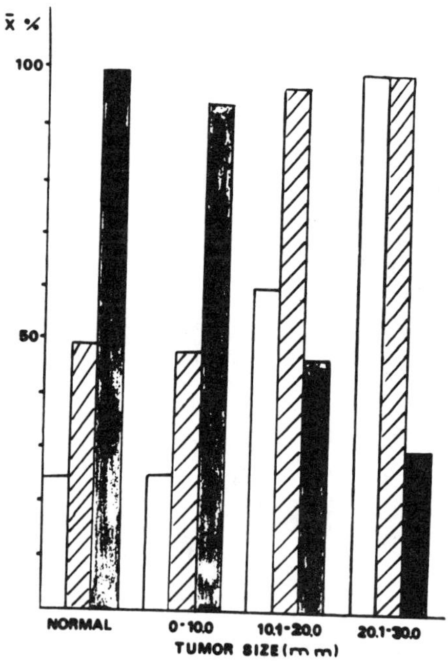

Serum Copper(□), Ceruloplasmin Oxidase(▨), and Total Liver SOD(■) Activity

		CONTROL %	1° %	2° %	3° %
☐	SERUM COPPER:	24.71 ± 9.67	24.92 ± 25.80	59.92 ± 69.63	100.00 ± 64.81
▨	CERULOPLASMIN	49.32 ± 21.78	47.95 ± 18.95	97.58 ± 42.79	100.00 ± 55.71
■	SOD ACTIVITY	100.00 ± 25.42	93.79 ± 18.64	46.89 ± 8.47	28.91 ± 7.91

The 100 % express the highest level.

Significant differences in these parameters were not found when the tumor was smaller than 10 mm in size. When the tumor was bigger, serum copper and ceruloplasmin increased ($p<0.001$) and liver SOD decreased ($p<0.01$). In the third stage serum copper increased, SOD decreased, but ceruloplasmin reached a plateau.

The correlation between ceruloplasmin oxidizing activity and total liver SOD activity is shown below. A negative correlation was found for this comparison, r= -0.71 (n=12).

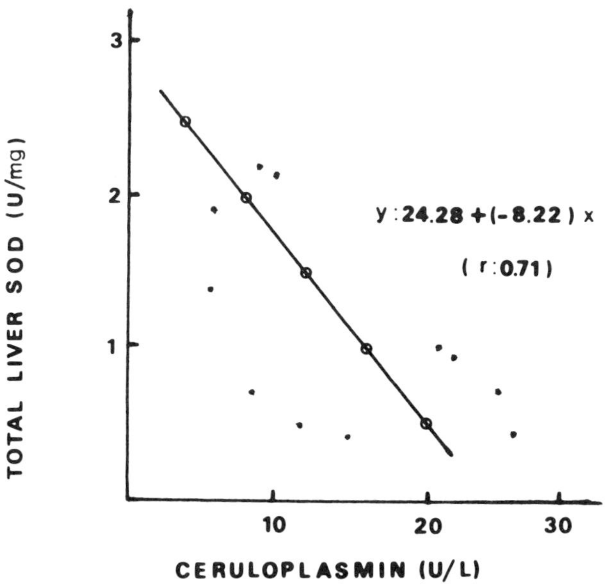

DISCUSSION

Positive copper cells were observed in both tumors using histochemistry. As the MM3 tumor has a greater percentage of positive cells, it is suggested that this greater expression is related to its higher angiogenic and metastastic capacity. The normal mammary gland presented a homogeneus distribution of stain in different locations. It is probable that this difference is related to the alteration of cellular components of the tumor cell. It has been pointed out that elements released by a tumor affect organs which are far from it. This might influence the movement of copper toward the blood stream. Apelgot's results suggest that the tumor should greedily seek copper. We have observed a progressive increase in copper and ceruloplasmin and a decrease in SOD in tumor bearing subjects along with tumor growth. It has also been observed that an increase in ceruloplasmin oxidase activity and a decrease in the total liver SOD activity are negatively correlated. We suggest that the redistribution of copper might partly affect the total liver SOD activity of the tumor bearing animal since liver SOD decreases in normal animals fed a hypocupric diet.[14] Though the oxidase activity of ceruloplasmin reaches a plateau in the last tumor stage, serum copper keeps on increasing progressively. This is probably due to its connection with another serum protein fraction.

ACKNOWLEDGMENT

This work was supported by CONICET (Consejo Nacional de Investigaciones Científicas y Técnicas. Argentina).

REFERENCES

1. McAuslan, BR; Reilly, W. Exp. Cell. Res. 1980, 130, 147.
2. Ziche, M; Jones, J; Gullino, PM. J.N.C.I. 1982, 69, 475.
3. Schwartz, AW; Ledicotte, GW; Fink, RW; Friedman, EW. Surgery. 1974, 76, 325.
4. Apelgot, S; Cooper, J; Grisvard, J; Guillé, E; Sissoef, I. Cáncer Res. 1981, 41, 1502.

5. Linder, MC; Bynt, RR; Lun, S; Scott, LE. Enzyme. 1979, 24, 85.
6. Miatto, O; Casail, M; Ganrielli, GB; Nicoli, N; Bellisola, C; Conocher, TR. Cancer. 1985, 55, 774.
7. Lamoreux, G; Mandeville, R; Poisson, R; Legault-Poisson, S; Jolicoeur, R. Cancer. 1982, 49, 502.
8. Boveris, A; Llesuy, SF; Fraga, CF. Journal of Free Radicals in Biology and Medicine. 1985, 1, 131.
9. Oberley, LW. In: Superoxide Dismutase, Vol.11 (LW Oberley ed.) C.S.C. Press. Boca Raton FL, 1982; p 127.
10. Lindquist, RR. Arch. Pathol. 1969, 87, 370.
11. Colombo, L; Bonaparte, Y; Klein, S; Stillitiani-D'Elía, I. Medicina (Argentina). 1980, 40, 827.
12. Schosinsky, KH; Lehman, HP; Beeler, FM. Clin. Chem. 1974, 20/12, 1556.
13. Misra, HP; Fridovich, IJ. Biol. Chem. 1972, 247, 3170.
14. Paynter, DI; Moir, RJ; Underwood, EJ. J. Nutr. 1979, 109, 1570.

STRUCTURE OPTIMIZATION OF AMIDES AND AMIDINES DERIVED FROM 1- AMINO- 3-(2-PYRIDYL)ISOQUINOLINE CONCERNING ANTIMYCOPLASMAL ACTIVITY

M. A. H. de Zwart, H. van der Goot, H. Timmerman

Department of Pharmacochemistry, Vrije Universiteit, de Boelelaan 1083, 1081 HV Amsterdam, The Netherlands

Introduction

It is known for many years that compounds containing a 2,2'-bipyridyl moiety are active against *M.gallisepticum.*
Previous investigations in our laboratory have revealed that in addition to this 2,2'-bipyridyl moiety, a largely planar structure, a certain extent of lipophilicity and the presence of non toxic amounts of copper are of primary neccesity for compounds to show antimycoplasmal activity[1].
Further investigations led to the conclusion that these compounds are able to chelate copper, resulting in the formation of lipophilic copper complexes.
In this way the compounds are able to serve as a carrier to transport the copper ions, which themselves are the ultimate toxic species, across the membrane into the cell[2].
Recently the high antimycoplasmal activity of amides and amidines derived from 4-amino-2-(2-pyridyl) quinazoline was discovered[3].
The increased activity of these compounds is probably due to the presence of a third coordination centre, which is possibly involved in the chelation of the copper ion.
Our aim in the present investigation was to develop highly active compounds by structure optimization within a series of amides and amidines derived from 1-amino-3-(2-pyridyl) isoquinoline.
For a fast acquirement of enhanced potency we used the method of J.G. Topliss[4], which is based on the Hansch approach.

SYNTHESIS.

I. Formation of the 2,2'-bipyridylskeleton

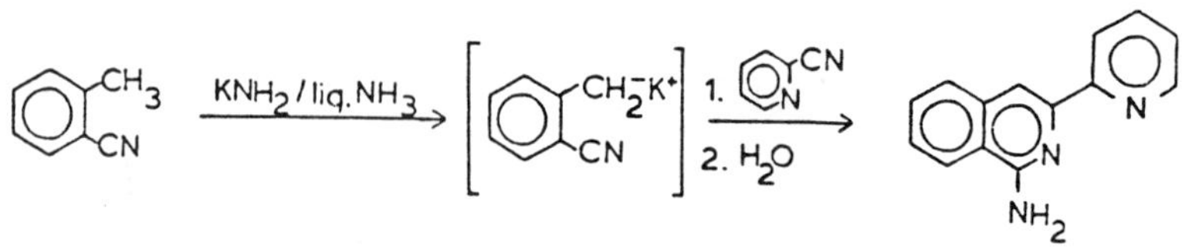

1

II. Amides

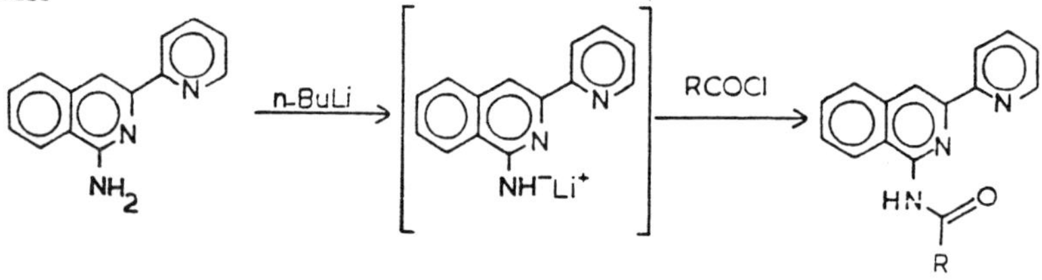

2,5

III. Amidines

a. From Amides

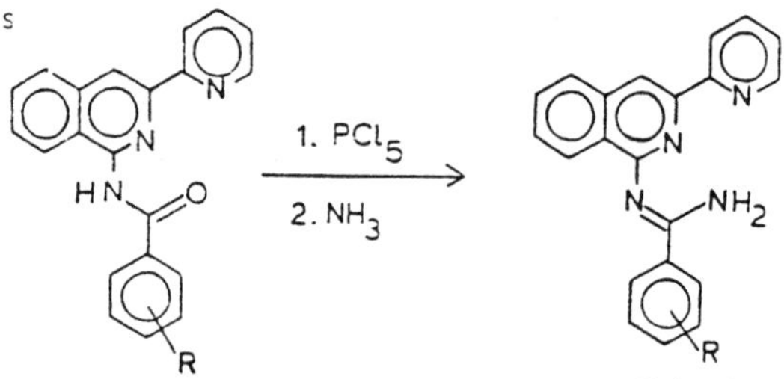

3b,c,f

b. From 1-amino-3-(2-pyridyl)isoquinoline and electron deficient aromatic nitriles

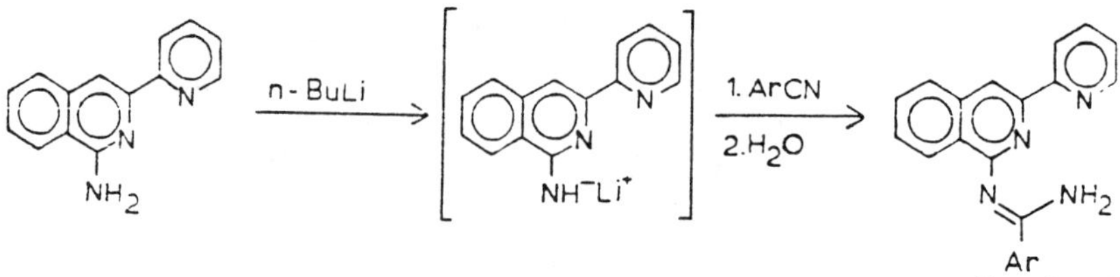

3a,d,e,g,h
4

Determination of biological activity.

For determination of antimycoplasmal activity we used *Mycoplasma gallisepticum* K514 as test organism.
M. gallisepticum is one of the fermentative mycoplasmas in which ATP is formed by glycolysis, ultimately leading to the formation and excretion of lactic acid.
So, it is possible to detect mycoplasmal growth through the change of colour of a properly chosen indicator.
The biological activity of the compounds was determined as the Minimal Inhibitory Concentration (**MIC**), which was indicated by the change of colour of phenol red.
Serial twofold dilutions of the test compounds in Adler medium were made. When the growth-inhibitory effect of the test compounds was determined in the presence of copper, 10 µg/ml $CuSO_4.5H_2O$ was added to the culture medium.
After inoculation the mixtures were incubated 24 hours at 37°C.

Table 1. Hydrophobic fragmental values[5]

Fragment	Σf
CH_3	0.701
C_2H_5	1.220
i- C_3H_7	1.739
t- C_4H_9	2.258
C_7H_{15}	3.815
C_8H_{17}	4.334
C_6H_5	1.840
3-OCH_3 (C_6H_4)	1.920
4-OCH_3 (C_6H_4)	1.920
3-CH_3 (C_6H_4)	2.359
4-CH_3 (C_6H_4)	2.359
3-Cl (C_6H_4)	2.582
4-Cl (C_6H_4)	2.582
3,4 di-Cl (C_6H_3)	3.324
pyridinyl	0.520

Results. Tables 2a-e: MIC-values (μg/ml) against M. gallisepticum K514 in Adler medium at 37°C.

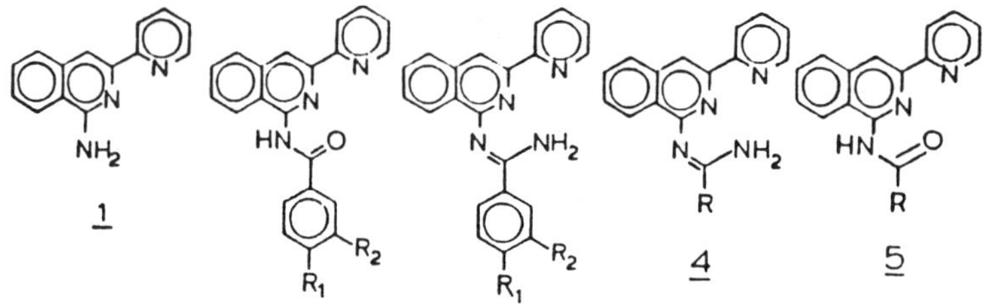

1 2 3 4 5

2a

Compound	without extra copper	extra copper[1] added
$CuSO_4$	100	-
Tylosin	0.1	0.1
1	100	0.1

2b

Compound	R_1	R_2	without extra copper	extra copper[1] added
2a	H	H	4	0.031
2b	CH_3	H	>64	0.063
2c	OCH_3	H	>64	0.063
2d	Cl	H	>64	0.125
2e	Cl	Cl	>64	0.25
2f	H	CH_3	>8	0.031
2g	H	OCH_3	>8	0.063
2h	H	Cl	>8	0.125

2c

Compound	R_1	R_2	without extra copper	extra copper[1] added
3a	H	H	0.125	0.002
3b	CH_3	H	0.125	0.002
3c	OCH_3	H	0.125	0.008
3d	Cl	H	0.5	0.016
3e	Cl	Cl	4	0.063
3f	H	CH_3	0.125	0.002
3g	H	OCH_3	0.25	0.002
3h	H	Cl	2	0.016

2d

Compound	R	without extra copper	extra copper[1] added
4a	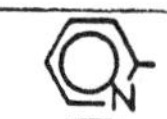	0.008	0.0005
4b	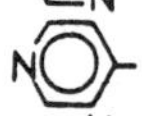	0.031	0.0005
4c	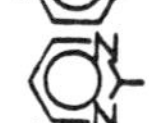	0.016	0.004
4d	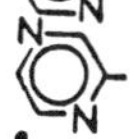	0.031	0.001
4e	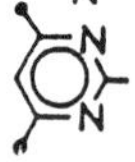	0.016	0.002

2e

Compound	R	without extra copper	extra copper[1] added
5a	CH_3	>8	0.500
5b	C_2H_5	>8	0.125
5c	$CH(CH_3)_2$	>8	0.063
5d	$C(CH_3)_3$	>8	0.063
5e	C_7H_{15}	>64	0.125
5f	C_8H_{17}	>64	0.125

1). 10 μg/ml $CuSO_4.5H_2O$

Discussion

In this investigation the molecular structure of amides and amidines derived from 1-amino-3-(2-pyridyl) isoquinoline (1) has been modified according to the structure optimization method proposed by Topliss.
Thus, we synthesized a variety of 3- and 4- substituted benzamides(2) and benzamidines(3).
Due to the dependency of the biological activity on copper we preferably compare the biological activity determined in the presence of 10 µg/ml $CuSO_4.5H_2O$.
The biological activity of the benzamides 2 was found to increase in the following way:

3,4-di-Cl<3-Cl,4-Cl<3-OCH_3,4-OCH_3,4-CH_3<3-CH_3,H.

According to Topliss the potency order of these compounds suggest a negative lipophilic dependency.

Comparison of the biological activity of the corresponding amidines 3 reveals the following order:

3,4-di-Cl<3-Cl,4-Cl<4-OCH_3<3-OCH_3,3-CH_3,4-CH_3,H.

As for the amides the biological activity shows an increasement when the hydrophobic parameter Σf of the substituent decreases (Table 1).

Therefore we decided to synthesize a series of N-heterocyclic substituted amidines 4.
A fourfold increase of the biological activity was observed in the case of pyridyl substituted amidines.
However, the less lipophilic pyrimidyl derivatives lead to a decrease of the biological activity which in combination with the previous results suggest the existence of an optimal lipophilicity.

Besides the benzamides 2 we synthesized a series of aliphatic amides 5, revealing the following order of biological activity:

$$CH_3 < C_2H_5 < CH(CH_3)_2,\ C(CH_3)_3 > C_7H_{15},\ C_8H_{17}$$

The aliphatic amides 5 show an increase of biological activity due to an increase of lipophilicity up to R= $CH(CH_3)_2$, $C(CH_3)_3$ However a further increasement of the lipophilicity (R= C_7H_{15}, C_8H_{17}) results in a decrease of biological activity.
When the biological activities of both the aliphatic and aromatic amides are compared in combination with their lipophilicity, the existence of an optimal lipophilicity necessary for maximal activity in the case of the amides is probable too.

Conclusions.

1. In the presence of copper amides and amidines of 1-amino-3-(2-pyridyl) isoquinoline are very potent antimycoplasmal agents.

2. The amidines are about ten times more active than the corresponding amides.

3. Copper complexes of the amidines and amides are involved in the growth inhibiting process.

4. A decrease of the hydrophobic parameter Σf resulted in an increase of the antimycoplasmal activity of benzamides 2 and benzamidines 3.

5. In a series N-heterocyclic substituted amidines 4 the biological activity is enhanced up to 0.5 ng/ml in the case of 2- and 4-pyridyl derivatives.
6. In a series of aliphatic amides 5 the hydrophobic parameter of group R shows an optimum for antimycoplasmal activity.

Acknowledgements. These investigations are supported by the Netherlands Technology Foundation (STW).

References.

1. Pijper, P.J.; van der Goot, H. ; Timmerman, H. ; Nauta,W. Th. ; Eur.J.Med. Chem.1984,19,389-404.
2. Gaisser, H.-D. ; van der Goot, H. ; Timmerman, H. ; Eur.J.Med.Chem.1985,20,513.
3. Linschoten, M.R. ; Gaisser, H.-D. ; van der Goot, H. ; Timmerman, H. ; Eur.J.Med.Chem.1984,19,137.
4. Topliss, J.G. ; J.Med.Chem.1977,20,463.
5. Rekker, R.F. ; de Kort, H.M. ; Eur.J.Med.Chem.1979,14,479.

THE ROLE OF CU(II) IN THE ADDITION OF ETHANOL TO THE CYANO GROUP AND THE PREPARATION OF NOVEL ORGANIC COMPOUNDS

Mohsen M. Mostafa, Rafaat M. El-Shazely and Abd El-Hammid M. Shallaby

Chemistry Department, Faculty of Science, Mansoura University, EGYPT

The addition of ethanol to the cyano group of salicylaldehyde cyanoacetylhydrazone (SCH) is catalyzed by Cu(II) chloride, bromide, and sulphate, while it remains unchangeable in complexes derived from Cu(II) acetate. The isolated Cu(II) complexes have been characterized by conventional physical and chemical methods. Infrared spectra indicate that the ligand coordinates in a tridentate and/or bidentate manner. The stereochemistry of these Cu(II) complexes is discussed and elucidated with the help of magnetic and spectral measurements. The novel organic compound, 1-salicylhydrazo-3-imino-3-ethoxypropionic acid hydrazide (SIEPH), was synthesized and its structure confirmed by elemental analysis and spectrophotometric techniques.

INTRODUCTION

Transition metal complexes of cyanoacetylhydrazine and their condensation products with salicylaldehyde as well as the addition of water or ethanol to the cyano group have been discussed earlier by us[1-3] and others.[4-8] In continuation of our work[2-3] on complexes containing the cyano group which are subject to addition of different solvents, we report herein the reaction of SCH in absolute ethanol with some Cu(II) salts and Zn(II) chloride. The imidate compound, SIEPH, has been extracted from the Zn(II) complex and characterized. Finally, the role of Cu(II) ion as well as the anion on the addition of ethanol to the cyano group has also been studied.

EXPERIMENTAL

All chemicals used are of BDH quality. SCH(H_2L) was prepared by slowly adding 10.6 ml of salicylaldehyde to 9.9 g of cyanoacetylhydrazine in absolute ethanol. The reaction mixture was refluxed on a water bath for 2 hours. The pale yellow product was crystallized from absolute ethanol and dried in a vacuum desiccator over anhydrous calcium chloride, m.p., 157°C. Cu(II) complexes of the general formulae Cu(HL*)$_2$, Cu(HL*) $(SO_4)_{0.5}(H_2O)_{0.5}$, (H_2L*=SIEPH) and Zn(HL*)$_2$ H_2O were prepared by refluxing the metal halide (Cl^- or Br^-) and/or sulphate with SCH in absolute ethanol for a time depending on the metal ion and the metal:ligand ratio. The complex, Cu(HL*)Ac(EtOH)$_2$, was prepared by mixing SCH in ethanol with Cu(II) chloride dihydrate in ethanol and in the presence of anhydrous sodium acetate as a buffering agent. The reaction mixture was refluxed on a water bath for 1 hour.

The intermediate compound, SIEPH (structure 1), was extracted by refluxing a suspension of Zn(II) complex, Zn(HL*)$_2$ H_2O, with excess disodium ethylenediaminetetra-acetate on a water bath for 1 hour. The formed pale yellow crystals were filtered, washed several times with EtOH and dried in a vacuum desiccator over anhydrous $CaCl_2$, m.p. 248°C.

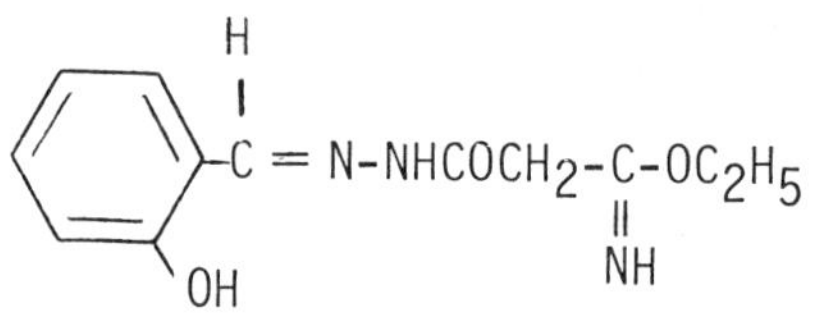

(1)

Analyses for copper and zinc were done by standard methods.[9] Carbon and hydrogen analyses were performed by the Microanalytical Unit at Mansoura University, Egypt. Molar conductivities in DMSO were measured using a Tacussel conductivity bridge type CD6NG at room temperature (25°C). IR (Nujol) and electronic (Solid and DMF) spectra were recorded on Pye Unicam SP 2000 and Perkin-Elmer 1430 spectrophotometers. Magnetic susceptibility measurements were done using Gouy method and Hg Co(SCN)$_4$ as calibrant at Alexandria University using Pascal's constants.[10] Proton magnetic resonance spectra were recorded on a Varian 360 (60MHz) using d_6-DMSO.

RESULTS AND DISCUSSION

Analytical data for ligands and their complexes are given in Table 1. Complexes having the following compositions Cu(HL)Ac(EtOH)$_{0.5}$, Cu(HL*)$_2$, and Cu(HL*)X·S (X=Ac$^-$ or (SO_4^{2-})$_{0.5}$ and EtOH or H_2O) have been isolated in pure form and characterized. Also, the imidate compound, SIEPH, was extracted from the Zn(II) complex as mentioned in the experimental section.

These metal complexes are quite stable in air and insoluble in common organic solvents but soluble in DMF and DMSO. Molar conductivities of Cu(II) complexes in DMSO at room temperature (25°C) are in the 1-18 ohm^{-1} cm^2 mole^{-1} range indicating non-electrolytic nature.[11]

Important i.r. assignments in the spectrum of the free ligand (SCH) as well as bonding sites have been determined by a careful comparison of SCH with its metal complexes and by considering our previous work.[2,3] The i.r. spectrum of the free ligand (SCH) shows bands at 3215, 3180, 2265, 1710, 1680 and 1630 cm^{-1} assignable to γ(OH), γ(NH), γ(C≡N), γ(CO) (hydrogen-bonded), γ(CO) (Free) and γ(C=N) vibrations, respectively. The existence of the OH (phenolic) band at lower wavenumber as well as the observation of two signals for protons of each OH and NH groups, in the ^{1}H n.m.r. spectrum of SCH in d$_6$-DMSO downfield from TMS, at 11.20 and 10.40 and at 9.70 and 9.55 ppm respectively, suggest that the free ligand (SCH) exists mainly in the keto form with intramolecular hydrogen bonding between the OH (phenolic) and the nitrogen of the azomethine group (structure 2). Moreover, the presence of intermolecular hydrogen bonding between the carbonyl oxygen and the NH groups is also proposed as shown below.

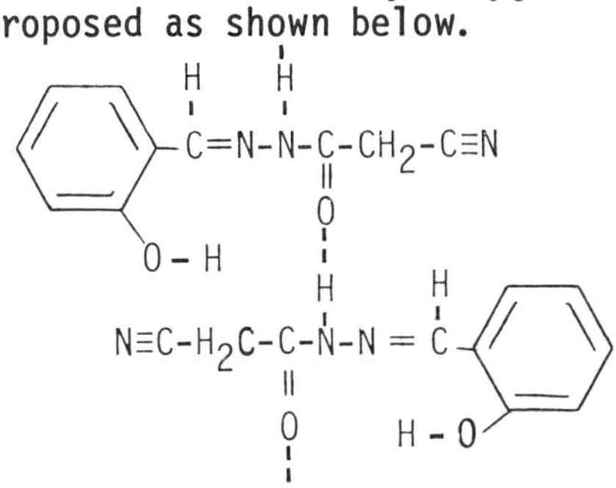

(2)

TABLE 1. ANALYTICAL AND OTHER PHYSICAL DATA

COMPOUND	M.P. (°C)	Colour	Calc. % C	H	Cu	Found % C	H	Cu	Λ[a] in DMSO	U_{eff} (B.M.)	d-d Bands
$SCH=H_2L$	157	Pale-Yellow	59.1	4.5	-	59.2	4.1	-	-	-	-
$SIEPH=H_2L*$	248	Yellow	57.8	6.1	-	58.1	5.8	-	-	-	-
$Cu(HL)Ac(EtOH)_{0.5}$	>300	Yellowish-green	44.9	4.1	18.3	44.9	3.2	20.4	6.0	2.20	16700
$Cu(HL*)_2$	>300	Yellowish-green	51.5	5.0	11.3	51.3	4.8	10.9	5.0	1.76	16400
$Cu(HL*)Ac(EtOH)_2$	263	Yellowish-brown	46.7	6.3	13.7	47.3	5.7	14.1	9.0	1.93	16200
$Cu(HL*)(SO_4)_{0.5}(H_2O)_{0.5}$	>300	Yellowish-green	39.1	4.1	17.2	40.2	3.4	17.1	18.0	1.77	16400
$Zn(HL*)_2\ H_2O$	266	Yellowish-orange	49.7	5.2	11.3	49.3	4.8	10.9	28.0	diamag	-

[a] $ohm^{-1}cm^2mole^{-1}$

In comparing the i.r. spectra of these metal complexes, we observed that both ligands gave the same complex. In other words, SCH containing a cyano group was converted to the imidate structure containing a C-O-C group due to the addition of ethanol to the cyano group during complex formation.

SCH coordinates in a tridentate manner via the azomethine nitrogen (C=N), OH (phenolic), and the enolized carbonyl oxygen with replacement of a hydrogen atom from the latter group. The observation of the band at 2260 cm^{-1}, assigned to (C≡N), in the spectrum of the Cu(II) complex, $Cu(HL)Ac(EtOH)_{0.5}$, suggests that no addition to the triple bond occurred and is supported by the following evidence.

The lack of any band at 1710 and 1680 cm^{-1} due to the carbonyl oxygen, the appearance of new bands at 1650 and 1240 cm^{-1}, assigned to the γ(C=N*) and γ(C-O)[12] vibrations respectively, the negative shift of the azomethine nitrogen (C=N) at 1630 cm^{-1} to lower wavenumbers, the observation of two strong bands at 1600 and 1490 cm^{-1}, assigned to γ_{as} (Ac)[13] and γ_s (Ac)[13] respectively, and the appearance of new bands at ca. 580, 515, 430, 385, and 330 cm^{-1}, assigned to γ(M-O) (H_2O)[13], γ(M-O) (phenolic), γ(M-O) (enolized carbonyl oxygen)[14], γ(M-O) (acetate), and γ(M-N)[15] respectively.

The electronic spectrum of the Cu(II) complex, $Cu(HL)Ac(EtOH)_{0.5}$, shows a broad band centered at 16700 cm^{-1}. This band together with the value of magnetic moment (U_{eff}=2.20 B.M.) support octahedral geometry around the Cu(II) ion as shown in structure 3.

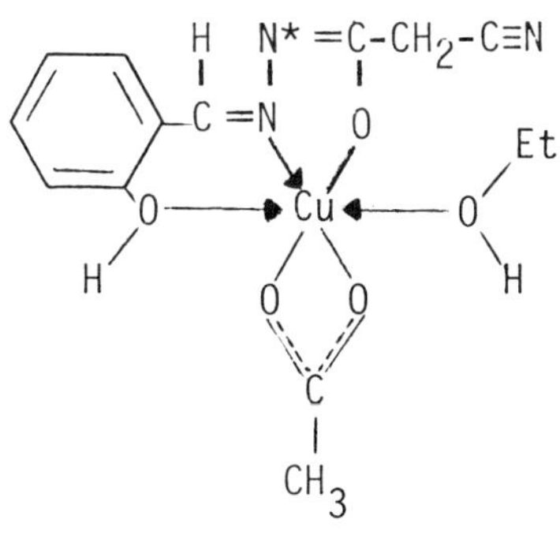

(3)

SIEPH coordinates in a tridentate and bidentate manner with Cu(II) and Zn(II) ions, respectively, during complex formation. This ligand coordinates in a tridentate manner via the carbonyl oxygen, azomethine nitrogen, and the OH (phenolic) groups with displacement of a hydrogen atom from the last group. The mode of complexation and structures 4 and 5 are suggested on the basis of the following evidence: The negative shift (20 cm^{-1}) of the azomethine nitrogen to lower wavenumber, the decrease of the intensity of the carbonyl oxygen band, the disappearance of the band at 1350 cm^{-1} assigned to the OH deformation[12] with the appearance of a new band at 1450 cm^{-1} assigned to γ(C-O)[16] phenolic, the appearance of new bands at 530, 430, and 340 cm^{-1} assigned to γ(M-O)(phenolic)[17], γ(M-O)(carbonyl)[14], and γ(M-N)[15] vibrations, respectively, and the disappearance of the cyano group (2260 cm^{-1}) together with the appearance of a new band at 1300 cm^{-1} assigned to γ(C-O-C)[18,19] vibration. The displacement of a hydrogen from the OH (phenolic) was checked qualitatively[20] using $FeCl_3$ solution. Also, the replacement of a proton from the OH (phenolic) group is determined quantitatively from the H^1 n.m.r. spectrum of the analogous uranyl complex $UO_2(HL^*)_2$, in d_6-DMSO. The lack of the signal at 11.20 ppm downfield, due to the proton of OH group, can be taken as additional evidence for the participation of the OH group in bonding with displacement of the proton (structures 4 and 5).

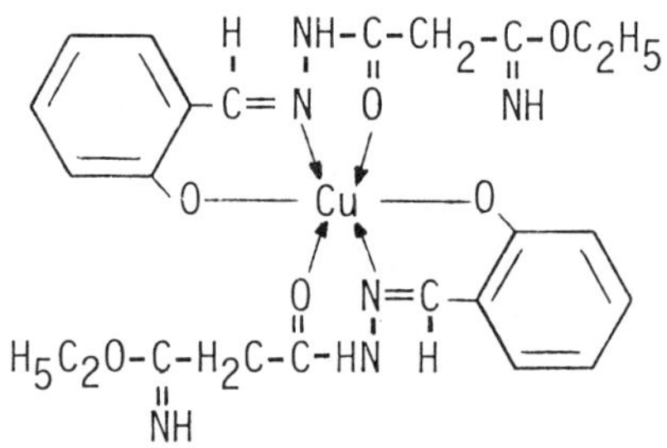

(4)

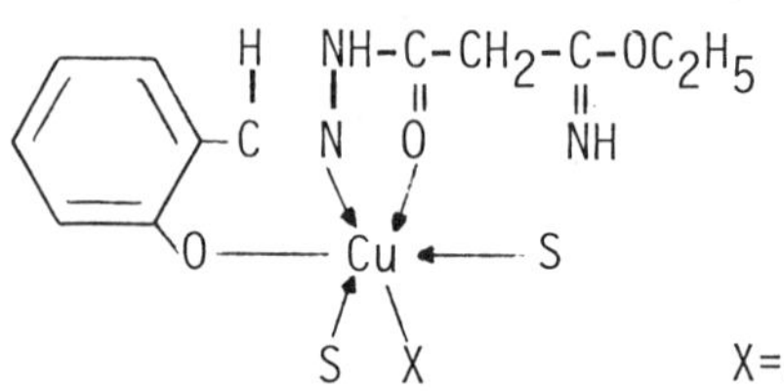

(5)

X=Ac^- or $(SO_4^{=})_{0.5}$
S=EtOH, $(H_2O)_{0.25}$

Electronic spectra of Cu(II) complexes, Cu(HL*)Ac$(EtOH)_2$ and Cu(HL*)$(SO_4)_{0.5}(H_2O)_{0.5}$, show a broad band centered at 16400 cm^{-1} indicating a distorted octahedral conformation around Cu(II).[21] Also, magnetic moment values (1.93 and 1.77 B.M.) lie in the permissible region for octahedral geometry with no copper-copper interaction.[22] The bidentate nature of the sulphate ion (C_{2V})[13] in Cu(II) sulphate complex is confirmed from the splitting of both γ_2 and γ_4 into three bands as well as the observation that both γ_2 and γ_4 are medium bands.[13]

The spectrum of the Zn(II) complex, Zn$(HL^*)_2$ H_2O, is more or less the same as those discussed above. The only difference is that the carbonyl group did not take part in coordination and the active centers are the azomethine nitrogen and phenolic OH groups with displacement of a proton from the latter group.

Finally, the i.r. spectrum of the novel organic compound, SIEPH, shows new bands which are absent in the spectrum of SCH. Bands located at 3200, 1590, and 1305 cm^{-1} are assigned to the NH of (C=NH), azomthine of (C=NH)[12], and γ_s(C-O-C) vibrations, respectively. Also, the observation of a strong band at 3460 cm^{-1}, assigned to γ(OH) vibration, as well as the positive shift of the carbonyl oxygen band to higher wavenumber (1730 cm^{-1}) indicates the absence of the hydrogen bonding which exists in SCH.

In conclusion, the cyano group remains in complexes prepared from Cu(II) acetate while it disappears on using Cu(II) chloride, bromide, or sulphate. This phenomenon can be explained as due to the need of a strong acid medium for the addition of ethanol which is achieved by the chloride, bromide, and sulphate salts of copper. These salts produce HCl, HBr, or H_2SO_4, respectively, while the acetate gives acetic acid which is not acidic enough to catalyze the addition process.

REFERENCES

1. Mostafa, M.M; Hassan, S.M; El-Asmy, A.A. J. Ind. Chem. Soc. 1978, 55, 529.
2. Shallaby, A.M; Soliman, M.S; El-Shazely, R.M; Mostafa, M.M. Polyhedron. Submitted for publication.

3. Soliman, M.S; Shallaby, A.M; El-Shazely, R.M; Mostafa, M.M. Polyhedron. Submitted for publication
4. Diana, G.D; Zalay, E.S; Cutler,Jr. R.A. J. Org. Chem. 1965, 30, 298.
5. Dulta, R.L; Ray, P.J. Ind. Chem. Soc. 1959, 36, 499.
6. Payne, G.B. J. Org. Chem. 1961, 26,668.
7. Dutta, R.L; Singh, A.M. J. Inorg. Nucl. Chem. 1978, 40, 417.
8. Barnard, P.F.B. J. Chem. Soc. 1969, A, 2140.
9. Vogel, A.I. In: Quantative Inorganic Analysis, 3rd ed.; Longman; London, 1961; p 434.
10. Lewis, J; Wilkins, R.G. In: Modern Coordination Chemistry, Interscience; New York, 1960; p 403.
11. Geary, W.J. Coord. Chem. Rev. 1971, 7, 81.
12. Bellamy, L.J. In: The Infrared Spectra of Complex Molecules, Methuen; London, 1958.
13. Nakamoto, K. In: Infrared Spectra of Inorganic and Coordination Compounds, Wiley; New York, 1970.
14. Teotia, M.P; Rastogi, D.K; Malik, W.U. Inorg. Chim. Acta. 1973, 7, 339.
15. Browall, K.W; Interranta, L.V. J. Coord. Chem. 1973, 3, 27.
16. Biradar, N.S; Patil, B.R; Kulkarni, V.H. J. Inorg. Nucl. Chem. 1975, 37, 1901.
17. Ferraro, J.R. In: Low Frequency Vibrations of Inorganic and Coordination Compounds, Plenum Press; New York, 1971.
18. Hadži, D; Prevoršvek, D. Spectrochim. Acta. 1957, 10, 38.
19. Figgins, P.E; Busch, D.H. J. Phys. Chem. 1961, 65, 2236.
20. Feigl, R. In: Spot Tests in Organic Analysis, Elsevier; Amsterdam, 1966.
21. Lever, A.B.P. In: Inorganic Electronic Spectroscopy, Elsevier; Amsterdam, 1968.
22. Kato, M; Jonassen, H.B; Fanning, J.C. Chem. Rev. 1964, 64, 99.

SUPEROXIDE DISMUTASE ACTIVITY OF HEART AND LIVER TISSUE FROM COPPER DEFICIENT RATS

Augusta A. Mylroie, Constance Umbles and Joseph Kyle. Department of Chemistry, Chicago State University, 95th Street at King Drive, Chicago, IL 60628

Superoxide dismutase (SOD) is considered to have an important role in protecting cells against direct and indirect oxidative damage[1-3]. Total SOD activity is the result of the activities of a copper dependent enzyme (CuSOD), present primarily in the cytoplasm, and a manganese dependent enzyme (MnSOD), localized in the mitochondrial matrix[4]. The catalytic activity of CuSOD is inhibited by cyanide and this property is used to distinguish it from MnSOD.[5]

Alterations in SOD activity have been reported to occur as a result of various experimental treatments such as nutritional deficiencies in Cu[6], Mn[7,8] and vitamins A or E[9], the intraperitoneal injection of glucose[10], hyperoxia[11,12] and hypoxia[13]. Recently we have shown that, in rats, Pb ingestion (250 or 500 ppm Pb in the drinking water) results in reduced levels of Cu and CuSOD in erythrocytes and reduced levels of Cu in tissues[14,15]. The Pb-induced reduction in tissue Cu levels was accompanied by a decrease in CuSOD activity in blood, liver, heart and kidney[16,17]. The decrease in CuSOD was reflected in a decrease in total SOD in liver, as expected. Unexpectedly, decreased CuSOD activity was not accompanied by a corresponding decrease in total SOD activity in heart or kidney homogenates-apparently as a result of a compensatory increase in MnSOD[17]. Similar observations were made in preliminary experiments in heart tissue of Cu deficient rats[18].

The present experiments were designed to determine whether MnSOD activity is enhanced in heart and liver of Cu depleted rats.

MATERIALS AND METHODS

In a series of experiments, weanling male Sprague-Dawley rats were fed AIN-'76 purified diet[19] containing either 6 ppm Cu (control) or 0.4 ppm Cu (Cu deficient) for 5-9 weeks. Animals were housed in individually suspended stainless steel cages in a temperature controlled room with a 12-hour light/dark cycle. Food and drinking water were provided ad libitum.

At the end of the 5-7 week treatment period, animals were anesthetized with Metofane and blood was collected by cardiac puncture into heparinized syringes. Aliquots of blood were allocated for hematological analysis, ceruloplasmin and SOD assays and atomic absorption spectrophotometry. Tissues were excised and either processed immediately or frozen at -70°C for further analysis. Weighed portions of excised organs and aliquots of blood were wet ashed prior to metal analysis by flame atomic absorption spectrophotometry[20]. Weighed heart tissue was combined with 9 volumes of Triton-X 100 (0.2% v/v) and homogenized at ice water temperature in glass-glass Potter-Elvehjem homogenizers with the pestle driven at half speed by a Con-Torque power unit. The crude homogenate was sonicated, using a Branson W-350 sonifier cell disruptor equipped with a microtip, for two 15 second periods with cooling. Heart homogenates were centrifuged at 800 g for 10 minutes and aliquots of the supernatant used for SOD assays. Liver was homogenized and sonicated as described above for heart tissue and the homogenate assayed for SOD activity. Mitochondrial fractions were prepared by homogenizing tissue in 0.25 M sucrose (1 g/9 mL) at pH 7.4.

The homogenate was centrifuged for 10 minutes at 800 g. The pellet was discarded and the supernatant centrifuged for 10 minutes at 9,800 g. The resultant pellet was washed in 9 volumes of sucrose solution using a Teflon-glass homogenizer to resuspend the pellet. The resuspended pellet (mitochondrial fraction) was centrifuged for 10 minutes and the washing repeated. The washed pellet was finally resuspended in cold 0.2% Triton-X 100 (1 g of original tissue: 1 mL of Triton X-100), sonicated and analyzed for SOD activity and protein content[21].

SOD activity was measured by its ability to increase the rate of the riboflavin sensitized photo-oxidation of o-dianisidine[22]. Superoxide dismutase activity was assayed in cuvettes containing a final (3.0 mL) reaction mixture of the following components: 45 mM potassium phosphate, pH 7.8; 0.013 mM riboflavin, 0.2 mM o-dianisidine and either 0 or 4 mM KCN. Cuvettes were illuminated using 20 W Sylvania Grow Lux Fluorescent tubes placed 5 cm above and to one side of the cuvettes and maintained at 37°C. CuSOD activity was calculated by substracting enzyme activity (MnSOD) obtained in the presence of 4 mM KCN from SOD activity (total SOD) obtained in the absence of KCN. CuSOD was also determined independently by measuring its activity in ethanol-chloroform treated homogenates[8]. A standard curve was prepared routinely using bovine superoxide dismutase (Sigma Chemical Co., 3000 units/mg) as a reference. One activity unit is defined as that activity equivalent to 1 µg of purified bovine CuSOD.

Ceruloplasmin activity was determined by measuring spectrophotometrically the rate of oxidation of p-phenylenediamine and values are expressed in terms of International Units (umoles Bandrowski's base/minute/ L of serum)[23]. Hemoglobin determinations were made by the cyanmethemoglobin method[24] and hematocrit values determined manually with a microcapillary centrifuge. Data were analyzed statistically[25].

RESULTS AND DISCUSSION

Compared to rats fed the Cu sufficient control diet, rats fed the Cu deficient diet exhibited the expected decreases in ceruloplasmin values (88%) and Cu levels (Table 1). The following decreases in Cu concentration resulted after a five week treatment period: blood Cu (93%), liver Cu (83%) and heart Cu (73%). Additional changes as a result of the Cu deficient diet included decreased hemoglobin (32%) and hematocrit (39%) values and an increase in liver Fe (93%).

In liver homogenates (Table 2), the decrease in tissue Cu levels was accompanied by a decrease in CuSOD (80%) and a corresponding decrease in total SOD (64%). There was a small, but not statistically significant (P>.05), increase in MnSOD activity. In heart homogenates (Table 2), the decrease in Cu levels was associated with a decrease in CuSOD activity (36%). Contrary to expectations, total SOD was not correspondingly decreased - the result of an apparent compensatory increase (25%) in MnSOD activity. Copper deficiency has been reported to result in cardiac hypertrophy[26-28] and in alterations of the size and/or number of mitochondria in heart[27,28] and liver[28]. Electron micrographs of heart muscle of Cu deficient rats were characterized by an enlargement of the area occupied by mitochondria relative to myofibrils, possibly as a result of an increase in the number of mitochondria[27,28].

In order to determine whether the increase in MnSOD reported here is not merely a reflection of an increase in the number of mitochondria or a nonspecific increase in mitochondrial proteins, MnSOD activity was assayed in mitochondrial fractions and calculated on the basis of units MnSOD/mg protein. There was a significant (P<.01) increase in MnSOD activity in both heart (38%) and liver (65%) mitochondrial fractions as

Table 1. Effect of Dietary Copper Deficiency on Parameters of Copper Status.[1]

Determination	Treatment Group Control	Cu deficient
Blood Cu (µg/ml)	0.84±.08	0.06±.01*
Liver Cu (µg/g)	3.58±.05	0.62±.06*
Liver Fe (µg/g)	68.7 ±3.1	133 ±8.7*
Heart Cu (µg/g)	5.20±.12	1.42±.05*
Heart Weight/ 100 g Body wt	0.35±.02	0.43±.02*
Body Wt Gain (g)	224 ±5	204 ±5*
Ceruloplasmin(I.U.)	95.0 ±4.3	11.1 ±1.9*
Hematocrit (%)	39.7 ± .6	28.2 ±.8*
Hemoglobin (g/100ml)	14.4 ±.3	9.8 ±.4*

[1]Values were obtained after a 5 week treatment period and are expressed as mean ± SEM (10-12 animals per treatment group).

*Significantly different from corresponding control group (P<.01).

a result of Cu deficiency (Table 2).

The increase in MnSOD activity reported here could be the result of enzyme induction or the activation of existing MnSOD; it could be due to a compensatory increase in MnSOD in response to a decrease in cytosolic CuSOD or in response to an increased $O_2^{\cdot -}$ production in the mitochondrial electron transport chain - an indirect result of a Cu deficiency.

A number of investigators have proposed that MnSOD activity is regulated by the intracellular concentration of $O_2^{\cdot -}$. Superoxide dismutase activity has been shown to be induced by increased exposure to molecular oxygen in adult rat lung[29] and rapid oxygen - induced synthesis of MnSOD has been demonstrated in pulmonary macrophages of neonatal rats in vivo and in vitro[30]. The

Table 2. Effect of Dietary Copper Deficiency in the Rat on Superoxide Dismutase Activities[1] of Liver and Heart.

Determination	Treatment Group[2] Control	Cu Deficient
<u>Heart Homogenate</u>		
Total SOD	4.3±.1	3.8±.2*
CuSOD	2.6±.2	1.7±.2*
MnSOD	1.7±.1	2.1±.1
<u>Liver Homogenate</u>		
Total SOD	13.9±.6	5.0±.3*
CuSOD	11.7±.6	2.5±.2*
MnSOD	2.2±.2	2.5±.3
<u>Heart Mitochondria</u>		
MnSOD	3.7±.2	5.1±.2*
<u>Liver Mitochondria</u>		
MnSOD	2.0±.2	3.3±.2*

[1]Units SOD/mg protein
[2]Values were obtained after a 5-7 week treatment period and are expressed as mean ± SEM of 10-12 animals per treatment group
*Significantly different from corresponding control group (P<.01)

increased MnSOD activity has been postulated to be due to enhanced levels of $O_2^{\cdot -}$ formed in the electron transport chain[30]. Superoxide dismutase activity was inhibited by intraperitoneal injection of glucose in mouse heart and brain[10]. Glucose was believed to reduce MnSOD activity because its metabolism is associated with lower $O_2^{\cdot -}$ production than other energy sources. The age-dependent increase of MnSOD in the cerebral cortex of rat has been attributed to an enhanced release of $O_2^{\cdot -}$ with age that corresponds to the increase in respiratory activity in mitochondria isolated from adult rats with increasing age[31]. An increase in MnSOD activity was also observed with the aging of human fibroblasts in vitro[32]. The authors suggested that this increase in MnSOD activity is related to the increase of oxygen radicals reported in aged membranes in vivo and in vitro and that sensitive regulatory mechanisms exist that can adjust MnSOD levels to the metabolic state of the cell[32].

It has been proposed that nutritional deficiencies of either Cu or Mn are reflected in a decrease in the corresponding SOD and accompanied by a compensatory increase in the other SOD and that MnSOD and CuSOD, although compartmentalized, are not independently regulated[7,33]. Results in support of this hypothesis, which emphasizes the common function of CuSOD and MnSOD, have been observed in a fungus, in peas, and in chickens (for a review, see 33), but not in mice or rats[7,8,34,35]. Chickens fed a Mn deficient diet had lower levels of MnSOD and higher levels of CuSOD in liver than the control group[7]. On the other hand, in Mn depleted rats no significant changes were observed in CuSOD activity, although MnSOD levels were significantly lower in heart and kidney[8]. In Cu deficient rats, liver CuSOD activity was reported to be decreased, but MnSOD activity remained unchanged compared to controls[34]. CuSOD activity in liver homogenates decreased in the cytosol and mitochondrial fractions of liver in Cu deficient swine to less than 15% of control values, with no concurrent increase in MnSOD[35]. The discrepancy between these results and the results we obtained may be due to the concentration of KCN (1 mM vs 4 mM) used in the SOD assay.

The increased MnSOD activity in liver and heart of Cu deficient rats reported here could be the result of an increased production of $O_2^{\cdot -}$ in the electron transport chain. The decrease in cytochrome oxidase reported by other investigators in heart and liver as a result of Cu depletion[26,36-39] and the increase in the activities of enzymes involved in the citric acid cycle[39] could result in a more reduced state of the electron transport chain and lead to an increased production of O_2^- and H_2O_2[40].

SUMMARY

Dietary-induced Cu deficiency in rats resulted in reduced levels of Cu in blood and tissues. This decrease in Cu content was accompanied by a significant reduction in cytosolic CuSOD in heart and liver homogenates and a reduction in total SOD activity in liver. However, total SOD activity was not significantly reduced in heart homogenate as a result of an apparent compensatory increase in MnSOD. This increase in MnSOD activity (units SOD/mg protein) was also observed in mitochondrial fractions prepared from both liver and heart of rats fed a Cu deficient diet. The enhanced MnSOD activity could be in response to an increased production of O_2^- in the electron transport chain and/or to the reduction in cytosolic CuSOD activity.

ACKNOWLEDGEMENTS

Supported by NIH Grant #RR08043.

REFERENCES

1. Halliwell, B. Cell Biol. Intern. Repts. 1978, 2, 113.
2. Fridovich, I. Ann. Rev. Biochem. 1979, 44, 147.
3. Freeman, B.A: Crapo, J.D. Lab. Invest. 1982, 47, 412.
4. Weisiger, A.R: Fridovich, I. J. Biol. Chem. 1973, 248, 3582.
5. Salin, M.L: Day, E.D: Crapo, J.D. Arch. Biochem. Biophys. 1978, 187, 223.
6. Paynter, D.I: Moir, R.J: Underwood, E.J. J. Nutr. 1979, 109, 1570.
7. deRosa, G: Keen, C.L: Leach, R.M: Hurley, L.S. J. Nutr. 1980, 110, 795.

8. Paynter, D.I. J. Nutr. 1980, 110, 437.
9. Sklan, D: Rabinowich, H.D: Donoghue, S. Nutr. Repts. Int. 1981, 24, 551.
10. Kasemset, D: Oberley, L.W. Biochem. Biophys. Res. Commun. 1984, 122, 682.
11. Crapo, J.D: Tierney, D.F. Am. J. Physiol. 1974, 226, 1401.
12. Bannister, W.H. In: The Biology and Chemistry of Active Oxygen; Bannister, J.V.; Bannister, W.H. eds. Elsevier, New York, NY 1984; p 208.
13. Sjostrom, K: Crapo, J.D. Bull. Eur. Physio path. Resp. 1981, 17, 111.
14. Mylroie, A.A. In: Trace Substances in Environ. Health, 14; Hemphill, D.D. ed. University of Missouri, Columbia, MO, 1980; p 195.
15. Mylroie, A.A: Collins, H: Umbles, C: Kyle, J. Toxicol. Appl. Pharmacol. 1986, 82, 512.
16. Mylroie, A.A: Ariyo, F: Kyle, J. In: Trace Substances in Environ. Health, 19, Hemphill, D.D. ed. University of Missouri, Columbia, MO, 1985; p 317.
17. Mylroie, A.A: Boseman, A: Kyle, J. In: Trace Substances in Environ. Health, 20, Hemphill,D.D. ed. University of Missouri, Columbia, MO 1986 (in press).
18. Umbles, C: Tucker, C: Kyle, J: Mylroie, A.A. MBRS Symposium, Abstr. 1986.
19. American Institute of Nutrition Ad Hoc Committee on Standards for Nutritional Studies. J. Nutr. 1977, 107, 1340.
20. Instrumentation Laboratory Operator's Manual, Lexington, MA, 1970; p 22.
21. Lowry, O.H: Rosebrough, N.J: Farr, A.L: Randall, R.L. J. Biol. Chem. 1951, 193, 265.
22. Misra, H.P: Fridovich, I. Arch. Biochem. Biophys. 1977, 183, 308.
23. Rice, E.W. Anal. Biochem. 162, 3, 452.
24. Drabkin, D: Austin, H. J. Biol. Chem. 1933, 112, 51.
25. Winer, B.J: Statistical Principles in Experimental Design, McGraw-Hill, New York, NY, 1971.

26. Abraham, P.A: Evans, J.L. In: Trace Substances in Environ. Health, 5; Hemphill, D.D. ed. University of Missouri, Columbia, MO, 1971; p 335.
27. Goodman, J.R: Warshaw, J.B: Dallman, P.R. Pediatr. Res. 1970, 4, 244.
28. Dallman, P.R: Goodman, J.R. Blood 1970, 35, 496.
29. Crapo, J.D: Tierney, D.F. Am. J. Physiol. 1974, 226, 1401.
30. Stevens, J.B: Autor, A.P. J. Biol. Chem. 1977, 252, 3509.
31. Vanella, A: Geremia, E: D'Urso, G: Tiriolo, P: Di Silvestro, I: Grimaldi, R: Pinturo, R. Gerontology 1982, 28, 108.
32. Crosti, N: Bajer, J: Rigo, A: Serra, A: Viglino, P: Argese, E: Mavelli, I. Acta Med. Rom. 1983, 21, 458.
33. Rabinowich, H.D: Fridovich, I. Photochem. Photobiol. 1983, 37, 679.
34. Balevska, P.S: Russanov, E.M: Kassabova, T.A. Int. J. Biochem. 1981, 13, 489.
35. Williams, D.M: Lynch, R.E: Lee, G.R: Cartwright, G.E. Proc. Soc. Exp. Biol. Med. 1975, 149, 534.
36. Elvehjem, C.A. Phys. Rev. 1935, 15, 471.
37. Dallman, P.R. J. Clin. Inv. 1967, 46, 1819.
38. Evans, J.L: Brar, B.S. J. Nutr. 1974, 104, 1285.
39. Kelly, W.A: Kesterson, J.W: Carlton, W.W. Exp. Molec. Pathol. 1974, 20, 40.
40. Boveris, A: Turrens, J.F. In: Chemical and Biochemical Aspects of Superoxide and Superoxide Dismutase. Bannister, W.H: Bannister, J.V. eds. Elsevier North Holland, New York, NY, 1980; p 84.

SYNTHESIS AND BIOCIDAL ACTIVITY OF SOME COPPER-TRIDENTATE SCHIFF BASES

R.K. PARASHAR, R.C. SHARMA

Department of Chemistry,

Agra University, Agra (India)

GOVIND MOHAN

Department of Pharmacology,

S.N.Medical College, Agra (India)

ABSTRACT

Copper(II) complexes with salicylidine anthranilic acid (SAA) salicylidine-o-aminophenol (SAP), salicylidine-o-aminothiphenol (SATP) and salicylidine-o-aminopyridine (SAPy) have been synthesized and characterised by their elemental analyses, TGA, IR and electronic spectral studies, molar conductance and magnetic susceptibility measurements. Antimicrobial activity of these ligands and their respective copper(II) complexes have been determined on gram positive(Staphylococous aureus)and gram negative (Escherichia coli) bacteria and on fungi like Aspergillus niger, Aspergillus nidulense and Candida albicans by serial dilution methods. A considerable increase in the biocidal activity of the ligands on being co-ordinated with the metal-ion has been observed in terms of their minimum inhibitory concentration (MIC) values.

INTRODUCTION

Schiff bases are well known for their biological activities[1,2]. The introduction of some functional groups at proper position of the suitably tailored potentially active molecule, further increases their biological activity[3].

Since the biocidal activity of an organic ligand can be enhanced several times on being coordinated with suitable metal ions[4-6], we have synthesized some biologically active tridentate schiff bases and their copper(II) complexes with a view to compare the biocidal properties of these ligands with their metal complexes to explore the possibility of their uses in the field of bio-inorganic relevance.

EXPERIMENTAL

All the chemicals used were of A.R.Grade.

Synthesis of Ligands

The general structure of ligands synthesized by condensing (I:I) molar amounts of salicylaldehyde with anthranilic acid/2-aminophenol or 2-aminopyridine in ethanol is depicted in Fig.I.

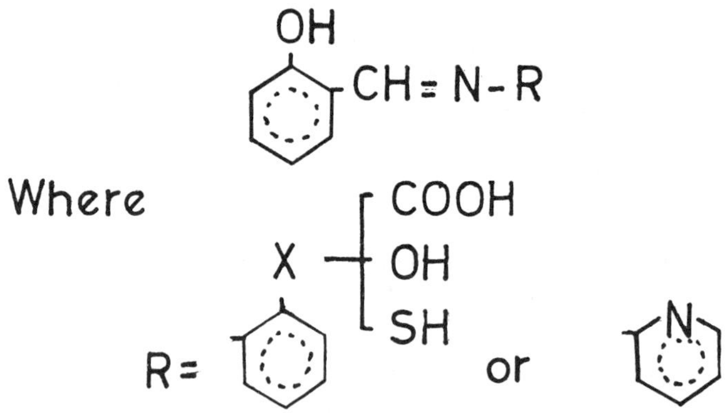

Fig.I General structure of ligands synthesized.

Preparation of Metal Complexes

Copper complexes were prepared by refluxing equimolar mixture of cupric acetate in methanol and the ligand either in benzene or methanol except SAPy, in which ratio of the metal to ligand was maintained as I:2 (M:L). The resultant green crystals were filtered and washed first with water and then with methanol, benzene and finally with ether and dried over P_4O_{10} in vaccum.

Elemental Analysis and Physical Measurements

All the synthesized ligands and their copper complexes were analysed for C,H,N, and S by micro-analytical techniques. Copper was estimated as thiocyanate. The IR spectra were recorded in KBr matrix using Perkin Elmer -521 spectrophotometer and Electronic spectra were taken in Cary-14 instrument. Magnetic susceptibility Measurements were done at room temperature by the Gyoy's method using mercury tetra-thiocyanato cobalt(II) as the calibrant and corrected for diamagnetism in the usual manner. Molar conductivity measurements were performed by Toshniwal digital conductivity meter in DMF solution of the complexes.

RESULTS AND DISCUSSION

The low values of molar conductance (0.8-2.20 $ohm^{-1}\ cm^2\ mol^{-1}$) of all the complexes indicated that ligands neutralised its charge during coordination to the metal(II) ion. Elemental analyses data (table-I) indicated that one molecule of all the ligands except SAPy were attached to a copper (II) ion. In case of SAPy, 1:2 M-L complex formation was observed. The presence of coordinated water molecule in all the copper complexes except SAPy was confirmed by their elemental analysis, IR and TGA data where loss in weight corresponds to one water molecule occured at 275°C.

TABLE-I Analytical Data of Compounds

Compound	Analysis %, Found (Calcd.) C	H	N	S	Cu	% Wt. Loss at 275°C	μ eff. in B.M.
SAA	70.06 (69.69)	3.84 (4.59)	5.32 (5.80)				
SAP	72.38 (73.22)	4.98 (5.19)	7.04 (6.56)				
SATP	67.64 (68.09)	5.03 (4.83)	7.18 (6.10)	14.36 (13.98)			
SAPy	73.43 (72.71)	6.07 (5.08)	13.78 (14.13)				
Cu(II)-SAA-H_2O	53.26 (52.41)	3.84 (3.43)	5.32 (4.32)		20.08 (19.81)	6.86 (5.61)	2.06
Cu(II)-SAP-H_2O	55.36 (54.40)	4.08 (3.76)	5.32 (4.78)		22.46 (21.71)	5.92 (6.15)	1.98
Cu(II)-SATP-H_2O	51.08 (50.55)	4.18 (3.59)	4.06 (4.53)	9.68 (10.38)	22.06 (20.57)	6.12 (5.83)	2.01
Cu(II)-$(SAPy)_2$	63.13 (62.94)	5.16 (4.40)	13.34 (12.23)		14.27 (13.87)		1.73

Infra Red Spectral Studies

The free schiff bases showed the characteristic azomethine and -OH frequencies at 1650 and 3500cm^{-1} respectively. The former shifted towards lower frequency region by 80-50cm^{-1} in the spectra of copper complexes due to involvement of N of-C=N-group[7,8], while the ϑOH band disappeared showing that it had been deprotonated during chelation. In the case of SAA and SAPy the carboxylic and pyridine nitrogen frequencies are observed at 1560 and 1170 cm^{-1}, which shifted lower region in the spectra of their copper complexes indicating the involvement of carboxylic oxygen and pyridine nitrogen in coordination[9]. A weak band around 2570 cm-I was probably due to ϑSH in SATP, also disappeared in the spectra of its complex showing its deprotonation during complexation. The presence of coordinated water molecule in all the complexes except copper-SAPy was confirmed by the appearance of bands around 3450, 850 and 700cm^{-1} probably due to -OH streching, rocking and wagging vibrations. The appearance of some new bands in the spectra of metal complexes in the region of 450, 370 and 310 cm^{-1} were probably due to the formation of M-O, M-N and M-S bonds respectively through complexation.[10]

Electronic Spectra and Magnetic Studies

The magnetic moment values(table-I) of all the complexes lies between I.73 to 2.06 BM indicates the square planar geometry of all the complexes except Cu(II)-SAPy, which was assigned to octahedral geometry. These geometries were further confirmed by their electronic spectra. In the case of SAA, SAP and SATP complexes, only a band seen around 16500 cm^{-1} may be due to the combination of the two transitions-

$$^2B_{Ig} \dashrightarrow {}^2A_{Ig} \qquad \text{and} \qquad ^2B_{Ig} \dashrightarrow {}^2E_g$$

confirms [11,12] the square planar geometry of these complexes whereas in the case of SAPy complex the band around 15450-14800 cm^{-1} probably due to following transition -

$$^2E_g \dashrightarrow {}^2T_{2g}$$

indicated a distorted octahedral geometry[13,14].

BIOCIDAL ACTIVITY

The synthesized ligands and their copper complexes were screened for their antimicrobial activity in slant and broth culture media by serial dilution method[15]. The antibacterial activity of the compounds was tested both on G+(S. aureus) and G⁻(E.coli) bacteria and on the three common fungi, (A.niger, A.nidulense and C.albicans). The incubation period for the bacteria was kept 24 hours at 37°C and for fungi 96 hours at 28°C. A comparative study of the MIC values (table-2) of the ligand to their copper complexes indicate that the activity of organic ligand was considerably increased on complexation with metal ion against the same microbe and under identical experimental conditions, probably due to their more lipophilic nature or due to combined activity effect.

The pharmacological specially the anti-inflammatory activities of these compounds are in progress in this laboratory.

ACKNOWLEDGEMENT

This investigation was supported by the Council of Scientific & Industrial Research, New Delhi (India) under the sponsored research project.

TABLE-2 Minimum Inhibitory Concentration (MIC) in µg/ml of the compound

Compound	BACTERIA		FUNGI		
	S.aureus	E.coli	A.niger	A.nidulense	C.albicans
$(CH_3COO)_2Cu.H_2O$	Nil	Nil	Nil	Nil	Nil
SAA	> 100	75	>100	>100	100
SAP	50	100	>100	100	75
SATP	100	75	100	75	50
SAPy	75	>100	>100	75	>100
Cu(II)-SAA-H_2O	6.25	25	75	>100	100
Cu(II)-SAP-H_2O	25	12.5	>100	50	25
Cu(II)-SATP-H_2O	50	12.5	25	6.25	50
Cu(II)-$(SAPy)_2$	12.5	50	50	12.5	75

REFERENCES

1. Hodnett E.M : Dunn W.J J.Med.Chem. 1970,13,768
2. Billmann H.J : Schmidgall R.L J Pharm.Sci. 1970,59, 1191
3. Sharma R.C. : Parashar R.K Unpublished results
4. Sharma R.C : Sharma R.S Tripathi S.P Curr. Sci.1983,52,410
5. Sorenson J.R.J : J.Med. Chem. 1976,19,135
6. Albert A. In : Selective Toxicity, 4th ed., Methuen, London, 1968
7. Csazar J : Kiss Acta Chem. Acad.Sci.1973,78,17
8. Davies M : Jones R.L. J. Chem. Soc.1954,121
9. Dutta R.L : Husain M.Md. Indian J.Chem.1984, 23A, 30
10. Ferraro J.R. In : Low Frequency Vibrations of Inorganic and Co-ordination Compounds; Plennum Press, New York, 1971
11. Byers W : Lever A.B.P. Parish R.V. Inorg. Chem.1968, 7, 1835
12. Bjerrum J : Ballhausen C.J : Jorgensen C.K. Acta.Chem.Scand. 1954, 8, 1275
13. Lever A.B.P. In : Inorganic Electronic Spectroscopy, Elsevier, Amsterdam, 1968
14. Bullock J.I : Jones S.L. J.Chem.Soc(A),1971, 2357
15. Donald C.G : Williams A.R. In : Assay Methods of Antibiotics, A Laboratory Manual, Medical Encyclopedia INC, 1955

DETERMINATION OF HUMAN MATERNAL SERUM COPPER LEVELS BY AAS AND DCP SPECTRA-SPAN VI™ TECHNIQUES.

Sophie Schmidt,[1] Martin Risk,[1] Jerry Simmons[2]

[1]OB/GYN, [2]Laboratory Medicine
University of South Dakota, School of Medicine
VA Medical Center, Sioux Falls, South Dakota

INTRODUCTION:

Previous work in our department showed maternal serum copper is elevated significantly during labor[1,2]. A preliminary study on 26 patients with threatened preterm labor indicated significant correlation between maternal serum copper levels, effect of tocolysis, and labor outcome. Those with levels < 2.5 ppm serum copper were arrested with or without tocolysis and their pregnancies were prolonged. Patients with levels > 2.5 ppm showed either arrest with tocolytic therapy and decreased copper levels, or continued elevated copper and progressive labor despite tocolytic therapy[3]. We plan to extend this study using a larger patient population and analyzing serial samples during labor.

This study examines several methods for copper determination in serum samples for sensitivity, ease, and accuracy. Flame Atomic Absorption (AA) and Direct Current Plasma Spectra-Span VI™ (DCP) were compared on a serum pool. Maternal serum copper levels were studied on a patient during the course of induced labor.

MATERIALS & METHODS:

Instruments: 1) Atomic Absorption: Perkin-Elmer model 2830 with multi-element hollow cathode lamp. 2) DCP: Beckman-ARL Spectra-Span VI™ Direct Current Plasma Atomic Emission instrument.

The Spectra-Span VI™ is a rapid scanning, high resolution spectrometer. It differs from the better known ICP multielement analyzer, to which it is a direct rival, primarily by the energy source.

The sample is aspirated through a nebulizer into an argon thermal plasma constrained at the juncture of three electrodes. Combustion occurs at this region, in which the excited plasma and sample react at ~8,000°C. During return to ground state, photons characteristic of each element are emitted; they are separated by wavelength by an Echelle grating and prism. By preselecting each element's wavelength, concentration is determined by emission intensity[4].

The model VI DCP can quantitate up to 60 elements per sample in the sequential mode, using characteristic wavelengths with computer control. It is important for routine analysis that this instrument is rapid & convient to operate.

There is little published on DCP in biological analysis; several application notes are available[4]. A recent review of DCP analysis provides operation principles of this technique[5].

Standards: Calibration standards were prepared from stock AA standards of 1000 ppm metal (Banco Scientific) or from ultra high purity standards (Sherba). Stock copper solutions at levels which bracket typical serum concentrations of 1 ppm were prepared in 0.2% nitric acid. Standards were prepared in both aqueous and 10% glycerol solutions.

Reagants: High purity nitric acid (Ultrex grade, J. T. Baker) was used in all cases. All labware was prewashed in 10 % nitric acid. Glycerol was obtained from MCB. Water was distilled and deionized by a four cartridge Millipore™ system (dd water).

Calibration Standards: WasteWatR™ trace metal standard (ERA Labs) was used for checking the aqueous standards . The copper concentration of the reconstituted standard was 0.286 - 0.30 ppm (manufacturer's value).

Cation-Cal™ (American Scientific-Dade) control simulates human serum using suitable concentration of bovine serum albumin with known amounts of trace elements added. The copper level for this standard is determined to be 1.93 ppm (manufacturer's value).

Serum: All blood was collected with the approval of the appropriate hospital & university committees; detailed consent forms were provided to the subjects prior to blood collection. A multidonor serum pool was collected and stored at 5° C for analysis for method verification.

Maternal blood was drawn from a 37 week gestation grand-multiparous woman who was induced by amniotomy after biochemical confirmation of fetal lung maturity. The woman had some difficulty with mild chronic hypertension during the pregnancy; this suggested the need for induction of labor. The first four samples were drawn by venapuncture, the remainder by IV site after flushing. Samples were collected into trace metal analysis Vacutainers™ (Becton-Dickenson) without additive.

Preparations:

Washings: All vessels were washed with 10% nitric acid and rinsed with dd water. .

Tubes: Falcon™ 2070 polyproplyene tubes were used for storing all samples and standards. Blood was collected in trace metal free Vacutainer™ tubes, centrufuged 2,000 g, 10 minutes, & the serum transfered to polyproplyene capped tubes for analysis.

Analysis: The serum pool was analyzed 1) undiluted 2) diluted 1:2, 1:2.5, 1:3, and 1:5 in 0.2% nitric acid. In addition, a pool aliquant was hydrolyzed thouroughly by preparation in 5 ml Teflon screw cap vessels (Savillex), diluted 1:2:1 serum/nitric acid/ water at room temperature or heated 130° C, 1 hour[6]. Maternal serum was diluted1:2.5 with 0.2% nitric acid.

Method of Standard Additions (MSA): serum was diluted 1:2 with either 1.0, 2.0, or 2.5 ppm copper standard and initial value calculated by formula[7].

Settings: AA wavelength was set at 324.8 nm. All readings were calibrated with a 2.0 ppm copper standard in acidified water or acidified 10% glycerol. The DCP was realigned at 324.728 nm at the beginning of each run. The samples were run in the multi-element setting useing a prelabeled file interspersing standards and references to recalibrate the standard curve and correct for drift (see graph 1).

RESULTS:

Pooled serum copper was determined by flame atomic absorption using one point calibration as recommended by the manufacturer. DCP analysis of whole, diluted, and hydrolyzed serum was performed by sample nebulization into the plasma combustion region. A computer generated five point calibration curve was used for sample determination. For both instruments, all determinations were the mean of three replicates.

Several methods were employed for sample preparation. Table 1 compares these methods on both instruments. Good results were obtained on the DCP with undiluted serum, serum diluted 1:2 and 1:3, acid digested serum at room temperature, and acid digested serum that was heated for one hour. The preparation procedure for flame AA can be critical; in our case, perhaps due to few interferences with copper, this was not the case. Acid digestion by the method of Passey et al[6] did not apear to confer advantages for our instrumental methods; this procedure requires extra effort & correction for vapor loss that is not warrented with DCP analysis. It is likely to be a helpful method for some AA analyses, especially for lipemic samples.

The results showing comparison of the different sample preparations on both the AA and DCP are given in table 1. All values are the mean of three replicates.

DCP drift can be a major factor for introducing errors. This problem can be obviated by the insertion of recalibration standards within the analysis sequence every few samples (at least every 10 minutes), as demonstrated by graph 1. The model VI DCP is much more stable than other versions since the optic system is equilibrated thermally.

Matrix matching is highly important for accurate analysis. Table 2 shows that the Cation-Cal™ and WasteWatR™ standards give accurate values only if there is proper matrix matching; that is, calibration standards must be prepared in a medium with viscosity corresponding to that of the sample. If the calibration standards medium is not matched to be consistant with the sample solution characteristics, misleading values will be obtained.

Samples	AAS results	DCP results
undiluted serum	1.00	1.088
serum 1:2 dilution	0.94	1.067
serum 1:3 dilution	1.04	0.968
serum/acid/H_2O 1:2:1	1.02 (1.00)	1.442 (1.440)
serum/acid/H_2O 1:2:1 with heat	1.26 (1.00)	1.214 (1.440)
MSA* calculation	1.07 (1.06)	1.220 (1.030)
Cation-Cal™	1.97	2.040
Waste-WatR™	0.33	0.292

*MSA = Method of Standard Additions; mean of three replicates

Table 1. Comparison of AAS and DCP copper values in serum pool. The numbers in parentheses are the values of a second serum pool run at the same time.

A time-course study of maternal serum copper levels was obtained from a volunteer patient. Samples were taken at early stages of contractions, during induced labor and after delivery, as shown in graph 2. It should be noted that throughout pregnancy, maternal serum copper levels show a continuous increase. This is accentuated during normal labor. However, it is not known how differing physiological states may affect copper transport & levels, such as in cases with induced labor (amniotomy), as was the situation with the patient we studied . Therefore, this study will be useful as a initial examination of cases of this nature.

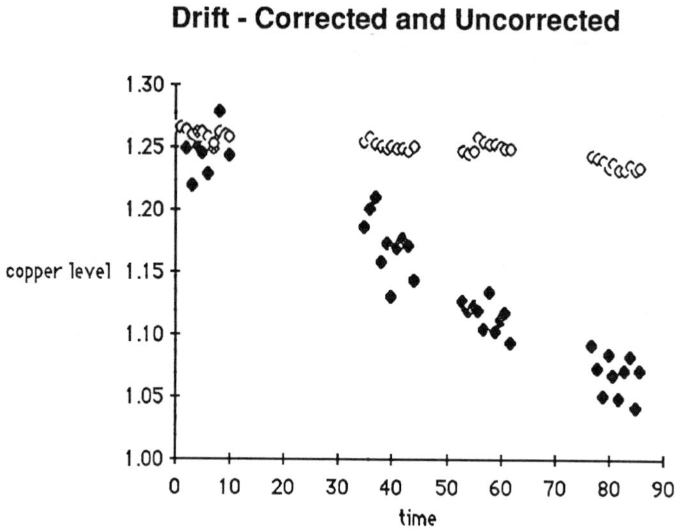

Graph 1. Drift with (o) and without (•) standard and reference recalibration during DCP analysis.

	Known value	10% glycerol standard	dd H2O standard
Cation-Cal™	1.93	2.08	4.420
Waste-WatR™	0.286- 0.330	0.2088	0.304

Table 2. Cation-Cal and WasteWatR™ checks on DCP in glycerol and water standards.

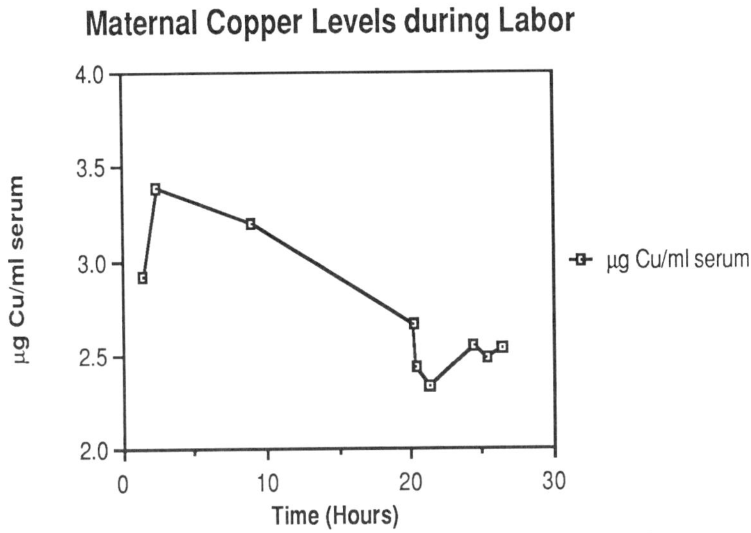

Graph 2. Patient serum copper levels during labor

pt.1 just before amniocentesis
pt.2 just after amniocentesis
pt.3 upon admission to hospital
pt.4 just prior to amniotomy
pt.5 1 hr post-amniotomy
pt.6 during latent labor
pt.7 during active labor
pt.8 at delivery
pt.9 1 hr post delivery

DISCUSSION:

This study is a preliminary step in our analysis of the function and medical significance of maternal copper levels in gestation and labor. This initial work is devoted to meticulous method development, including the comparison of several techniques to determine the most suitable analysis conditions for further studies. We have compared AA and DCP techniques; work in preparation will describe HPLC ion chromatography for serum trace elements.

It is important for our work in trace metals to determine the significance of manipulative parameters. For example, it is clear that matrix matching (by viscosity and perhaps total organic content) is essential in both AA & DCP (see table 2). Since AA is recognized as the standard method for trace metal analysis, alternate techniques should be compared with it on the same samples to demonstrate validity of the rival procedure. Our results on pooled serum and known standards demonstrate that the less well known DCP instrument is as accurate and reproducible as conventional flame AA. DCP has sensitivity intermediate between flame and graphite furnace AA (data not shown - determined for copper). It has several advantages: it is at least as convenient to operate as flame AA, it is quite sensitive, and is a more robust instrument. The most important point in its consideration is that DCP is amenable to rapid multielement determination.

The amount of sample required for DCP in the standard configuration is about 2.0 ml; several elements can be quantitated from this. Flame AA determination requires around 0.2 ml for each element to be measured. It should be noted that our results are obtained on a 1:2.5 dilution, so that limited volume samples are still ameable to DCP multielement analysis. We are in the process of introducing minor hardware modifications to permit the determination of 40 or more elements from one ml total volume.

Prior work with the DCP in biological analysis has used slightly different sample preparation methods[8,9]. It should be noted that most investigators use other models (especially III or IIIB DCP) less well stabilized than our model VI. This & minor preparative factors may influence the choice of methodological details. It must be noted that in all trace element studies, minute aspects of sample preparation are of the greatest importance for validity of results[7,10].

With regard to the function of copper in gestation, it is well known that there is a three to four fold increase in copper levels from beginning to end of pregnancy[2]. This was demonstrated in our maternal samples, taken around delivery. In standard physiolgical states, humans have ~1 μg/ml copper; our sample was elevated three-fold. There was elevation of copper levels around the time of amniocentesis, but a reduction before delivery. This patient appeared tense at the time of the amniocentesis, but once labor began she relaxed (probably correlated with being multiparous). Pitocin was not used. It remains to be investigated if stress or a specific labor related mechanism is responsible for elevation & variation of copper levels during gestation & the final hours of labor.

Acknowledgements:
To the maternal patient for her interest and cooperation.
USD School of Medicine Office of Student Affairs, & Aberdeen Area Indian Health Service for support.
Dr. Jim Thomas, Barb Dudgeon, and staff at Sioux Valley Hospital for sample collection.
Particular thanks to Prof. Loren Petersen for support & suggestions.

REFERENCES:

1. Kundu, N.; Wachs, M., Iverson, G. and Petersen, L. Fetal and Maternal Serum Copper Levels Before and During Labor in Normal and Complicated Pregnancies. Obstet. Gynecol. 1981, 57:313-7
2. Kundu, N.; Parke, P., Palmer, I., Olsen, O., & Petersen, L. P. Distribution of Serum Selenium, Copper, & Zinc in Normal Human Pregnancy. Arch. Env. Health 1985, 40:268-73
3. Petersen, L. and Kundu, N. Maternal Serum Copper in the Management of Threatened Preterm Labor. Proc. Fifth Annual Conference of the Society of Perinatal Obstetricians, 1985, Abs. #232, P.247
4. Kacsir, J. M.; Kinsey, W. J., & Seelt, R. C. Determination of Trace Metals in Human Whole Blood, Urine, & Serum by DC Plasma Emission Spectroscopy. Spectra-Span VI™ Instructions & Application notes. ARL-Beckman Instrument, Irvine, CA.
5. Frank, A. & Petersson, L. Direct Current Plasma - Atomic Emission Spectrometer as a Simultaneious Multi-Element Tool for Analysis of Biological Materials. Kemia-Kemi 1985, 5: 426-30
6. Passey, R.; Maluf, K. and Fuller, R. Quantitation of Zinc in Nitric Acid Digested Plasma by Atomic Absorption Spectrophotometry. Anal. Biochem. 1985, 151:462-5
7. Van Loon, J. C. Sélected Methods of Trace Metal Analysis: Biological and Environmental Samples. John Wiley and Sons, New York. 1985
8. Roberts, N. B.; Fairchough, D., McLoughlin, S., & Taylor, W. H. Measurement of Cu, Zn, & Mg in Serum & Urine by DCP. Ann. Clin. Biochem. 1985, 22:533-8
9. Gardiner, P. E.; Bratter, P., Negretti, V. E., & Schulze, G. Use of DCP as a Detector in Gel Filtration Chromatography of Biological Fluids. Spectrochemica Acta 1983, 38B:427-36
10. Ericson, S.P.; McHalsky, M. L., Rabinow, B. E., Kronholm, K. G., Arceo, C. S., Weltzer, J. A., & Ayd, S. W. Sampling & Analysis Techniques for Monitoring Serum for Trace Elements. Clin. Chem. 1986, 32:1350-6

MODULATION OF IMMUNE FUNCTION AS A CONSEQUENCE OF COPPER DEPRIVATION

D.W. Roberts[1], V. Kishore[2], J.B. Barnett[3], R.W. Benson[1], and J.R.J. Sorenson[4]

National Center for Toxicological Research[1], FDA, Jefferson, AR; Departments of Microbiology and Immunology[3], College of Medicine, and Biopharmaceutical Sciences[4], College of Pharmacy, University of Arkansas for Medical Sciences, Little Rock, AR; and Division of Basic Pharmaceutical Sciences[2], College of Pharmacy, Xavier University of Louisiana, New Orleans, LA.

INTRODUCTION

Considerable evidence has been presented regarding the complex relationships between nutrition, metabolism, responses to inflammation,[1,2] susceptibility to neoplasia, and immunocompetence.[3] These relationships, coupled with the potential effects of copper bioavailability on the outcome of toxicology studies,[4] led us to evaluate the effects of diet and copper deprivation on immunocompetence.

Reports have indicated an association between copper deficiency, diminished host-resistance and compromised immune function.[1,5-7] However, these results have not been consistent and the biochemical basis of immunosuppression has not been established; probably as a result of complex interactions between copper, zinc, and iron.[8]

Interpretation of animal studies where dietary copper is withheld is confounded by concomitant effects on the bioavailabilty of other metalloelements, such as iron, which are known to be important for normal immune function. In light of the known variation in the quantity and quality of nutrients in many dietary sources and complex dietary interactions affecting copper absorption, metabolism and mobilization, we elected to study immune responses in rats fed a standard copper deficient semipurified diet supplemented with drinking water of known copper content. Since differences in protein source can alter immune function,[9,10] we included a conventional diet of defined copper content but mixed protein source.

MATERIALS AND METHODS

Experimental Design

Equal numbers of three-week-old male Sprague-Dawley rats were randomly assigned to four diet groups. All animals were housed, 1 per cage, in stainless-steel drop-through cages and provided with 12-hr light and dark cycles, an ambient temperature of approximately 22°C, and either distilled deionized water (DDW) or tap water ad libitum. The copper deficient (CuD) group received semipurified diet and DDW; the marginally copper deficient (MCuD) group received semipurified diet and DDW containing 0.4 μg Cu/ml (as cupric acetate, E. Merck Co., USA); the copper sufficient (CuS) group received semi-purifed diet plus DDW containing 4 μg Cu/ml; and the control rodent chow (CRC) group received Purina Certified Rodent Chow® plus tap water. Rats were observed daily and weighed weekly. Immune function was evaluated between 12 and 20 weeks of age (days 65-125 on diet). Rats were sacrificed after 140 days on their respective diets and copper, zinc, and iron levels were measured in tissues. Comparisons of interest included effect of semipurified diet, CuS vs CRC; and effects of copper sufficiency within the semipurified diet groups (CuD, MCuD, and CuS).

Diets

Copper-deficient, semipurified diet purchased from ICN Nutritional Biochemicals, Cleveland, Ohio, had a copper content of 0.5 μg of copper per gram of diet by atomic absorption spectrophotometry. The composition of the semipurified diet was identical to the AIN 76 mixture[11] except that the biotin level was increased to 4 mg/kg diet to avoid avidin-induced biotin deficiency and copper was left out of the AIN mineral mix. This diet provided Zn and Fe levels of 32.5 μg/g and 34.6 μg/g diet, respectively.

The control diet, Certified Rodent Chow® (#5002, Ralston Purina Co., St. Louis, MO) had a guaranteed analysis of 13.3 ppm Cu, 52.4 ppm Zn, 180.0 ppm Fe, a mixed protein source which included soybean meal, fish meal, and casein, and an unspecified amount of cane molasses. DDW used for drinking water did not contain any detectable copper, zinc, or iron when analyzed by flame atomic absorption spectrophotometry. Tap water supplied to animals fed the CRC diet contained 1.7 μg copper/ml.

Evaluation of Immune Function

The capacity to respond to the contact sensitizing agent, 4-ethoxymethylene-2-phenyloxazol-5-one (oxazolone; BDH Chemicals, Poole, England), was evaluated in 15-week-old rats as a correlate of cell-mediated immunity.[12] T-Cell competence was assessed after 65 days on diet by sensitizing rats on a shaved flank with 50 μl of 10% (w/v) oxazolone in acetone and then challenging them nine days later with 10 μl of 1% oxazolone in acetone (w/v) applied to the dorsal surface of the right ear and measuring swelling of the ear.

The T-cell independent (TI) antibody response to Type III Pneumococcal Polysaccharide (S3) was evaluated after animals were fed their respective diets for 83 days, when rats were immunized by intraperitoneal (ip) injection of 25 μg S3 and 5 lime flocculating (Lf) units of tetanus toxoid (TT, Connaught Lab. Ltd.). The antibody response to S3 was measured by RIA as described by Benson and Roberts.[13]

The T-cell dependent (TD) antibody response to TT was evaluated in rats given a secondary immunization consisting of 5 Lf TT injected 28 days after combined S3 and TT primary immunization. Serum was obtained 14 days after secondary immunization (day 125 on diet) and levels of IgG anti-TT were quantitated in a modified micro-ELISA assay.[14]

Tissue Metal Concentrations

Levels of Cu, Fe, and Zn in plasma, liver, kidney, and spleen were determined using atomic adsorption spectrophotometry (IL 157, Instrumentation Laboratory, Inc., USA). Tissue samples were analyzed following wet ashing in a 2:1 mixture of concentrated HNO_3 using methods found to be interference free.[15]

Statistical Evaluation

A one-way analysis of variance (ANOVA), with group as the effect of interest, was the primary method of statistical analysis. Standard Pearson product moment correlations were calculated between variables (plasma and tissue metalloelement levels, and immune response data).

RESULTS AND DISCUSSION

Our initial intent was to examine the effect of copper deprivation on immunocompetence with the Certified Rodent Chow® control diet serving to show, by comparison with the semipurified diet, that copper was the major important variable. Cu deficiency was verified and levels of the related micro-nutrients zinc and iron were determined by measuring metalloelement levels in plasma, liver, kidney and spleen (Fig. 1). By the criteria of plasma, liver, kidney, and spleen copper, the CRC and CuS diets provided the intended comparable levels of copper nutriture.

As expected, animals fed either CuD or MCuD diets had significantly decreased levels of Cu in every tissue examined compared to animals fed either CuS or CRC (all p values < 0.0005; Fig. 1). Animals fed the CuS or CRC diet had 17X greater concentrations of plasma Cu than animals fed the CuD diet. Animals fed CuS or CRC diets had similar levels of Cu in all organs tested. Liver, kidney, and spleen Zn concentrations were not affected by changes in the level of dietary copper and were similar for the CuS and CRC diet groups.

The concentration of Fe in tissues and plasma was affected both by diet type and level of dietary Cu. In plasma, kidney, and spleen, the pattern was similar and characterized by a decrease in Fe in rats fed the semipurified diet compared to those rats fed CRC. Within the three semipurified-diet groups, the plasma, kidney, and spleen Fe decreased as levels of dietary Cu decreased ($p < 0.001$). Liver Fe concentrations showed a unique pattern which suggested they were affected by multiple factors. Liver Fe was decreased in two of the semipurified diet groups, MCuD and CuS, as compared to the CRC group

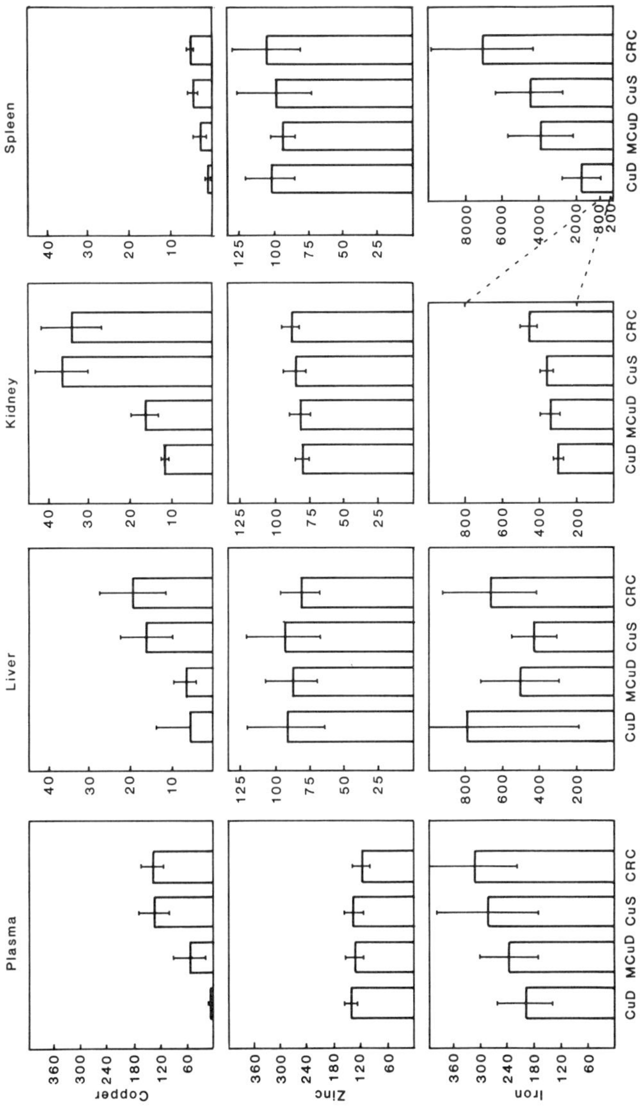

Fig. 1. Effect of diet and dietary copper content on plasma and organ levels of copper, zinc, and iron. Plasma metalloelement concentrations are expressed as μg/dl ± S.D. and tissue metalloelement contents are expressed as μg/g dry weight ± S.D. Note the 10-fold scale change for spleen iron.

consistent with the semipurified-diet having decreased levels of Fe. This was similar to the pattern seen for plasma, kidney, and spleen Fe. In contrast, the CuD group showed a marked increase not only in liver Fe as compared to MCuD and CuS groups, but also in the variability associated with these data.

Rats maintained on CuD, MCuD, or semipurified CuS diets from 3 weeks of age until sacrifice at 24 weeks-of-age grew normally at similar rates and, with the exception of increased mortality in the CuD and MCuD groups, did not exhibit overt signs of toxicity. Animals fed the CRC diet weighed slightly less than rats fed semipurified diet; and within the semipurified diet groups, copper sufficiency did not significantly affect weight. Mean group weights for rats fed CuD, MCuD, and CuS diets were 104%, 113%, and 109% of the CRC (control) group, respectively (Fig. 2A). Decreasing levels of dietary copper in the semipurified diet groups were associated with increased mortality (Fig. 2B). Final survival by group was CuD, 60%; MCuD, 80%; CuS, 90%; and CRC, 100%. Whereas copper deficiency had a dose-dependent effect on survival, there was no apparent effect of copper deficiency on the weight of surviving animals (Fig. 2A).

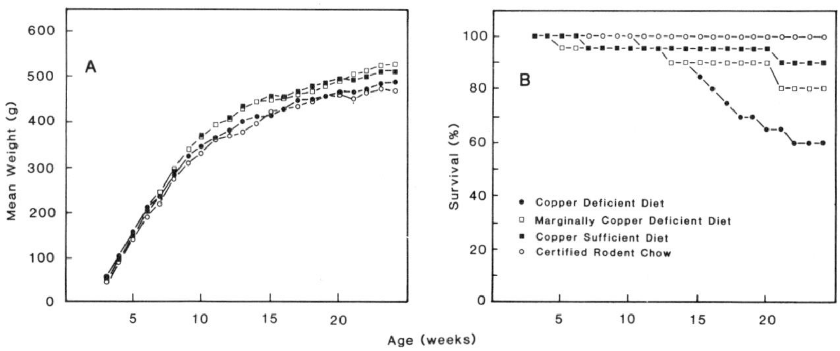

Fig. 2. Effect of diet and dietary copper content on weight gain and mortality. For each group, initial n = 20 at 3-weeks-of-age. At 24 weeks of age the number of surviving animals by group was: CuD, 12; MCuD, 16; CuS, 18; and CRC, 20.

Under the conditions described, immunization and the manipulations necessary to evaluate immunocompetence did not result in persistent alterations of tissue or plasma levels of Cu, Zn, or Fe (data not presented).

When the ear-swelling response was measured at 24 hours post-challenge in oxazolone-sensitized rats (Fig. 3A), the semipurified diet, but not dietary Cu content, was associated with a significantly enhanced CMI response ($p < 0.025$). Oxazolone-sensitized rats fed the CuS diet had greater ear-swelling responses at 24, 48, 72, and 96 hours after oxazolone challenge than similarly treated rats fed the CRC diet, and these differences were significant at the 24- and 48-hour measurements ($p<0.025$). In order to control for possible effects of diet and dietary copper content on non-specific inflammation, separate animals received a single application of the irritant croton oil on the ear. Rats fed the CuD diet had smaller, immunologically nonspecific, ear-swelling responses to croton oil at 24, 48, 72, and 96 hours than rats fed MCuD, CuS, or CRC diets, suggesting an impaired inflammatory response in the CuD fed animals (Fig. 3B).

Both copper deficiency and the semipurified diet reduced the capacity of rats to respond to the T-cell independent antigen S3. Fig. 4 shows results obtained from rats in each diet group 5 days after ip immunization with 25 μg of S3. Within the semi-purified diet groups, rats deprived of copper (CuD) had smaller responses than either MCuD or CuS rats ($p < 0.005$). When the capacity to respond to S3 was compared between rats receiving the CuS diet and the CRC diet, the CRC group was found to produce a higher response ($p < 0.001$). The capacity to respond to S3 showed a strong positive correlation with levels of spleen iron ($p < 0.005$), and this relationship is also shown in Fig. 4.

Capacity to make an IgG anamnestic response to the T-cell dependent antigen TT was evaluated by ELISA 14 days after secondary immunization. Rats receiving the semipurified diets, regardless of copper content, made higher responses to TT than rats fed CRC (Figure 5). There was a negative correlation between capacity to respond to TT and levels of spleen copper ($p < 0.01$).

Dietary copper deficiency beginning at 3-weeks-of-age

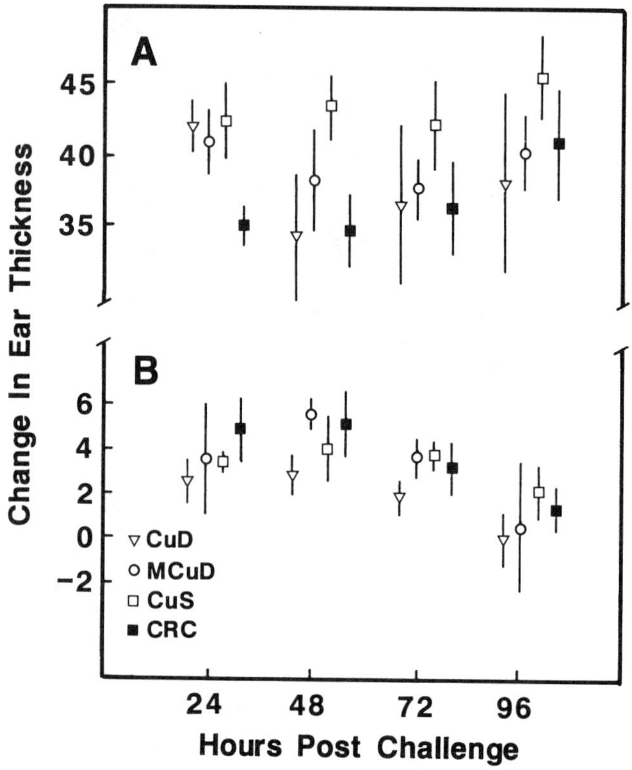

Fig. 3. Effect of diet and dietary copper content on response to the contact allergen oxazolone (Panel A) and the irritant croton oil (Panel B). Data expressed on mean change in ear thickness (10^{-2} cm $\pm$ SEM) for 9-10 animals per point (Panel A) and 4-5 animals per point (Panel B).

was associated with decreased capacity to respond to the TI antigen S3, and decreased levels of plasma, kidney, and spleen iron. Levels of liver iron were markedly increased and variable in rats fed the CuD diet, and this result was similar to results reported previously for mice chronically deprived of copper.[16] Dietary copper deficiency beginning at 3 weeks of age had no effect on spleen weight (not presented), caused a minimal increase in plasma zinc,

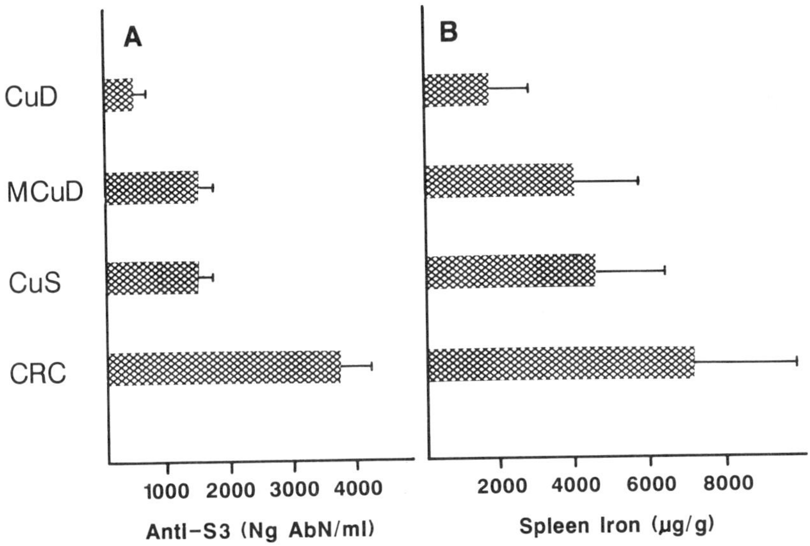

Fig. 4. Correlation between capacity to respond to the T-independent immunogen S3, expressed as nanograms of antibody nitrogen (NgAbN/ml ± SD), and level of spleen iron, expressed as μg Fe/g dry spleen ± S.D.

had no effect on liver, kidney, or spleen zinc, and had no effect on the TD antibody response to TT or on the TD cell-mediated response to oxazolone. These findings contrast somewhat with results of previous studies employing C58 mice deprived of copper from birth.[6,7] In addition to the possibility that rats and mice may respond differently to copper deficiency, it is likely that differences in the temporal relationship of copper deprivation to the ontogeny of immunologic systems is critical. Prohaska and Lukasewycz[6,7] reported suppression of both humoral and CMI in mice deprived of copper from birth. In a similar study, Prohaska et al.[16] found decreased growth, thymic atrophy, splenomegally, and morphologic changes in lymphoid tissues as a result of copper deprivation during the early postnatal period. Together, these observations suggest that capacity to make

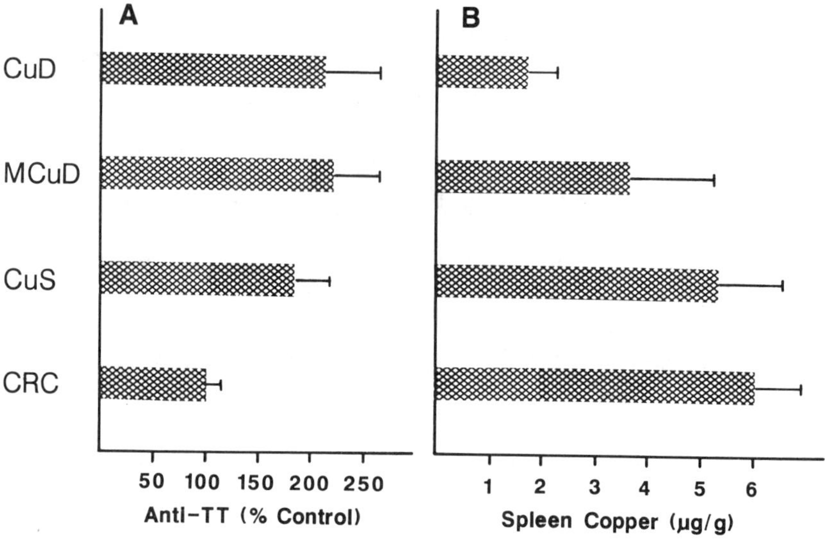

Fig. 5. Correlation between capacity to respond to the T-dependent immunogen TT (mean ± SEM) and levels of spleen copper (mean ± S.D).

TD immune responses is relatively vulnerable to copper deprivation during the early postnatal period, and relatively resistant to copper deprivation in adult animals. The fact that capacity to respond to TI antigens develops before the capacity to respond to TD antigens,[17] coupled with the observation in the current study that responses to TD antigens were unchanged by copper deficiency after three weeks of age, indicates that the effects of copper deficiency on capacity to respond to the TI antigen S3 are due to effects on the acquisition or expression of TI immunity, not to effects on immunologic development.

In addition to the immunologic differences associated with copper insufficiency within the semipurified diet group, there were differences in immune function between rats fed the two copper sufficient diets. Compared to rats fed the CRC diet, rats fed the CuS semipurified diet had increased responses to the TD immunogens TT and oxazolone, and decreased responses to the TI immunogen S3, accompanied by decreased levels of plasma, liver, kidney, and spleen iron. Recent reports suggest that the immunologic differences between groups fed the CuS semipurified diet and the CRC diet are likely attributable to differences in iron content,[18-21] carbohydrate source,[22-25] or protein source.[9,10]

When copper intake is restricted, plasma levels of ceruloplasmin drop, and ceruloplasmin is not available to reoxidize stored iron so that it may be bound by transferrin.[26] This mechanism could explain the variable accumulation of iron in livers of rats fed the CuD diet and in mice chronically deprived of copper.[16] When ceruloplasmin is absent, iron cannot be recycled from macrophages and a hypoferremia develops.[27] In the present study, splenic hypoferremia in individual animals had a close positive correlation to TI but not to TD immune function.

The semipurified diet contained 31% sucrose as the carbohydrate source and 20% eggwhite solid as protein source. In contrast, the CRC diet contained an unspecified amount of cane molasses, and a mixed protein source including soybean meal, fish meal, and casein. Semipurified diets containing large amounts of reducing sugars as the carbohydrate source have been reported to exacerbate symptoms of copper deficiency.[18-21]

Differences in type of dietary protein have been shown to markedly affect the capacity of mice to make both TD and TI immune responses. The findings that the CuS diet as compared to the CRC diet caused a decrease in TI antibody responses and a concomitant increase in TD antibody and CMI responses suggests that dietary differences other than copper are affecting a regulatory mechanism common to both T- and B-dependent responses.

Adequate dietary zinc is known to be essential for development and subsequent function of immunologic systems.[28] The similarity of plasma and organ levels of zinc (Fig. 1) regardless of dietary copper content indicate that immunologic differences observed in rats fed the different diets were probably not the result of changes in the bioavailability of zinc.

Data reported here emphasize the importance of metalloelements and other dietary components as factors to be considered in toxicological testing, especially where the endpoints being measured could be affected directly or indirectly by metalloenzymes or altered immunologic competence. The complex interactions among metalloelements and components of diet, coupled with the difficulty in replicating diets, suggest the use of functional tests to monitor the adequacy of diet. In this context, we propose that the capacity to respond to the TI antigen S3 may serve as a valuable adjunct in the evaluation of the capacity of diet to support immune function and in the evaluation of nutritional adequacy. In combination with previous work,[6,7,16] these data suggest that copper deprivation has different effects on developed, as compared to developing, immune systems and that the temporal relationship of metalloelement deprivation to ontogeny and to the evaluation of immune function are of critical importance in the interpretation of results.

ACKNOWLEDGEMENTS

This work was supported by the International Copper Research Association, the Kroc Foundation, the Max and Victoria Dryfus Foundation, Elsa U. Pardee Foundation, and Arthur Armburst Cancer Research Fund. V.K. is also supported by NIH Grant 08008-1451, and J.B.B. and J.R.J.S are supported by P.H.S. grant number CA40380 awarded by the National Cancer Institute, D.H.S.S. We thank Cindy Hartwick for her help in preparing this manuscript.

REFERENCES

1. Kishore, V.; Latman, N.; Roberts, D.; Barnet, J.; Sorenson, J. Agents and Actions 1984, 14, 274.
2. Sorenson, J.; Inorg. Perspect. Biol. Med. 1978, 2, 1.
3. Gross, R.; Newberne, P. Physiol Rev. 1982, 60, 188.
4. Farkas, C. Sci. of the Total Environment 1978, 9, 149.
5. Anon. Nutr. Revs. 1982, 40, 107.
6. Prohaska, J.; Lukasewycz, O. Science 1981, 213, 559.
7. Lukasewycz, O.; Prohaska, J. J. Natl. Cancer Inst. 1982, 69, 489.
8. Newberne, P. Lab Animal 1975, 4, 20.
9. Bounous, G.; Kongshavn, P. J. Nutr. 1985, 115, 1403.

10. Bounous, G.; Letourneau, L.; Kongshavn, P. J. Nutr. 1983, 113, 1415.
11. American Institute of Nutrition. J. Nutr. 1977, 107, 1340.
12. Asherson, G.; Ptak, W. Immunology 1968, 15, 405.
13. Benson, R.; Roberts, D. J. Toxicol. Environ. Health 1982, 10, 859.
14. Roberts, D.; Wolff, G.; Campbell, W. Proc. Natl. Acad. Sci. 1984, 81, 2152.
15. Sorenson, J.; Melby, E.; Nord, P.; Petering, H. Arch. Envir. Health 1973, 27, 36.
16. Prohaska, J.; Downing, S.; Lukasewycz, O. J. Nutr. 1983, 113, 1583.
17. Roberts, D.; Chapman, J. In: Developmental Toxicology - Target Organ Toxicology Series; Raven Press, New York, 1983; p 123.
18. Martson, H.; Allen, S.; Swaby, S. Br. J. Nutr. 1971, 25, 15.
19. Alfaro, B.; Heaton, F. Br. J. Nutr. 1973, 29, 73.
20. Alfaro, B.; Heaton, F. Br. J. Nutr. 1974, 32, 435.
21. Gross, R.; Newberne, P. Physiol. Rev. 1982, 60, 188.
22. Landes, D. Proc. Soc. Exptl. Biol. Med. 1975, 150, 686.
23. Cohen, A.; Teitelbaus, A.; Miller, E.; Bentor, V.; Hirk, R.; Fields, M. Israel J. Med. Sci. 1982, 18, 840.
24. Fields, M.; Ferretti, R.; Smith, Jr., J.; Reiser, S. J. Nutr. 1983, 113, 1335.
25. Fields, M.; Ferretti, R.; Reiser, S.; Smith, Jr., J. Proc. Soc. Exptl. Bio. Med. 1984, 175, 530.
26. Frieden, E. In: Copper in the Environment, John Wiley and Sons, New York, 1979; p 241.
27. Weinberg, E. Physiol. Rev. 1984, 64, 65.
28. Leuke, R.; Fraker, P. J. Nutr. 1979, 109, 137.

ACCELERATED RECOVERY OF HEMATOPOIESIS WITH RADIOPROTECTIVE CU-DIPS

L.S.F. Soderberg,[1] J.B. Barnett,[1]
M.L. Baker,[2] H. Salari,[3]
and J.R.J. Sorenson[3]

Departments of Microbiology and Immunology[1] and Radiology,[2] College of Medicine, and the Department of Biopharmaceutical Sciences,[3] College of Pharmacy, University of Arkansas for Medical Sciences, Little Rock, Arkansas

INTRODUCTION

Copper (II)(3, 5-diisopropylsalicylate)$_2$ (Cu-DIPS), administered to mice 3 hr prior to exposure to 8.0 Gy irradiation, increased the 42 day survival rate from 40% to 85%.[1] Treatment with Cu-DIPS significantly accelerated spleen repopulation and the recovery of immune responsiveness following irradiation.[2] In the present study the effects of Cu-DIPS on the recovery of hematopoietic activity in bone marrow and spleen following irradiation were investigated.

METHODS

Experimental Design

Mice (C57BL/6) were injected subcutaneously with 80 mg/kg Cu-DIPS or vehicle (4.0% propyleneglycol, used as a wetting agent, and 1.4% polyvinyl alcohol in saline). Three hours later some of the mice from each group were exposed to 8.0 Gy (1.7 Gy/min) in a ^{137}Cs irradiator. Irradiated mice in the Cu-DIPS group had a 42 day survival rate of 85% while the vehicle group had a 40% survival rate. Colony-forming unit (CFU) assays were performed 7, 14, 24, and 42 days after treatment.

Colony-Forming Cell Assays

Mice were sacrificed and their femoral bones removed and flushed with medium using a syringe with a 26 g needle. The resulting bone marrow cell suspensions were washed and placed in standard semi-solid agar cultures[3] with granulocyte/macrophage-colony stimulating factor (GM-CSF) or 6.5u/ml interleukin-3 (IL 3, Genzyme Corp., Boston, MA). A standardized dilution of endotoxin-mouse lung conditioned medium was used as a source of GM-CSF.[4] Colonies of 50 or more cells were counted on days, 3, 7, and 14 of incubation. Semi-solid agar cultures of spleen cells were similarly prepared and stimulated with IL 3.

RESULTS

Femoral bone marrow was extracted from the mouse bones as completely as possible 7, 14, 24, or 42 days after irradiation. As shown in Figure 1, the number of recoverable, viable bone marrow cells per femur was reduced by 99.5% 7 days after irradiation, whether or not the animals were treated with Cu-DIPS. A fairly large number of femurs was required to provide sufficient viable cells for CFU assays at 7 days. At subsequent time periods bone marrow cellularity recovered at a significantly greater rate in Cu-DIPS-treated mice than in vehicle-treated mice. The increase in bone marrow cellularity was not as great as the Cu-DIPS-mediated increases in spleen cellularity following irradiation.[2] By 42 days bone marrow cellularity approached normal levels.

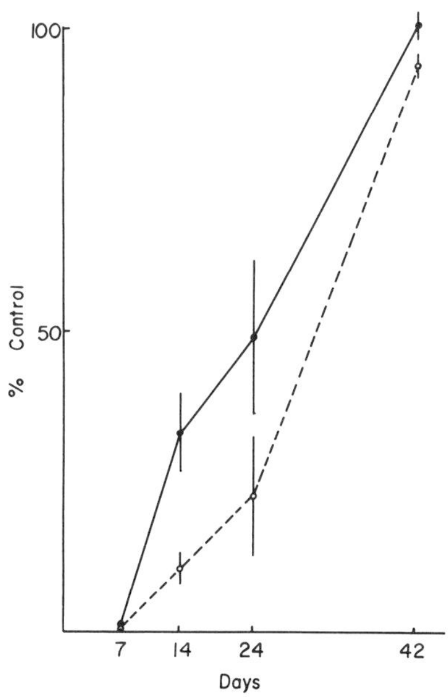

Figure 1. Bone marrow cellularity at various times after irradiation. Solid line: Cu-DIPS; dashed line: vehicle. Unirradiated control values are presented in Table 1.

The recovered bone marrow cells were tested for regenerational capacity in semi-solid agar cultures. Figure 2 shows the results of CFU assays in which the bone marrow cells were stimulated with GM-CSF. Irradiation reduced the number of GM-CFU to zero at 7 days. Significant numbers of colonies did not appear until 24 days after irradiation, but approached normal levels by 42 days. No differences in the regeneration of GM-CFU in irradiated bone marrow were found due to Cu-DIPS treatment.

Interleukin 3 stimulates a more primitive stem cell than GM-CSF and IL 3-induced colonies contain cells of mixed lineage, including granulocyte, macrophage, lymphoid, erythroid, megakaryocyte, and other lineages. IL 3 is believed to stimulate the same precursors as measured by the *in vivo* focus-forming units (CFU-S) assay.[5] Figure 3 shows that irradiation destroyed IL 3-CFU, when measured at 7 days. Very low levels of IL 3-CFU persisted until 42 days after

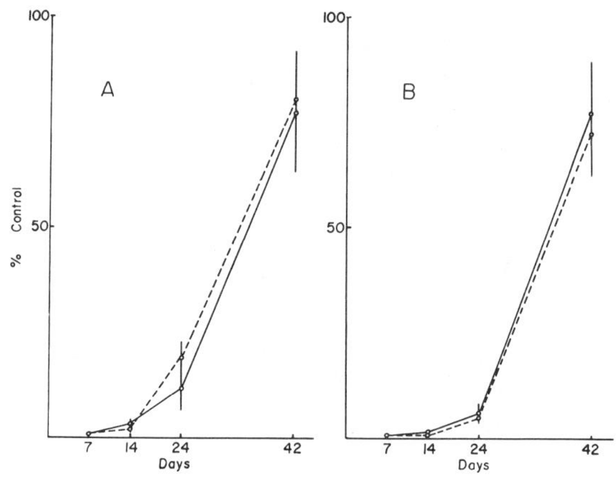

Figure 2. Bone marrow GM-CFU at various times after irradiation. A: CFU/culture; B: CFU/femur. Solid lines: Cu-DIPS; dashed lines: vehicle. Control values are presented in Table 1.

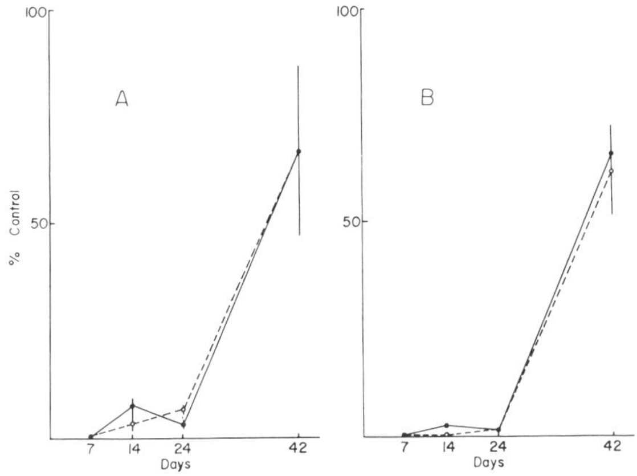

Figure 3. Bone marrow IL 3-CFU at various times after irradiation. A: IL 3-CFU/culture; B: IL 3-CFU/femur. Solid lines: Cu-DIPS; dashed lines: vehicle. Control values are shown in Table 1.

irradiation, when they had recovered to about 65% of control levels. No differences due to Cu-DIPS treatment were seen in the regeneration of IL 3-CFU in bone marrow after irradiation.

We have reported that Cu-DIPS significantly enhanced the repopulation of murine spleens and the regeneration of lymphocyte activity following irradiation.[2] Since Cu-DIPS had no obvious enhancing effects on bone marrow hematopoiesis, we examined IL 3-CFU in spleen cells following irradiation (Figure 4).

Spleen cells, isolated from Cu-DIPS or vehicle groups 7 days after irradiation, were unresponsive to IL 3. However, at 14 days, while vehicle-treated spleen cells still produced no colonies, spleen cells from Cu-DIPS-treated mice had 660% of control levels of IL3-CFU. This level dropped off and

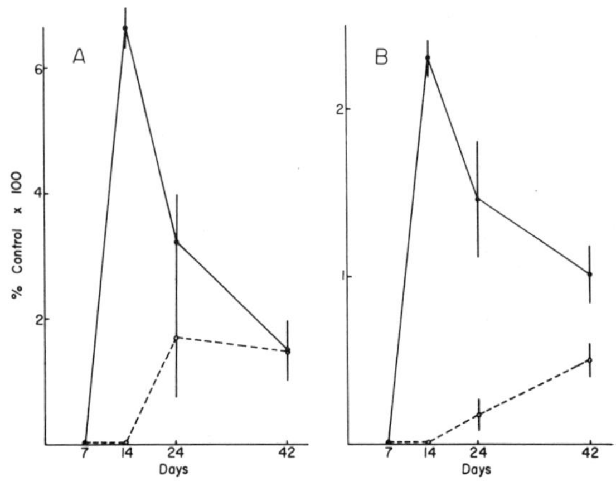

Figure 4. Splenic IL 3-CFU at various times after irradiation. A: IL 3-CFU/culture; B: IL 3-CFU/spleen. Solid lines: Cu-DIPS; dashed lines: vehicle. Control values are shown in Table 1.

Table 1

Control Values

	(1)	(2)	(3)
Cells X 10^6 Femur	22.1±1.7	29.1±0.7	19.1±3.5
GM-CFU/10^6 Marrow Cells	363±12	310±90	247±35
GM-CFU/Femur	8022±256	9021±2619	6867±973
IL 3-CFU/10^6 Marrow Cells	190±15	303±20	300±12
IL 3-CFU/Femur	2540±3255	3249±420	2322±369
IL 3-CFU/10^7 Spleen Cells	40±12	87±7	47±7
IL 3-CFU/Spleen	290±87	398±84	200±100

converged with a late and much lower increase in IL3-CFU in the vehicle-treated mice. The total number of IL3-CFU/spleen (Fig. 4B) peaked at 230% at 14 days in Cu-DIPS-treated mice and dropped off to control levels by day 42. In vehicle-treated mice, total IL3-CFU increased slowly to reach only 50% of control values by day 42.

DISCUSSION

Treatment of mice with radioprotective levels of Cu-DIPS was found to accelerate the recovery of spleen cellularity and spleen weight.[2] Independently of increases in cell numbers, Cu-DIPS was found to accelerate the recovery of lymphocyte reactivity in irradiated mice.[2] In the adult, spleen cells are constantly replenished by hematopoietic activity in bone marrow. It was expected that Cu-DIPS would accelerate bone marrow hematopoietic activity. However, no effects of Cu-DIPS on bone marrow GM-CFU or IL 3-CFU were found. In contrast, the spleens of Cu-DIPS-treated mice had a dramatic early (14 days) burst in IL 3-CFU activity. While the vehicle-treated mice developed increased splenic IL 3-CFU activity, such activity occurred later and was much more limited. It is likely that Cu-DIPS enhancement of recovery of spleen cellularity and immune activity after irradiation was due to spleen, not bone marrow, hematopoietic activity. Although we only tested two measures of hematopoiesis, it could be speculated that the observed Cu-DIPS increases in bone marrow cellularity may have been due to cells migrating from the spleen. The specificity of Cu-DIPS action on spleen and not bone marrow hematopoiesis, might be due to preferential deposition of the Cu-DIPS in the spleen. This possibility is currently under investigation.

ACKNOWLEDGEMENTS

We are indebted to the Arthur Ambrust Cancer Research Fund, International Copper Association, Max and Victoria Dreyfus Foundation, Elisa U. Pardee Foundation, and Denver Roller Corp. for financial support. Mice were kindly provided by the National Center for Toxicological Research, Jefferson, Arkansas.

REFERENCES

1. Salari, H., Baker, M. L., Barnett, J. B., Soderberg, L. S. F., Willingham, W. M. and Sorenson, J. R. J. In: Biology of Copper Complexes; Ed. J. Sorenson; Humana Press, NJ, in press. (this proceedings)
2. Soderberg, L. S. F., Barnett, J. B., Baker, M. L., Salari, H., and Sorenson, J. R. J. In: Biology of Copper Complexes; Ed. J. Sorenson; Humana Press, NJ, in press. (this proceedings)
3. Nicola, N. A., Burgess, A. W., Metcalf, D. J. Biol. Chem. 1979. 254, 5290.
4. Nicola, N. A., and Metcalf, D. J. Cell. Physiol. 1981, 109, 253.
5. Hopel, A. J., Fung, M. C., Johnson, R. M., Young, I. G., Johnson, G., and Metcalf, D. Blood 1985, 65, 1453.

ANTIASTHMATIC ACTIVITY OF SOME COPPER COMPLEXES

Abdel Fattah M. Badawi,[1] Abdul Hadi M. Summan,[2] H.M. Sarau,[3] J.J. Foly,[3] Fadia Ali,[3] John R.J. Sorenson[4]
[1]Department of Chemistry, Centre of Science and Mathematics, Taif, Saudi Arabia; [2]Faculty of Science, Umm-Al-Qura University, Holy Makkah, Saudi Arabia; [3]Smith Kline and French Laboratories, Swedeland, Pennsylvania 19479 U.S.A; and [4]Biopharmaceutical Sciences, College of Pharmacy, University of Arkansas for Medical Sciences, Little Rock, Arkansas 72201 U.S.A

INTRODUCTION

Normal lung development and function requires a number of Cu-dependent enzymes:[1] cytochrome c oxidase, superoxide dismutase(Cu-Zn SOD), tyrosinase, dopamine-β-hydroxylase, amine oxidases, and lysyl oxidase. A variety of pathological lung disorders are accompanied by obstructive pulmonary disease(OPD) which is associated with abnormal Cu metabolism. Serum copper values are significantly elevated in chronic OPD[2] and in bronchial asthma patients.[3]

The lung of the copper-deficient rat is proposed as a model for developmental pulmonary emphysema.[4] D-penicillamine, which complexes copper, produces mesal tracheobronchial cystic lesions in lungs of rats.[5]

5-Lipoxygenase catalyses the oxygenation of arachidonic acid (AA) at the C-5 position in producing various leukotrienes.[6] The biological activity of slow reacting substances in anaphylaxes(SRS-A) is now ascribed predominantly to leukotrienes C4 and D4.[7] Compounds which inhibit the action of lipoxygenase also block the synthesis and release of SRS-A and may be of therapeutic use in asthma.

We investigated the possibility that copper, like certain other metalloelements, is necessary for the function of the respiratory system. Since divalent copper complexes are characterized as inhibiting hydroperoxide-dependent oxidations involving liver microsomes[8] and copper-dependent SOD inhibited peroxidation by lipoxygenase,[9] it occurred to us that copper complexes might also inhibit the action of 5-LO, thus exerting anti-asthmatic activity.

To investigate the possibility that bronchial asthma might be suppressed by treatment with Cu complexes with SOD activity we submitted Cu(II)(3,5-diisopropylsalicylate)$_2$ (CuDIPS)[A] and Cu(II)bis(chromone-2-carboxylate)(CuBCC)[B] to Smith Kline & French Laboratories, Philadelphia, PA. for evaluation as inhibitors of 5-LO. Both Cu complexes were found to significantly inhibit 5-HETE synthesis suggesting 5-LO inhibitory activity.

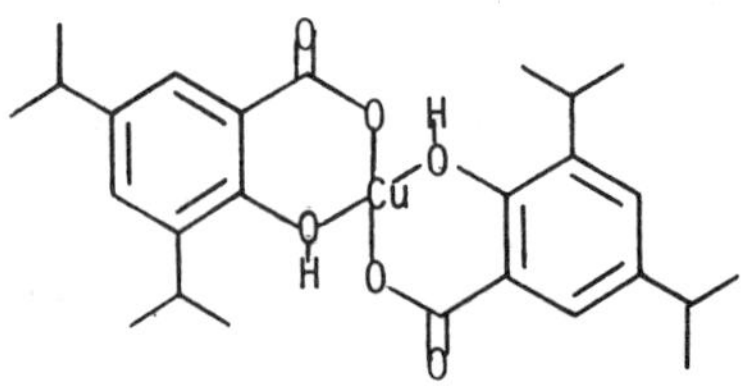

A

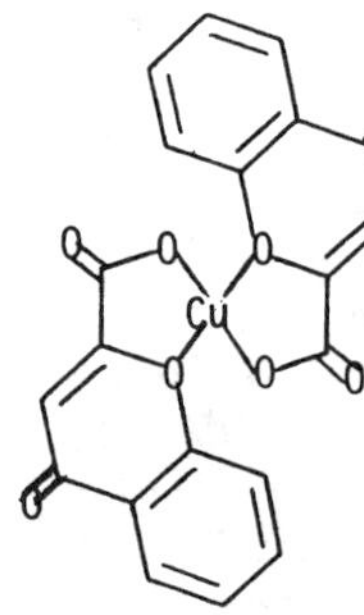

B

MATERIALS AND METHODS

CuDIPS and CuBCC were synthesized using reported methods.[10] Details of the CuBCC synthesis will be described elsewhere. These test compounds were suspended in DMSO at 10 mM concentration and diluted to assay concentration in assay buffer. Activities of 5-lipoxygenase and LTA_4 synthetase were determined in high speed supernatants of RBL-1 cells using a modification of the method of Jakschik and Lee.[11] Rat basophilic leukemia(RBL-1) cells were obtained from American Type Culture Collection and grown in spinner culture at $37^{o}C$ in Eagle's minimum essential medium supplemented with 10% heat inactivated fetal calf serum in 5% CO_2/95% air. RBL-1 cells were harvested by centrifugation at 400 x g for 10 min. and washed once with phosphate buffer containing 50 mM sodium phosphate, pH 7.0, 0.1 percent gelatin, and 1 mM EDTA. These cells were then resuspended in buffer at 5 x 10^7 cells/ml and disrupted in a Parr nitrogen bomb at 750 psi for 10 min. The disrupted cell suspension was centrifuged at 35,000 x g for 10 min. and the supernatant further centrifuged for 1 hour at 100,000 x g and the final supernatant was stored in aliquots at -80^oC.

To 250 ul aliquots of the supernatant, maintained at 5^oC, was added 2.5 ul of test compound in DMSO and 10 ul of 50 mM $CaCl_2$ to give a final concentration of 2 mM. The reaction was initiated by the addition of 2.5 ul of 1-^{14}C -arachidonic acid (specific activity 7.5 uCi/umole) to achieve a final concentration of 25 uM. Incubation continued for 5 min. at 5^oC. The reaction was terminated by addition of 2 volumes of ice cold acetone and allowed to deproteinize on ice for 15 min. prior to centrifugation. DMSO concentration in the assay was $\leq$ 1% which had no significant effect on the activity.

To quantitate metabolites the supernatant was extracted with two volumes of ice cold ethyl acetate. The organic phase was dried under argon and the metabolites redissolved in a small volume of ethyl acetate. Whatman LK6D thin layer plates were pre-developed in toluene: diethylether:ethanol:acetic acid (50:40:2:0.5) and dried in air. Samples of the ethyl acetate solution were rapidly spotted and developed in the thin layer solvent system. Chromatogram radioactivity was quantitated using a Berthold LB2832 automatic TLC linear analyzer.

Data presented in the Figure show inhibition of 5-HETE synthesis at 10^{-5} and 10^{-4} M concentrations, which suggests that these two Cu complexes have 5-LO inhibitory activity. CuDips was found to produce 26% and 33% inhibition of 5-HETE synthesis at 10^{-5} and 10^{-4}M concentrations, respectively, while CuBCC produced 14% and 33% inhibition at these same concentrations. The ID_{50} calculated by linear regression was 3.2 x 10^{-4} for CuDIPS and 1.8 x 10^{-4} for CuBCC. CuDIPS and CuBCC were found to produce 14% and 35% inhibition of Di-HETE synthesis respectively at 10^4 M concentration. Because of the limited solubility of the compounds in aqueous solution, data presented is the minimal response. More soluble Cu complexes will probably show better activity. Neither 3,5-diisopropylsalicylic acid nor chromone-2-carboxylic acid affected 5-HETE and Di-HETE synthesis in this test system.

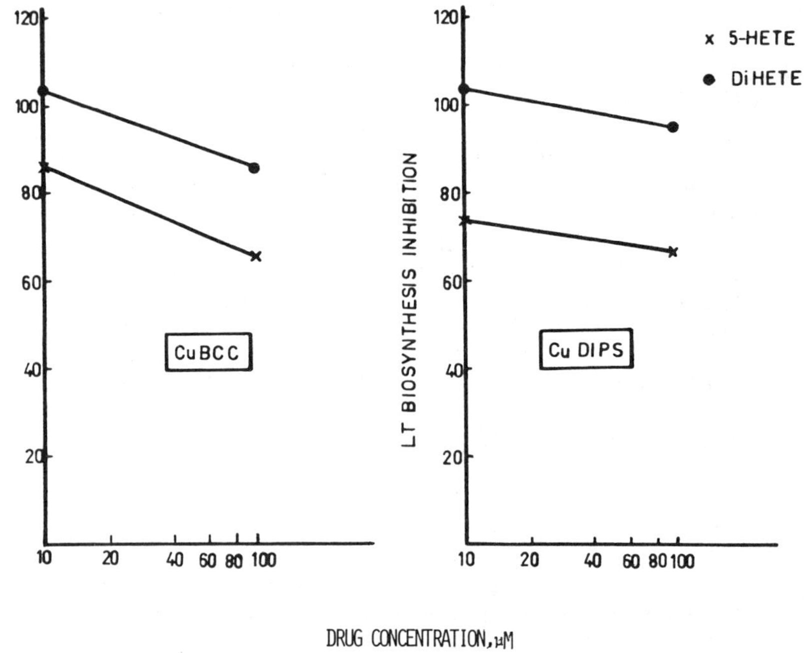

To evaluate the possibility that inorganic forms of copper might have potential antiasthmatic activity $Cu(II)Cl_2$ was previously studied with human platelet lipoxygenase (12-LO). The inorganic copper salt caused an artificial and dramatic increase in oxygen consumption which was apparently due to Cu^{2+}-catalysed auto-oxidation of the substrate.[12] This did not occur in the HETE synthesizing system when complexes of copper were used. $Cu(II)Cl_2$ in the absence of Ca in the present system had no effect on arachidonate oxidation.

DISCUSSION

This is the first published report supporting the possibility that SOD-like Cu complexes may have potentially useful prophylactic activity against bronchial asthma and chronic OPD.

Since bronchial asthma and chronic OPD are accompanied by a fall in lung copper concentration, it seems reasonable to suggest that this fall may be associated with increased glutathione-S-transferase activity and lipid peroxidation as a result of a lack of copper-dependent SOD activity. Thus enhanced superoxide formation by neutrophils[13] may have a role in pulmonary pathology associated with asthma and chronic OPD.

During irreversible inhibition of 5-LO from RBL-1 cells, it was postulated that the AA hydroperoxide intermediate could undergo facile oxygen-oxygen bound homolysis to form radicals which may inactivate the enzyme.[14] Many of the reported 5-LO inhibitors are free radical scavengers.[15] The fact that small molecular weight copper complexes disproportionate superoxide might explain the inhibitory effect of CuDIPS and CuBCC on 5-LO catalyzed hydroperoxidation.

Lipoxygenases belong to the class of enzymes known as dioxygenases which characteristically incorporate two atoms of molecular oxygen into a substrate, e.g., $R + O_2 \longrightarrow R{-}O_2$. Current evidence points to a requirement of ferric iron for lipoxygenase activity.[16] The proposed mechanism of action of copper complexes in inhibiting AA peroxidation can be described by the simplified scheme:

Step (1) $Fe^{3+} + S \longrightarrow SFe^{3+}$

(2) $SFe^{3+} + e^- \longrightarrow SFe^{2+}$

(3) $SFe^{2+} + O_2 \longrightarrow SFe^{2+}O_2 \rightleftharpoons SFe^{3+} + O_2^-$

(4) $SFe^{2+}O_2 + e^- \longrightarrow SFe^{2+}O_2^-$

(5) $SFe^{2+}O_2^- + H^+ \longrightarrow SOOH + Fe^{3+}$

Where S is the substrate, Fe^{3+} is the oxidized form of lipoxygenase and SOOH is the reaction product.

An inhibitory role for copper (II) complexes is suggested based upon their disproportionation of superoxide: $Cu^{2+} + 2O_2^- + 2H^+ \longrightarrow H_2O_2 + O_2$. This reaction may accelerate the dissociation of $SFe^{2+}O_2$ via step 3. The higher the concentration of copper complexes, the higher their inhibitory efficiency. CuDIPS and CuBCC complexes may react with O_2^- but may also complex with the substrate and prevent bonding at the active site of 5-LO in step 1 to cause mixed-type inhibition of AA peroxidation.

Apparent inhibition of AA oxidation by copper complexes might also be explained as consumption of hydroperoxide via a catalase or peroxidase (Haber-Weiss) mechanism:

$$2SOOH + 2Cu^{2+} \longrightarrow 2SOH + O_2 + 2Cu^{+}$$

$$2Cu^{+} + SOOH + 2H^{+} \longrightarrow 2Cu^{2+} + SOH + H_2O$$

Thus, the inhibitory effect of copper complexes on AA peroxidation might be associated with consumption of hydroperoxide, disproportionating superoxide and/or competing with the substrate for bonding to 5-LO.

Also, copper has an action which is similar, at the receptor level, to that of prostacyclin (PGI_2) which is characterized by inhibiting the anaphylactic release of SRS-A from guinea pig lung tissue.[17]

Bronchial asthma and chronic OPD are clearly associated with inflammation wherein superoxide release or accumulation may have an etiologic role. If superoxide accumulation is a cause of asthma and chronic OPD then superoxide disproportionating reactivity of copper complexes or their possible role in activation of Cu-Zn SOD in the respiratory system merit consideration as possible mechanism of action.

Copper chelates might also be considered as the active forms of antiallergic drugs in the respiratory system and their value in therapy warrant in depth study.

REFERENCES

1. Underwood, EJ. In: Trace ELements in Human & Animal Nutrition, 4th ed.; Academic Press; New York, 1977; p 56.
2. Sullivan, JF; Blotcky, AJ; Jetton, MM; Hahn, HK; Burch, RE. J. Nutr. 1979, 109, 1432.
3. Gorski, M; Kaszewska-Jablonska, I; Przezdziak J. Pol. Arch. Med. 1966 Wewn 37 (12), 649
4. O'Dell, BL; Kilburn, KH; McKenzie, WN; Thurston, RJ. Am. J. Pathol. 1978, 91, 413.
5. Kilburn, KH; Hess, RA. Tetratology. 1982, 26, 1.
6. Samuelsson, B. Science 1983, 220, 568.
7. Lewis, RA; Austen, KF; Drazen, JM; Clark, JM; Marfat, DA; Corey, EJ. Proc. Natl. Acad. Sci. U.S.A. 1980, 77, 3710.
8. Kurchenko, VP; Usanor, SA; Metelitsa, DI. React. Kinet. Catal Lett. 1979, 12(1), 31.
9. Matsuda, Y; Beppu, T; Arima, K. Agric. Biol. Chem. 1979, 43, 1179.
10. Sorenson, JRJ. J. Med. Chem. 1976, 19, 135.
11. Jakschik, BA; Lee, LH. Nature. 1980, 287, 51.
12. Donald, PW; Brown, VR. Biochem. Biophys. Acta 1981, 663, 361.
13. Simchowitz, L. J. Lab. Clin. Med. 1979, 93, 583
14. Corey, EJ; Lansbury, PT; Cashman, JR; Kantner, SS. J. Am. Chem. Soc. 1984, 106, 1501.
15. Koshihara, Y; Neichi, T; Murota, S; Lao, A; Fujimoto, Y; Tatsuno, T. Biochem. Biophys. Acta. 1984, 792, 92.
16. Greewald, JE; Wong, LK; Alexander, MS; Bianchine, JR; Fertel, RH. Fed. Proc. 1980, 39, 1124.
17. Cunnane, SC; Zinner, H; Horrobin, DF; Manku, MS; Morgan, RO; Karmali, RA; Ally, RI; Karmazyn, M; Barnette, WE; Nicolaou, KC. Can.J. Physiol. Pharmacol. 1979, 57, 35.

THE EFFECT OF COPPER ON THE POLYAMINE OXIDASE REACTION

Vardan K. Gasparian and Robert M. Nalbandyan
Institute of Biochemistry of Armenian Academy of Sciences, 375044, Yerevan-44, USSR

It is well known that FAD-containing polyamine oxidase (PAO) is the main polyamine degradating enzyme which determines the level of polyamines in tissues.

Using PAO isolated from bovine liver mitochondria,[1] we have tested the effect of copper on the activity of PAO. Experiments were conducted according to the following scheme. The substrate (spermidine, 1 mM) was preliminarily incubated with 1 mM Cu^{+2} and then 0.1 uM PAO was added to the mixture. In control experiments no copper was added. The reaction was followed by aldehyde formation.[2] Results obtained showed that in the presence of copper the reaction is activated by a factor of 2.5 to 3. It is of interest that other metals such as Zn, Fe, and Hg had no activation effects. In the case of Hg, a strong inhibition of the reaction was observed.

Thus, copper should be considered as a stimulator of PAO and hence as a modulator of physiological effects of polyamines which are know to be universal tissue growth factors.[3]

REFERENCES

1. Hölttä, E. Biochemistry, 1977, 16, 91-100.
2. Padmanaban, R; Kim, K. Biochem. Biophys. Res. Commun. 1965, 19, 1-5.
3. Jänne, J; Rösö, H; Raina, A. Biochim. Biophys. Acta, 1978, 473, 241-293.

SYNOVIAL FLUID COPPER, ZINC AND SELENIUM IN RELATION TO INFLAMMATORY PARAMETERS IN RHEUMATIC DISEASES.

A. PERETZ[1], J. NEVE[2], F. VERTONGEN[3], J.P. FAMAEY[1]

Departments of Rheumatology[1] and of Medical Chemistry[3]

Hôpital Universitaire Saint-Pierre and Laboratory of Analitycal Chemistry and Toxicology, Institute of Pharmacy[2], University of Brussels, Belgium.

ABSTRACT : Trace elements concentrations in plasma are altered in inflammation. An increase in copper and a decrease in zinc and selenium levels have been reported in rheumatoid arthritis and other inflammatory arthritis. As the joints are primarily affected by the inflammatory process, we have investigated copper, zinc, selenium and caeruloplasmin levels as well as the activity of the seleno-enzyme glutathione-peroxidase (GPx) both in synovial fluid (SF) and in plasma (or the erythrocytes for GPx) from 27 patients with inflammatory arthritis (including 6 rheumatoid arthritis) and 27 osteoarthrotic or traumatic controls without inflammation. A significant increase in SF trace elements concentrations was observed in inflammatory arthritis but within different ranges according to the element : 1.3 fold increase for selenium, 2.1 fold increase for copper and 3.6 fold increase for zinc. Positive correlations were disclosed between SF proteins and trace elements in the non-inflammatory but not in the inflammatory group. Trace elements concentrations were not significantly correlated as well as SF Cu and caeruloplasmin. The increased diffusion of protein-bound trace elements cannot exclusively account for the modified concentrations in trace elements (except for Se) observed in inflammatory SF. A release from damaged tissues and cells might also play a role in these changes.

INTRODUCTION : The interest in trace elements such as copper (Cu), zinc (Zn) and selenium (Se) in rheumatic diseases was recently stimulated by the reports of their modified status during inflammation (1,2,3). The site of rheumatic inflammation is mainly the joint. It was thus interesting to measure trace elements concentrations in the synovial fluid (SF). Increase in SF trace elements especially SF-Cu (4) and SF-Zn (5) were already reported in inflammation. These modifications could be interpreted as physiological responses to prevent hyaluronic acid depolymerization and other tissues destructions mediated by superoxide and other toxic oxygen radicals (6). Indeed Se contributes to this protective mechanism through the selenium-containing enzyme glutathione-peroxidase (GPx) (7), Cu and Zn through the Cu-Zn cytosolic superoxide dismutase (SOD) (8) and Cu through the Cu-bound-α_2-globulin caeruloplasmin (Cp) (9). Few authors have determined the levels of more than one trace element in the same SF sample. In this contribution we have measured Cu, Zn, Se and Cp levels as well as the activity of GPx both in SF and plasma (or erythrocytes for GPx)of patients with various inflammatory rheumatic diseases, including rheumatoid arthritis (RA), and with non inflammatory osteoarthrosis or traumatism as controls.

MATERIALS AND METHODS : 27 inflammatory rheumatic patients with recent knee effusion (less than 4 days) were included in the study : 6 RA, 11 crystal-induced arthritis and 10 seronegative reactive arthritis. They were compared with 27 non-inflammatory controls with similar recent knee effusions : 22 osteoarthrotic and 5 traumatic joints. All hemorragic SF were discarded and the inflammatory character of the SF was accepted when more than 500 polymorphonuclear cells/mm^3 were present and the proteins level was above 3.5.g/dl.

The blood and the SF samples for trace elements determination were collected in acid-washed tubes containing heparin, between 8 and 10 a.m. when the subjects were still fasting. Plasma and SF were used after centrifugation.

Trace elements determinations were performed by atomic absorption spectroscopy (AAS). A Varian AA5 instrument was used in the flame mode for Zn and Cu. Zn was measured after dilution with 0.1% HCl and Cu after dilution and deproteinization with 20% trichloracetic acid (10). Se was determined by a direct graphite furnace AAS procedure using Zeeman-effect background correction (spectrophotometer Perkin-Elmer model Z 3030) (11,12). Accuracy of procedures was previously established (11,12).

Caeruloplasmin was determined by turbidimetry. SF and erythrocyte GPx activities were measured at 30°C according to the modified method of Paglia and Valentine and their values expressed in µmol of NADPH oxidised/min (U) per either ml or g Hb (13).

The following laboratory tests were performed using classical procedures: Westergreen erythrocyte sedimentation rate (ESR), hemoglobin concentration (Hb), serum and synovial proteins concentrations, blood and SF white blood cells (WBC) counting and determination.

Results were expressed as mean ± standard deviation (S.D.)

Statistical analysis were performed using regression analysis and Student's t test.

RESULTS : A. Plasma. A non significant increase in Cu concentration and a non significant decrease in Zn concentration were observed in the inflammatory patients group whereas Se concentration remained unchanged (Table 1)

Table 1

Cu, Zn, Se and related parameters in blood of inflammatory (IA) and non-inflammatory (NIA) arthritis.

	IA	NIA
- p Cu, µg/dl	170 ± 85	136 ± 23
- p Zn, µg/dl	76 ± 16	88 ± 12
- p Se, ng/ml	86 ± 14	85 ± 16
- E GPx, U/gHb	7.2 ± 1.6	7.5 ± 2.5
- p caeruloplasmin, mg/dl	41 ± 12	30 ± 6
- p Proteins, g/l	71 ± 7	71 ± 8
- ESR, mm/h	44 ± 31***	8 ± 5
(range)	(5-120)	(2-20)

*** $p < 0.001$ (difference between IA and NIA)

p : plasma

E : erythrocyte

B. Synovial fluid. Significant increases in the concentration of the three trace elements were observed but within different ranges : 1.3 fold for Se, 2.1 fold for Cu and 3.6 fold for Zn (Table 2).

A positive linear correlation was disclosed between SF Zn, Cu and Se, and SF protein levels in the non-inflammatory control group ($r = 0.63$, 0.59 and 0.52, $P < 0.05$) but not in the inflammatory group. A positive correlation with total SF WBC number and trace elements SF concentrations was found only for Zn inflammatory fluids ($r = 0.55$).

Trace elements concentrations were not correlated except positively for Cu and Zn in controls SF ($r = 0.84$; $p < 0.01$). SF GPx and Se concentrations were not correlated as well as SF Cu and Cp in inflammatory as well as in non-inflammatory groups. Assuming that the Cu content of Cp is 0.32% W/W (14), it could be calculated that all the Cu is Cp-bound in the non-inflammatory group. In the inflammatory SF, the calculated Cp-bound Cu was insufficient to account for all the Cu measured, which is therefore in excess regarding to the detected Cp.

Table 2

Synovial Cu, Zn, Se and related parameters in inflammatory (IA) and non-inflammatory (NIA) arthritis.

	IA	NIA
Cu, μg/dl	97 ± 46***	46 ± 19
Zn, μg/dl	69 ± 40***	19 ± 13
Se, ng/ml	50 ± 14**	39 ± 10
GPx, U/ml	3.7 ± 2.0	2.0 ± 1.0
Caeruloplasmin, mg/dl	20.5 ± 4.1**	14.1 ± 6.0
Proteins, g/l	46.5 ± 9.0***	33.1 ± 8.0
WBC/mm^3	14,700***	220
(range)	(3,200 - 58,800)	(32 - 340)
PMN/mm^3	12,600***	40
(range)	(650 - 53,360)	(2 - 120)

*** P < 0.001)
) difference between IA and NIA
*** P < 0.01)

WBC = white blood cells

PMN = Polymorphonuclear cells.

DISCUSSION : The present results confirm the various modifications for trace elements plasma levels already described in the literature (15-17). In addition to the classical rise in protein level, due to changes in synovial membrane permability related to inflammation, an increase in Cu, Zn and Se levels was observed in SF, as previously reported for Cu (4) and Zn (5).

The synovial membrane permeability mechanisms are still poorly understood. Synovial inflammation induces several anatomical changes resulting in a reduction of the interstitial diffusion of small solutes and an increase of vessel wall permeability responsible for an increased penetration of proteins. The diffusion of small solutes such as non-protein-bound trace elements is bidirectional and their concentration in SF is the result of a diffusion equilibrium while that of the large solutes such as proteins is unidirectional from the plasma to the SF (18).

With a decrease in plasma Zn, an absence of change in plasma Se and an increase in plasma Cu which is only due to a Cp-bound Cu increase (6), the increase in SF trace elements may not be interpreted as the consequence of a diffusion equilibrium of unbound elements between plasma and SF.

A 1.4 fold rise in SF proteins due to transsynovial unidirectional diffusion from the plasma was observed in the inflammatory group. In these conditions the 1.3 fold elevation of SF Se observed in inflammation could be explained by the diffusion of protein-bound Se from the plasma. This mechanism is insufficient to explain the observed 2.1 fold change in SF Cu and 3.6 fold change in SF Zn. A release from cartilage, synoviocytes or SF WBC due to increased rates of turnover and associated lysis of these tissues or cells during periods of inflammation could largely contribute to the elevated SF concentrations of these two trace elements (19). The difference in cellular concentrations might be the explanation for the different pattern for SF Zn and Cu modifications.

The importance of these changes in the pathogenesis of inflammation remains poorly understood. They are probably the consequence of physiologic responses of inflamed tissues related to an increased activity of various trace-elements dependent enzymes. Further investigations on trace elements metabolism in patients with rheumatic diseases will be necessary to elucidate their role in synovial inflammation.

ACKNOWLEDGEMENTS : We thank the "Fonds de la Recherche Scientifique Médicale" Belgium (grants n° 9.418.84, 15.419.86F, 15.006.87F and 3.4510.87) as well as Labcatal S.A. for support of this research.

REFERENCES.

1. MEACOCK, SCR, SWANN, BP, DAWSON, W.
The dynamics and possible roles of metal complexes in inflammation.
Agents Actions, Suppl. 8 (Eds Rainsford, KD, Brune, K, Whitehouse, MW) Trace elements in the pathogenesis and treatment of inflammation, Birkhäuser Verlag, 1981 pp 145-164.
2. KLASIG, K.
Effect of inflammatory agents and interleukin-I on iron and zinc metabolism.
Am. J. Physiol. 1984, 247 : R-901-R-904.
3. SORENSON, JRJ.
An evaluation of altered copper, iron, magnesium and zinc concentrations in Rheumatoid Arthritis.
Inorganic Perspectives in Biology and Medicine, 1978 ; 2 : 1-16.
4. SCUDDE, PR, AL-TIMINI, D, Mc MURRAY, W, WHITE, AG, ZOOB, BC, DORMANDY, TL.
Synovial fluid copper and related variables in rheumatoid arthritis and degenerative arthritis.
Ann. Rheum. Dis. 1978 ; 37 : 71-72.
6. BONNEBRAKE, RA, Mc CALL, JT, HUNDER, GG, POLLEY, HF.

Zinc accumulation in synovial fluid.
May. Clin. Proc. 1972 ; 47 : 746-750.

6. GREENWALD, RA, MOY, WW.
Effect of oxygen-derived free radicals on hyaluronic acid.
Arthritis Rheum. 1980 ; 23 : 455-463.

7. SUNDE, RA.
The Biochemistry of selenoproteins.
J. Am. Oil. Chemist. Soc., 1984 ; 61 : 1891-1900.

8. Mc CORD, JM.
Free radicals and inflammation : protection of synovial fluids by superoxide dismutase.
Science 1974 ; 185 : 529-531.

9. FRIEDEN, E.
Caeruloplasmin : a multifunctional metalloprotein of vertebrate plasma.
in Metal ions in Biological Systems (Ed. Siegl, H) vol. 13
Copper Proteins, Marcel Dekker , New-York, Basel 19, pp 117-142.

10. NEVE, J., MOLLE, L., HANOCQ, M., SINET, P.M., VAN GEFFEL, R.
Erythrocyte and plasma trace element levels in clinical assessments.
Biol. Tr. Elem. Res., 1983 ; 5 : 75-79.

11. NEVE, J, MOLLE, L.
Direct determination of selenium in human serum by graphite furnace atomic absorption spectroscopy.
Acta Pharmacol. Toxicol., 1986 ; 59 : 606-609.

12. NEVE, J, CHAMART, S, MOLLE, L.
Optimization of a direct procedure for the determination of selenium in plasma and erythrocytes using Zeeman-effect atomic absorption spectroscopy.
In Trace Element Analytical Chemistry in Medicine and Biology (Eds. P. Bratter, P. Schramel) vol. 4, W. de Gruyter, Berlin, in press.

13. GUNZLER, W, KREMERS, H, FLOHE, L.
An improved coupled test procedure for glutathione peroxidase in blood.
A. Klin. Chem. Klin. Biochem. 1974 ; 12 : 444-448.

14. KASPER, CB, DEUTSCH, HF.
Physiochemical studies on human caeruloplasmin.
J. Biol. Chem. 1963 ; 238 : 2325-2338.

15. SCUDDER, PR, AL-TIMINI, D, McMURRAY, W, WHITE, AC, ZOOB, BC, DORMANDY, TL.
Serum copper and related variables in rheumatoid arthritis.
Ann. Rheum. Dis. 1978 ; 37 : 67-70.

16. BALOGH, Z, EL-GHOBAREY, AF, FELL, GS, BROWN, DH, DUNLOP, J, DICK,WC.
Plasma zinc in relation to clinical symptoms and drug treatment in rheumatoid arthritis.
Ann. Rheum. Dis. 1980 ; 39 : 329-332.

17. TARP, U, OVERVAD, K, HANSEN, JC, THORLING, EB.
Low selenium in severe rheumatoid arthritis.
Scand. J. Rheumatol. 1985 ; 14 : 97-101.

18. SIMKIN, PA, NILSON, KL.
Transsynovial exchange of large and small molecules.
Clin. Rheum. Dis. 1981 ; 7 : 99-129.
19. NIEDERMEIER, W, GRIGGS, JH.
Trace metal composition of synovial fluid and blood serum of patients with rheumatoid arthritis.
J. Chron. Dis. 1971 ; 23 : 527-536.

INDEX